# The Quantum Mechanics Solver

Jean-Louis Basdevant • Jean Dalibard

# The Quantum Mechanics Solver

## How to Apply Quantum Theory to Modern Physics

**Third Edition**

Jean-Louis Basdevant
Département de Physique
Ecole Polytechnique
Palaiseau, France

Jean Dalibard
Collège de France
Paris, France

ISBN 978-3-030-13726-7     ISBN 978-3-030-13724-3    (eBook)
https://doi.org/10.1007/978-3-030-13724-3

Library of Congress Control Number: 2019934101

This Springer imprint is published by the registered company Springer Nature Switzerland AG
The registered company address is: Gewerbestrasse 11, 6330 Cham, Switzerland

# Preface to the Third Edition

Quantum mechanics is an endless source of new questions and fascinating observations. Examples can be found in fundamental physics and in applied physics, in mathematical questions as well as in the currently popular debates on the interpretation of quantum mechanics and its philosophical implications.

Teaching quantum mechanics relies mostly on theoretical courses, which are illustrated by simple exercises often of a mathematical character. Reducing quantum physics to this type of problem is somewhat frustrating since very few, if any, experimental quantities are available to compare the results with. For a long time, however, from the 1950s to the 1970s, the only alternative to these basic exercises seemed to be restricted to questions originating from atomic and nuclear physics, which were transformed into exactly soluble problems and related to known higher transcendental functions.

In the past 20 years or so, things have changed radically. The development of high technologies is a good example. The one-dimensional square-well potential used to be a rather academic exercise for beginners. The emergence of quantum dots and quantum wells in semiconductor technologies has changed things radically. Optronics and the associated developments in infrared semiconductor and laser technologies have considerably elevated the social rank of the square-well model. As a consequence, more and more emphasis is given to the physical aspects of the phenomena rather than to analytical or computational considerations.

Many fundamental questions raised since the very beginnings of quantum theory have received experimental answers in recent years. A good example is the neutron interference experiments of the 1980s, which gave experimental answers to 50-year-old questions related to the measurability of the phase of the wave function. Perhaps the most fundamental example is the experimental proof of the violation of Bell's inequality and the properties of entangled states, which have been established in decisive experiments since the late 1970s. More recently, the experiments carried out to quantitatively verify decoherence effects and "Schrödinger-cat" situations have raised considerable interest with respect to the foundations and the interpretation of quantum mechanics. Needless to say, the development of quantum information theory and quantum technologies is truly a new revolution of the field.

This book consists of a series of problems concerning present-day experimental or theoretical questions on quantum mechanics. All of these problems are based on actual physical examples, even if sometimes the mathematical structure of the models under consideration is simplified intentionally in order to get hold of the physics more rapidly.

The problems have all been given to our students in the École Polytechnique and in the École Normale Supérieure in the past 25 years or so. A special feature of the École Polytechnique comes from a tradition which has been kept for more than two centuries and which explains why it is necessary to devise original problems each year. The exams have a double purpose. On one hand, they are a means to test the knowledge and ability of the students. On the other hand, however, they are also taken into account as part of the entrance examinations to public office jobs in engineering, administrative and military careers. Therefore, the traditional character of stiff competitive examinations and strict meritocracy forbids us to make use of problems which can be found in the existing literature. We must therefore seek them among the forefront of present research. This work, which we have done in collaboration with many colleagues, turned out to be an amazing source of discussions between us. We all actually learned very many things, by putting together our knowledge in our respective fields of interest.

Compared to the second version of this book, which was published by Springer-Verlag in 2006, we have made several modifications. First of all, we have included new themes, such as quantum correlations in a multiparticle system, vortices in rotating Bose-Einstein condensates, neutrino transformations in the sun, etc. Second, we have updated the discussions and the references that are provided at the end of each chapter. We hope this will allow our readers to continue to explore all these topics at the forefront of current research in quantum physics.

Palaiseau, France                                                   Jean-Louis Basdevant
Paris, France                                                             Jean Dalibard
December 2018

# Acknowledgements

We are indebted to many colleagues who either gave us driving ideas or wrote first drafts of some of the problems presented here. We want to pay a tribute to the memory of Gilbert Grynberg, who wrote the first versions of "The Hydrogen Atom in Crossed Fields", "Hidden Variables and Bell's Inequalities" and "Spectroscopic Measurement on a Neutron Beam". We are particularly grateful to André Rougé and Jim Rich for illuminating discussions on "Neutrino Oscillations", and we pay a tribute to the memory of François Jacquet who wrote the first version of this topic. Finally we thank Philippe Grangier, who actually conceived many problems among which the "Schrödinger's Cat", the "Ideal Quantum Measurement" and the "Quantum Thermometer", Gérald Bastard for "Quantum Boxes", Jean-Noël Chazalviel for "Hyperfine Structure in Electron Spin Resonance", Thierry Jolicoeur for "Magnetic Excitons", Bernard Equer for "Probing Matter with Positive Muons", Vincent Gillet for "Energy Loss of Ions in Matter" and Yvan Castin, Jean-Michel Courty and Dominique Delande for "Quantum Reflection of Atoms on a Surface".

# Contents

# Part I
# Elementary Particles, Nuclei and Atoms

# Matter-Wave Interferences with Molecules

1

In this chapter we study the diffraction and interference phenomena which can be observed with molecules that are transmitted by a double or a multiple slit setup. We first show that this type of experiment allows one to reveal the existence of the dimer $He_2$ and the trimer $He_3$, whose existence had remained controversial for a long time. We then turn to the case of larger objects and set a bound on the maximal size of molecules that can be studied in this kind of setup.

## 1.1 Helium Dimers and Trimers

We consider the diffraction of a beam of helium atoms arriving on a periodic grating (Fig. 1.1). From the measured diffraction pattern, we wish to show the presence of a few dimers $He_2$ and trimers $He_3$ in the beam. For simplicity we assume that all atoms of the beam have the same velocity $v = 430$ m/s. The mass of a helium atom is $M = 6.69 \times 10^{-27}$ kg.

**1.1.1** Give the wavelength $\lambda_1$ associated to the helium atoms. What are the wavelengths $\lambda_2$ and $\lambda_3$ associated to the molecules $He_2$ and $He_3$, assuming that they propagate at the same velocity as the atoms?

**1.1.2** The beam crosses a grating with equally spaced slits. The distance between slits is $d = 200$ nm.

(a) Explain briefly why coherent diffraction (Bragg) can occur in particular directions of space.

(b) Calculate the angle $\theta$ between the incident beam axis and the first diffracted direction for helium atoms. Express $\theta$ in terms of $\lambda_1$ and $d$, using the fact that $\lambda_1 \ll d$.

© Springer Nature Switzerland AG 2019
J.-L. Basdevant, J. Dalibard, *The Quantum Mechanics Solver*,
https://doi.org/10.1007/978-3-030-13724-3_1

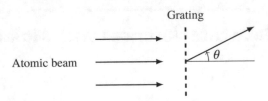

**Fig. 1.1** Diffraction of a beam of helium atoms by a periodic grating

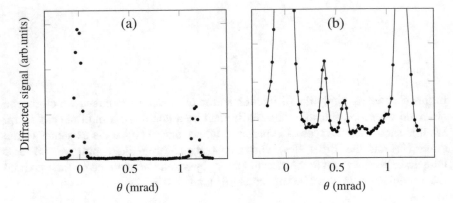

**Fig. 1.2** (**a**) Measured signal as function of the deflection angle $\theta$. (**b**) Same data as in (**a**), with a vertical scale multiplied by 60

**1.1.3** The measured angular distribution is shown in Fig. 1.2a. The peak at $\theta = 0$ corresponds to the non-diffracted beam. Does the position of the diffracted peak correspond to the prediction?

**1.1.4** Figure 1.2b shows a magnification of the signal of Fig. 1.2a. Explain why it allows one to conclude that the dimers $He_2$ and the trimers $He_3$ do exist.

## 1.2     Interferences of Large Material Particles

We now consider a Young double-slit experiment performed with material particles. These particles are assumed to have a spherical shape with radius $R$ and volume mass density $\rho = 1000 \, \text{kg/m}^3$ (Fig. 1.3). They are emitted by a source at temperature $T = 300 \, \text{K}$ and propagate along the axis of the interferometer with a velocity $v$ such that $\frac{1}{2}mv^2 = \frac{1}{2}k_BT$, where $m$ is the mass of a particle and $k_B$ the Boltzmann constant.

**1.2.1** Give the wavelength $\lambda$ of the particle in terms of $\rho$, $R$, $T$ and Planck's constant $h$.

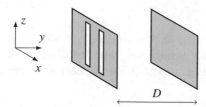
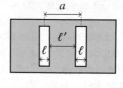

**Fig. 1.3** Left : Young double-slit interferometer. Right: Slit geometry. Each slit has a width $\ell$ and the distance between the slit centers is $a$. Hence the opaque zone between the slits has a width $\ell' = a - \ell$

**1.2.2** We denote by $D$ the distance between the slits and the detection screen. Give the expression of the fringe spacing in terms of $\lambda$, $D$ and $a$, assuming that the slit width $\ell$ is small compared to the distance between the slits $a$. One will also assume for simplicity that $a \ll D$. In the following we will extend this result to the case $\ell \approx a$.

**1.2.3** In order to observe an interference signal we require the two following conditions to be satisfied.

- $R \leq x_i/2$. This condition expresses the fact that the impact zone of a particle on the detection screen should not overlap with several interference fringes.
- $R \leq a/2$. This condition expresses the fact that some particles should be able to go through a slit without touching its edges. More precisely, one should have $R \leq \ell/2$, knowing that $\ell \leq a = \ell + \ell'$.

Let us fix the value of $D$ and choose the optimal value of $a$ to observe an interference signal with the largest possible particles. What is the maximal size $R_{max}$ for which interferences can be observed? Express $R_{max}$ in terms of $D$, $\rho$, $T$, $k_B$ and $h$. Which value should one choose for the slit spacing $a$?

**1.2.4** The interferometer length is $D = 1$ m. What is the value of $R_{max}$? What are the corresponding velocity and wavelength? What are the main obstacles to the realization of such an experiment?

## 1.3  Solutions

### 1.3.1  Helium Dimers and Trimers

**1.3.1.1** The de Broglie relation $\lambda = h/(mv)$ gives $\lambda_1 = 2.30 \times 10^{-10}$ m. For particles with mass $2m$ or $3m$ propagating with the same speed, we find $\lambda_2 = 1.15 \times 10^{-10}$ m and $\lambda_2 = 0.77 \times 10^{-10}$ m.

**1.3.1.2** Bragg diffraction.

**(a)** A constructive interference will occur in the direction $\theta$ when the path difference $d \sin \theta$ between the paths going through neighbouring slits is a multiple of the wavelength $\lambda$. The angles for which coherent diffraction occurs are therefore given by $\sin \theta = n\lambda/d$.

**(b)** The first diffraction order corresponds to $\theta \simeq \lambda/d = 1.15 \, \text{mrad}$.

**1.3.1.3** The diffracted peak is found at the deflection angle $\theta = 1.16 \, \text{mrad}$, in agreement with the prediction above.

**1.3.1.4** In the magnified picture, one sees additional peaks at $\theta = 0.58 \, \text{mrad}$ and $\theta = 0.40 \, \text{mrad}$, i.e. one half and one third of the angle found above. If we assume that all components of the beams propagate at the same velocity (which can be checked using a complementary time-of-flight experiment), it means that the mass of the objects at the origin of these peaks is twice or three times as large as the helium atom mass. These correspond to the molecules $He_2$ and $He_3$.

## 1.3.2 Interferences of Large Material Particles

**1.3.2.1** Using $\lambda = h/(mv)$ and $v = \sqrt{k_B T/m}$, we find $\lambda = h/\sqrt{m k_B T}$. The mass is $m = \rho (4\pi/3) R^3$, which leads to

$$\lambda = \sqrt{\frac{3}{4\pi}} \, \frac{h}{\sqrt{\rho R^3 k_B T}}. \tag{1.1}$$

**1.3.2.2** When $\ell \ll a$ and $a \ll D$, the fringe spacing in a Young double-slit experiment is given by $x_i = \lambda D/a$.

**1.3.2.3** The two necessary conditions stated in the text can be simultaneously satisfied only if $R^2 \leq a x_i/4$, which can also be written $R^2 \leq \lambda D/4$. Replacing $\lambda$ by its expression, we get

$$R^{7/2} \leq \frac{1}{8} \sqrt{\frac{3}{\pi}} \, \frac{hD}{\sqrt{\rho k_B T}}, \tag{1.2}$$

which can be written

$$R_{\text{max}} = \left( \frac{1}{8} \sqrt{\frac{3}{\pi}} \, \frac{hD}{\sqrt{\rho k_B T}} \right)^{2/7}. \tag{1.3}$$

We now choose $a = 2R_{\text{max}}$ and obtain an interference pattern with the fringe spacing $x_i = 2R_{\text{max}}$.

**1.3.2.4** For $D = 1$ m, the largest value of $R$ is 55 nm, the mass $m = 7.1 \times 10^{-19}$ kg, the velocity $v = 7.6$ cm/s and the wavelength $\lambda = 1.2 \times 10^{-14}$ m. For such large particles, it is extremely difficult to maintain the coherence between the two arms of the interferometer. Uncontrolled dephasing can be caused by collisions between the particles and the molecules of the residual gas in the vacuum chamber where the experiment is performed. Residual electric fields can also alter the interference signal if the particles carry some electric charge. In addition note that the velocity of the particles is low, therefore gravitational effects will be important in the time of flight (13 s) between the plane of the double slits and the detection screen.

### 1.3.3  References

The experiments on helium atoms and molecules that are presented in the first part of this chapter were performed in Göttingen and described in W. Schöllkopf and J.P. Toennies, *Nondestructive Mass Selection of Small van der Waals Clusters*, Science **266**, 1345 (1994).

The exploration of the wave nature of material objects with a large mass has made spectacular progress over the last two decades, and interferences of particles with a mass larger than $10^4$ atomic mass unit have been observed. See e.g. M. Arndt, *De Broglie's meter stick: Making measurements with matter waves*, Phys. Today **67**, 30 (2014). It is impressive to see that De Broglie's relation, which was initially formulated for electrons, applies equally well for objects that are more than $10^7$ times heavier.

# Neutron Interferometry

**2**

In the late 1970s, Overhauser and his collaborators performed several neutron interference experiments which are of fundamental importance in quantum mechanics,
and which settled debates that had started in the 1930s. We study in this chapter two
of these experiments, aiming to measure the influence on the interference pattern (i)
of the gravitational field and (ii) of a $2\pi$ rotation of the neutron wave function.

We consider an interferometer made of three parallel, equally spaced crystalline
silicon strips, as shown in Fig. 2.1. The incident neutron beam is assumed to be
monochromatic.

For a particular value of the angle of incidence $\theta$, called the Bragg angle, a
plane wave $\psi_{\text{inc}} = e^{i(\boldsymbol{p}\cdot\boldsymbol{r}-Et)/\hbar}$, where $E$ is the energy of the neutrons and $\boldsymbol{p}$ their
momentum, is split by the crystal into two outgoing waves which are symmetric
with respect to the perpendicular direction to the crystal, as shown in Fig. 2.2.

The transmitted wave and the reflected wave have complex amplitudes which can
be written respectively as $\alpha = \cos\chi$ and $\beta = i\,\sin\chi$, where the angle $\chi$ is real:

$$\psi_{\text{I}} = \alpha\,e^{i(\boldsymbol{p}\cdot\boldsymbol{r}-Et)/\hbar} \qquad \psi_{\text{II}} = \beta\,e^{i(\boldsymbol{p}'\cdot\boldsymbol{r}-Et)/\hbar}, \tag{2.1}$$

where $|\boldsymbol{p}| = |\boldsymbol{p}'|$ since the neutrons scatter elastically on the nuclei of the crystal.
The transmission and reflection coefficients are $T = |\alpha|^2$ and $R = |\beta|^2$, with of
course $T + R = 1$.

In the interferometer shown in Fig. 2.1, the incident neutron beam is horizontal.
It is split by the interferometer into a variety of beams, two of which recombine and
interfere at point $D$. The detectors $C_2$ and $C_3$ count the outgoing neutron fluxes.
The neutron beam velocity corresponds to a de Broglie wavelength $\lambda = 1.445$ Å.
We recall the value of neutron mass $M = 1.675 \times 10^{-27}$ kg.

The neutron beam actually corresponds to wave functions which are quasi-
monochromatic and which have a finite extension in the transverse directions. In
order to simplify the writing of the equations, we only deal with pure monochromatic plane waves, as in (2.1).

© Springer Nature Switzerland AG 2019
J.-L. Basdevant, J. Dalibard, *The Quantum Mechanics Solver*,
https://doi.org/10.1007/978-3-030-13724-3_2

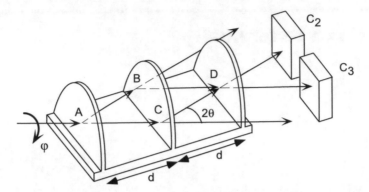

**Fig. 2.1** The neutron interferometer: The three "ears" are cut in a silicon monocrystal; $C_2$ and $C_3$ are neutron counters

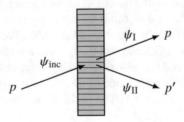

**Fig. 2.2** Splitting of an incident plane wave satisfying the Bragg condition

## 2.1   Neutron Interferences

**2.1.1** The measured neutron fluxes are proportional to the intensities of the waves that reach the counters. Defining the intensity of the incoming beam to be 1 (the units are arbitrary), write the amplitudes $A_2$ and $A_3$ of the wave functions which reach the counters $C_2$ and $C_3$, in terms of $\alpha$ and $\beta$. Calculate the measured intensities $I_2$ and $I_3$ in terms of the coefficients $T$ and $R$.

**2.1.2** Suppose that we create a phase shift $\delta$ of the wave propagating along $AC$, i.e. the wave function is multiplied by $e^{i\delta}$ in $C$.

(a) Calculate the new amplitudes $A_2$ and $A_3$ in terms of $\alpha$, $\beta$ and $\delta$.
(b) Show that the new measured intensities $I_2$ and $I_3$ are of the form

$$I_2 = \mu - \nu(1 + \cos\delta), \qquad I_3 = \nu(1 + \cos\delta), \tag{2.2}$$

and express $\mu$ and $\nu$ in terms of $T$ and $R$.
(c) Comment on the result for the sum $I_2 + I_3$.

## 2.2    The Gravitational Effect

The phase difference $\delta$ between the beams $ACD$ and $ABD$ is created by rotating the interferometer by an angle $\phi$ around the direction of incidence. This creates a difference in the altitudes of $BD$ and $AC$, which both remain horizontal, as shown in Fig. 2.3. The difference in the gravitational potential energies induces a gravitational phase difference.

**2.2.1** Let $d$ be the distance between the silicon strips, whose thickness is neglected here. Show that the side $L$ of the lozenge $ABCD$ and its height $H$, shown in Fig. 2.3, are related to $d$ and to the Bragg angle $\theta$ by $L = d/\cos\theta$ and $H = 2d\sin\theta$. Experimentally the values of $d$ and $\theta$ are $d = 3.6\,\text{cm}$ and $\theta = 22.1°$.

**2.2.2** For an angle $\phi$, we define the gravitational potential $V$ to be $V = 0$ along $AC$ and $V = V_0$ along $BD$.

(a) Calculate the difference $\Delta p$ of the neutron momenta in the beams $AC$ and $BD$ (use the approximation $\Delta p \ll p$). Express the result in terms of the momentum $p$ along $AC$, the height $H$, $\sin\phi$, $M$, and the acceleration of gravity $g$.
(b) Evaluate numerically the velocity $\sqrt{2gH}$. How good is the approximation $\Delta p \ll p$?

**2.2.3** Evaluate the phase difference $\delta$ between the paths $ABD$ and $ACD$. One can proceed in two steps:

(a) Compare the path difference between the segments $AB$ and $CD$.
(b) Compare the path difference between the segments $BD$ and $AC$.

**2.2.4** The variation with $\phi$ of the experimentally measured intensity $I_2$ in the counter $C_2$ is represented in Fig. 2.4. (*The data does not display a minimum exactly at $\phi = 0$ because of calibration difficulties.*)
    Deduce from these data the value of the acceleration due to gravity $g$.

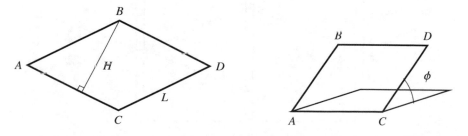

**Fig. 2.3** Turning the interferometer around the incident direction, in order to observe gravitational effects

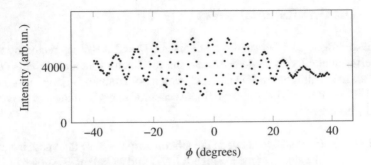

**Fig. 2.4** Measured neutron intensity in counter $C_2$ as the angle $\phi$ is varied

## 2.3    Rotating a Spin 1/2 by 360 Degrees

The plane of the setup is now horizontal. The phase difference arises by placing along $AC$ a magnet of length $l$ which produces a constant uniform magnetic field $\boldsymbol{B}_0$ directed along the $z$ axis, as shown in Fig. 2.5.

The neutrons are spin 1/2 particles, and have an intrinsic magnetic moment $\hat{\boldsymbol{\mu}} = \gamma_n \hat{\boldsymbol{S}} = \mu_0 \hat{\boldsymbol{\sigma}}$ where $\hat{\boldsymbol{S}}$ is the neutron spin operator, and the $\hat{\sigma}_i (i = x, y, z)$ are the usual $2 \times 2$ Pauli matrices. The axes are represented in Fig. 2.5: the beam is along the $y$ axis, the $z$ axis is in the $ABCD$ plane, and the $x$ axis is perpendicular to this plane.

We assume that the spin variables and the space variables are uncorrelated, i.e. at any point in space the wave function factorizes as

$$\begin{pmatrix} \psi_+(\boldsymbol{r}, t) \\ \psi_-(\boldsymbol{r}, t) \end{pmatrix} = e^{i(\boldsymbol{p}\cdot\boldsymbol{r} - Et)/\hbar} \begin{pmatrix} a_+(t) \\ a_-(t) \end{pmatrix}. \tag{2.3}$$

We neglect any transient effect due to the entrance and the exit of the field zone.

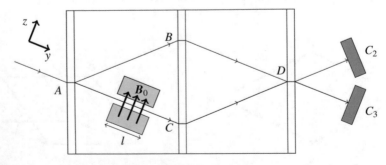

**Fig. 2.5** Experimental setup for observing the neutron spin Larmor precession. The plane $ABCD$ of the interferometer is now horizontal (the vertical is along $x$ and perpendicular to the figure) and a magnetic field $\boldsymbol{B}_0$ parallel to $z$ is added along the segment $AC$

The incident neutrons are prepared in the spin state

$$| + x \rangle = \frac{1}{\sqrt{2}} \begin{pmatrix} 1 \\ 1 \end{pmatrix}, \tag{2.4}$$

which is the eigenstate of $\hat{\mu}_x$ with eigenvalue $+\mu_0$. The spin state is not modified when the neutrons cross the crystal strips.

### 2.3.1 Evolution in the magnet.

(a) Write the magnetic interaction Hamiltonian of the spin with the magnetic field.
(b) What is the time evolution of the spin state of a neutron in the magnet?
(c) Setting $\omega = -2\mu_0 B_0/\hbar$, calculate the three components of the expectation value $\langle \hat{\mu} \rangle$ in this state, and describe the time evolution of $\langle \hat{\mu} \rangle$ in the magnet.

**2.3.2** When the neutron leaves the magnet, what is the probability $P_x(+\mu_0)$ of finding $\mu_x = +\mu_0$ when measuring the $x$ component of the neutron magnetic moment? For simplicity, one can set $T = Ml\lambda/(2\pi\hbar)$ and express the result in terms of the angle $\delta = -\omega T/2$.

**2.3.3** For which values $b_n = nb_1$ ($n$ integer) of the field $B_0$ is this probability equal to 1? To what motion of the average magnetic moment do these values $b_n$ correspond?

Calculate $b_1$ with $\mu_0 = -9.65 \times 10^{-27}$ JT$^{-1}$, $l = 2.8$ cm, $\lambda = 1.445$ Å.

**2.3.4** Write the state of the neutrons when they arrive on $C_2$ and $C_3$ (note $p_2$ and $p_3$ the respective momenta).

**2.3.5** The counters $C_2$ and $C_3$ measure the neutron fluxes $I_2$ and $I_3$. They are not sensitive to spin variables. Express the difference of intensities $I_2 - I_3$ in terms of $\delta$ and of the coefficients $T$ and $R$.

**2.3.6** The experimental measurement of $I_2 - I_3$ as a function of the applied field $B_0$ is given in Fig. 2.6. A numerical fit of the curve shows that the distance between two maxima is $\Delta B = (64 \pm 2) \times 10^{-4}$ T.

Comparing the values $b_n$ of question 2.3.3 with this experimental result, and recalling the result of a measurement of $\mu_x$ for these values, explain why this proves that the state vector of a spin-1/2 particle changes sign under a rotation by an *odd* multiple of $2\pi$.

**Fig. 2.6** Difference of counting rates ($I_2 - I_3$) as a function of the applied field

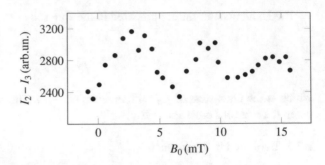

## 2.4    Solutions

### 2.4.1    Neutron Interferences

**2.4.1.1** The beams $ABDC_2$ and $ACDC_2$ interfere. Omitting the propagation factors, the amplitude in $C_2$ reads

$$A_2 = \alpha^2\beta + \beta^3 = \beta(\alpha^2 + \beta^2). \tag{2.5}$$

Similarly, for $ABDC_3$ and $ACDC_3$,

$$A_3 = 2\alpha\beta^2. \tag{2.6}$$

The intensities at the two counters are

$$I_2 = R - 4R^2T \quad I_3 = 4R^2T. \tag{2.7}$$

**2.4.1.2** When there is a phase shift $\delta$ in C, the above expressions get modified as follows:

$$A_2 = \alpha^2\beta e^{i\delta} + \beta^3 = \beta(\alpha^2 e^{i\delta} + \beta^2), \quad A_3 = \alpha\beta^2(1 + e^{i\delta}). \tag{2.8}$$

The intensities become

$$I_2 = R - 2R^2T\,(1 + \cos\delta) \quad I_3 = 2R^2T\,(1 + \cos\delta). \tag{2.9}$$

The fact that $I_2 + I_3$ does not depend on the phase shift $\delta$ is a consequence of the conservation of the total number of particles arriving at $D$.

## 2.4.2 The Gravitational Effect

**2.4.2.1** This results from elementary trigonometry.

**2.4.2.2 (a)** Since there is no recoil energy of the silicon atoms to be taken care of, the neutron total energy (kinetic+potential) is a constant of the motion in all the process. This energy is given by $E_{AC} = p^2/2M$ and $E_{BD} = (p - \Delta p)^2/2M + MgH \sin\phi$, hence

$$\Delta p \simeq M^2 gH \sin\phi/p. \tag{2.10}$$

**(b)** The velocity $\sqrt{2gH}$ is of the order of $0.5 \, \text{m/s}$, and the neutron velocity is $v = h/M\lambda \simeq 2700 \, \text{m/s}$. The change in velocity $\Delta v$ is therefore very small: $\Delta v = gH/v \simeq 2 \times 10^{-4} \, \text{m/s}$ for $\phi = \pi/2$.

**2.4.2.3 (a)** The gravitational potential varies in exactly the same way along $AB$ and $CD$. The neutron state in both cases is a plane wave with momentum $p = h/\lambda$ just before $A$ or $C$. The same Schrödinger equation is used to determine the wave function at the end of the segments. This implies that the phases accumulated along the two segments $AB$ and $CD$ are equal.

**(b)** When comparing the segments $AC$ and $BD$, the previous reasoning does not apply, since the initial state of the neutron is not the same for the two segments. The initial state is $\exp(ipz/\hbar)$ for $AC$, and $\exp[i(p - \Delta p)z/\hbar]$ for $BD$. After travelling over a distance $L = \overline{AC} = \overline{BD}$, the phase difference between the two paths is

$$\delta = \frac{\Delta p \, L}{\hbar} = \frac{M^2 g\lambda d^2}{\pi \hbar^2} \tan\theta \sin\phi. \tag{2.11}$$

**2.4.2.4** From the previous result, one has $\delta_2 - \delta_1 = Ag(\sin\phi_2 - \sin\phi_1)$, where $A = M^2\lambda d^2 \tan\theta/(\pi\hbar^2)$. Therefore,

$$g = \frac{\delta_2 - \delta_1}{A \, (\sin\phi_2 - \sin\phi_1)}. \tag{2.12}$$

There are nine oscillations, i.e. $(\delta_2 - \delta_1) = 18\pi$, between $\phi_1 = -32°$ and $\phi_2 = +24°$, which gives $g \simeq 9.8 \, \text{ms}^{-2}$. The relative precision of the experiment was actually of the order of $10^{-3}$.

### 2.4.3   Rotating a Spin 1/2 by 360 Degrees

**2.4.3.1** Since $B$ is along the $z$ axis, the magnetic Hamiltonian is:

$$\hat{H}_M = -\boldsymbol{\mu} \cdot \boldsymbol{B}_0 = \frac{\hbar\,\omega}{2} \begin{pmatrix} 1 & 0 \\ 0 & -1 \end{pmatrix}. \qquad (2.13)$$

At time $t$, the spin state is

$$|\Sigma(t)\rangle = \frac{1}{\sqrt{2}} \begin{pmatrix} e^{-i\omega t/2} \\ e^{+i\omega t/2} \end{pmatrix}. \qquad (2.14)$$

By a direct calculation of $\langle\boldsymbol{\mu}\rangle$ or by using Ehrenfest theorem ($\frac{d}{dt}\langle\boldsymbol{\mu}\rangle = \frac{1}{i\hbar}\langle[\hat{\boldsymbol{\mu}}, \hat{H}]\rangle$), we obtain:

$$\frac{d\langle\mu_x\rangle}{dt} = \omega\langle\mu_y\rangle \qquad \frac{d\langle\mu_y\rangle}{dt} = -\omega\langle\mu_x\rangle \qquad \frac{d\langle\mu_z\rangle}{dt} = 0. \qquad (2.15)$$

Initially $\langle\mu_x\rangle = \mu_0$ and $\langle\mu_y\rangle = \langle\mu_z\rangle = 0$; therefore,

$$\langle\boldsymbol{\mu}\rangle = \mu_0\,(\cos\omega t\,\boldsymbol{u}_x + \sin\omega t\,\boldsymbol{u}_y). \qquad (2.16)$$

**2.4.3.2** When the neutrons leave the field zone, the probability of finding $\mu_x = +\mu_0$ is

$$P_x(+\mu_0) = |\langle +x|\Sigma(T)\rangle|^2 = \cos^2\frac{\omega T}{2} = \cos^2\delta \qquad (2.17)$$

with $T = l/v = lM\lambda/h$.

**2.4.3.3** The above probability is equal to 1 if $\delta = n\pi$ ($\omega T = 2n\pi$), or $B_0 = nb_1$ with

$$b_1 = \frac{2\pi^2\hbar^2}{\mu_0 M l\lambda} = 34.5 \times 10^{-4}\ \text{T}. \qquad (2.18)$$

For $\delta = n\pi$ the magnetic moment has rotated by $2n\pi$ around the $z$ axis by Larmor precession.

**2.4.3.4** The formulas are similar to those found in question 2.4.1.2. The phase of the upper component of the spinor written in the $\{|+\rangle_z, |-\rangle_z\}$ basis, is shifted by $+\delta$, that of the lower component by $-\delta$:

Amplitude at the counter $C_2$ :     $e^{i(p_2\cdot r - Et)/\hbar} \dfrac{\beta}{\sqrt{2}} \left( \dfrac{\beta^2 + \alpha^2 e^{i\delta}}{\beta^2 + \alpha^2 e^{-i\delta}} \right),$     (2.19)

Amplitude at the counter $C_3$ :     $e^{i(p_3\cdot r - Et)/\hbar} \dfrac{\alpha\beta^2}{\sqrt{2}} \left( \dfrac{1 + e^{i\delta}}{1 + e^{-i\delta}} \right).$     (2.20)

**2.4.3.5** Since the measuring apparatus is insensitive to spin variables, we must add the probabilities corresponding to $S_z = \pm 1$, each of which is the modulus squared of a sum of amplitudes. Altogether, we obtain the following intensities of the total neutron flux in the two counters:

$$I_2 = R - 2R^2 T (1 + \cos\delta), \quad I_3 = 2R^2 T (1 + \cos\delta) \qquad (2.21)$$

and

$$I_2 - I_3 = R - 4R^2 T (1 + \cos\delta). \qquad (2.22)$$

**2.4.3.6** There will be a *minimum* of $I_2 - I_3$ each time $\cos\delta = +1$, i.e. $\delta = 2n\pi$. This corresponds to a constructive interference in channel 3. On the other hand, there appears a *maximum* if $\cos\delta = -1$, i.e. $\delta = (2n+1)\pi$, and this corresponds to a destructive interference in channel 3 ($I_3 = 0$).

If $\delta = n\pi$, whatever the integer $n$, one is *sure* to find the neutrons in the *same* spin state as in the initial beam. However, the interference pattern depends on the parity of $n$.

The experimental result $\Delta B = (64 \pm 2) \times 10^{-4}$ T confirms that if the spin has rotated by $4n\pi$, one recovers a constructive interference in channel 3 as in the absence of rotation, while if it has rotated by $(4n + 2)\pi$, the interference in $C_3$ is destructive. The probability amplitude for the path $ACD$ has changed sign in this latter case, although a spin measurement in this path after the magnet will give exactly the same result as on the incoming beam.

## 2.4.4 References

The experiments presented in this chapter have been described by A. W. Overhauser and A. R. Collela, *Experimental Test of Gravitationally Induced Quantum Interference*, Phys. Rev. Lett. **33**, 1237 (1974); R. Colella, A. W. Overhauser, and S. A. Werner, *Observation of Gravitationally Induced Quantum Interference*, Phys. Rev. Lett. **34**, 1472 (1975); S. A. Werner, R. Colella, A. W. Overhauser, and C. F. Eagen, *Observation of the Phase Shift of a Neutron Due to Precession in a Magnetic Field*, Phys. Rev. Lett. **35**, 1053 (1975). See also D. Greenberger and A.W. Overhauser, *The role of gravity in quantum theory*, Scientific American **242**, 54 (May 1980).

More information on neutron optics and neutron interferometry can be found in A. Zeilinger, R. Gähler, C. G. Shull, W. Treimer, and W. Mampe, *Single- and double-slit diffraction of neutrons*, Rev. Mod. Phys. **60**, 1067 (1988); A. Zeilinger, *Experiments and the foundations of quantum physics*, Rev. Mod. Phys. **71**, S288 (1999) and H. Rauch and S. A. Werner, *Neutron Interferometry: Lessons in Experimental Quantum Mechanics* (Oxford University Press, 2002).

# Analysis of a Stern–Gerlach Experiment

**3**

We analyze a Stern–Gerlach experiment, both experimentally and from the theoretical point of view. In the setup considered here, a monochromatic beam of neutrons crosses a region of strongly inhomogeneous magnetic field and one observes the outgoing beam.

## 3.1 Preparation of the Neutron Beam

Neutrons produced in a reactor are first "cooled", i.e. slowed down by crossing liquid hydrogen at 20 K. They are incident on a monocrystal, for instance graphite, from which they are diffracted. To each outgoing direction, there corresponds a well-defined wavelength, and therefore a well-defined momentum. A beryllium crystal acts as a filter to eliminate harmonics, and the vertical extension of the beam is controlled by two gadolinium blocks, which are opaque to neutrons, separated by a thin sheet of (transparent) aluminum of thickness $a$, which constitutes the collimating slit, as shown in Fig. 3.1.

**3.1.1** The de Broglie wavelength of these monochromatic neutrons is $\lambda = 4.32$ Å. What are their velocity and their kinetic energy?

**3.1.2** One observes the impacts of the neutrons on a detector at a distance $L = 1$ m from the slit. The vertical extension of the beam at the detector is determined by two factors, first the width $a$ of the slit, and second the diffraction of the neutron beam by the slit. We recall that the angular width $\theta$ of the diffraction peak from a slit of width $a$ is related to the wavelength $\lambda$ by $\sin \theta = \lambda/a$. For simplicity, we assume that the neutron beam is well collimated before the slit, and that the vertical extension $\delta$ of the beam on the detector is the sum of the width $a$ of the slit and the width of the diffraction peak.

© Springer Nature Switzerland AG 2019
J.-L. Basdevant, J. Dalibard, *The Quantum Mechanics Solver*,
https://doi.org/10.1007/978-3-030-13724-3_3

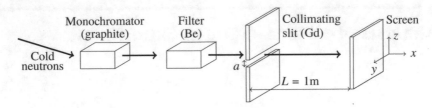

**Fig. 3.1** Preparation of the neutron beam

Show that one can choose $a$ in an optimal way in order to make $\delta$ as small as possible. What is the corresponding width of the beam on the detector?

**3.1.3** In the actual experiment, the chosen value is $a = 5\ \mu m$. What is the observed width of the beam at the detector?

Comment on the respective effects of the slit width $a$ and of diffraction, on the vertical shape of the observed beam on the detector.

The extension of the beam corresponds to the distribution of neutron impacts along the $z$ axis. Since the purpose of the experiment is not only to observe the beam, but also to measure its "position" as defined by the maximum of the distribution, justify the choice $a = 5\ \mu m$.

Figure 3.2 is an example of the neutron counting rate as a function of $z$. The horizontal error bars, or bins, come from the resolution of the measuring apparatus, the vertical error bars from the statistical fluctuations of the number of neutrons in each bin. The curve is a best fit to the experimental points. Its maximum is determined with an accuracy $\delta z \sim 5\ \mu m$.

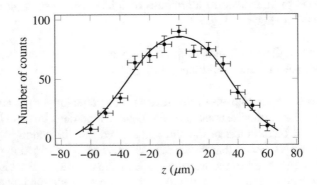

**Fig. 3.2** Measurement of the beam profile on the detector

## 3.2      Spin State of the Neutrons

In order to completely describe the state of a neutron, i.e. both its spin state and its spatial state, we consider the eigenbasis of the spin projection along the $z$ axis, $\hat{S}_z$, and we represent the neutron state as

$$|\psi(t)\rangle \; : \quad \begin{pmatrix} \psi_+(r,t) \\ \psi_-(r,t) \end{pmatrix}, \tag{3.1}$$

where the respective probabilities of finding the neutron in the vicinity of point $r$ with its spin component $S_z = \pm\hbar/2$ are

$$\mathrm{d}^3 P(r, S_z = \pm\hbar/2, t) = |\psi_\pm(r,\ t)|^2 \, \mathrm{d}^3 r. \tag{3.2}$$

**3.2.1**  What are the probabilities $P_\pm(t)$ of finding, at time $t$, the values $\pm\hbar/2$ when measuring $S_z$ irrespective of the position $r$?

**3.2.2**  What is, in terms of $\psi_+$ and $\psi_-$, the expectation value of the $x$ component of the neutron spin $\langle S_x \rangle$ in the state $|\psi(t)\rangle$?

**3.2.3**  What are the expectation values of the neutron's position $\langle r \rangle$ and momentum $\langle p \rangle$ in the state $|\psi(t)\rangle$?

**3.2.4**  We assume that the state of the neutron can be written:

$$|\psi(t)\rangle \quad : \quad \psi(r,t) \begin{pmatrix} \alpha_+ \\ \alpha_- \end{pmatrix}, \tag{3.3}$$

where the two complex numbers $\alpha_\pm$ are such that $|\alpha_+|^2 + |\alpha_-|^2 = 1$. How do the results of questions 3.2.2 and 3.2.3 simplify in that case?

## 3.3      The Stern–Gerlach Experiment

Between the slit, whose center is located at the origin ($x = y = z - 0$), and the detector, located in the plane $x = L$, we place a magnet of length $L$ whose field $B$ is directed along the $z$ axis. The magnetic field varies strongly with $z$, see Fig. 3.3.

**Fig. 3.3** Magnetic field
setup in the Stern–Gerlach
experiment

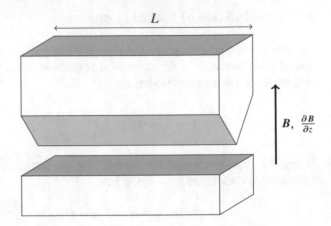

We assume that the components of the magnetic field are

$$B_x = B_y = 0 \quad B_z = B_0 + b'z. \tag{3.4}$$

In what follows we choose[1] $B_0 = 1$ T and $b' = 100$ T/m.

The magnetic moment of the neutron $\hat{\boldsymbol{\mu}}$ in the matrix representation that we have chosen for $|\psi\rangle$ is

$$\hat{\boldsymbol{\mu}} = \mu_0 \, \hat{\boldsymbol{\sigma}}, \tag{3.5}$$

where $\hat{\boldsymbol{\sigma}}$ are the Pauli matrices, and $\mu_0 = 1.913 \, \mu_N$, where $\mu_N$ is the nuclear magneton $\mu_N = q\hbar/2M_p = 5.051 \times 10^{-27}$ J·T$^{-1}$. Hereafter, we denote the neutron mass by $m$.

**3.3.1** What is the form of the Hamiltonian for a neutron moving in this magnetic field?

Write the time-dependent Schrödinger equation for the state $|\psi(t)\rangle$.

Show that the Schrödinger equation decouples into two equations of the Schrödinger type, for $\psi_+$ and $\psi_-$ respectively.

**3.3.2** Show that one has

$$\frac{\mathrm{d}}{\mathrm{d}t} \int |\psi_\pm(\boldsymbol{r}, t)|^2 \, \mathrm{d}^3 r = 0. \tag{3.6}$$

What does one conclude as to the probabilities of measuring $\mu_z = \pm\mu_0$?

---

[1]This form violates Maxwell's equation $\nabla \cdot \boldsymbol{B} = 0$, but it simplifies the following calculation. With a little modification (e.g. $B_x = 0$, $B_y = -b'y$, $B_z = B_0 + b'z$ and $B_y \ll B_z$ over the region of space crossed by the neutron beam), one can settle this matter, and arrive at the same conclusions.

**3.3.3** We assume that, at $t = 0$, at the entrance of the field zone, one has

$$|\psi(0)\rangle \quad : \quad \psi(r, 0) \begin{pmatrix} \alpha_+ \\ \alpha_- \end{pmatrix} \tag{3.7}$$

and that $\langle r \rangle = 0$, $\langle p_y \rangle = \langle p_z \rangle = 0$ and $\langle p_x \rangle = p_0 = h/\lambda$, where the value of the wavelength $\lambda$ has been given above. The above conditions correspond to the experimental preparation of the neutron beam discussed in Sect. 3.1.

Let $\hat{A}$ be an observable depending on the position operator $\hat{r}$ and the momentum operator $\hat{p}$. We define the numbers $\langle A_+ \rangle$ and $\langle A_- \rangle$ by

$$\langle A_\pm \rangle = \frac{1}{|\alpha_\pm|^2} \int \psi_\pm^*(r, t) \, \hat{A} \psi_\pm^*(r, t) \, \mathrm{d}^3 r. \tag{3.8}$$

What is the physical interpretation of $\langle A_+ \rangle$ and $\langle A_- \rangle$? Show in particular that $|\psi_+|^2/|\alpha_+|^2$ and $|\psi_-|^2/|\alpha_-|^2$ are probability laws.

**3.3.4** Apply Ehrenfest's theorem to calculate the following quantities:

$$\frac{\mathrm{d}}{\mathrm{d}t}\langle r_+ \rangle, \quad \frac{\mathrm{d}}{\mathrm{d}t}\langle p_\pm \rangle. \tag{3.9}$$

Solve the resulting equations and give the time evolution of $\langle r_\pm \rangle$ and $\langle p_\pm \rangle$. Give the physical interpretation of the result, and explain why one observes a splitting of the initial beam into two beams of relative intensities $|\alpha_+|^2$ and $|\alpha_-|^2$.

**3.3.5** Calculate the splitting between the two beams when they leave the magnet. Express the result in terms of the kinetic energy of the incident neutrons.

Given the experimental error $\delta z$ in the measurement of the position of the maximum intensity of a beam, i.e. $\delta z = 5 \times 10^{-6}$ m as discussed in question 3.1.3, what is the accuracy on the measurement of the neutron magnetic moment in such an experiment, assuming that the determination of the magnetic field and the neutron energy is not a limitation? Compare with the result of magnetic resonance experiments:

$$\mu_0 = (-1.91304184 \pm 8.8 \times 10^{-7}) \, \mu_N. \tag{3.10}$$

**3.3.6** In the same experimental setup, what would be the splitting of a beam of silver atoms (in the original experiment of Stern and Gerlach, the atomic beam came from an oven at 1000 K) of energy $E \simeq 1.38 \times 10^{-20}$ J? The magnetic moment of a silver atom is the same as that of the valence electron $|\mu_e| = q\hbar/2m_e \simeq 9.3 \times 10^{-24}$ J.T$^{-1}$.

**3.3.7** Show that, quite generally, in order to be able to separate the two outgoing beams, the condition to be satisfied is of the form

$$E_\perp t \geq \hbar/2, \tag{3.11}$$

where $E_\perp$ is the transverse kinetic energy acquired by the neutrons in the process, and $t$ is the time they spend in the magnetic field. Comment and conclude.

## 3.4 Solutions

### 3.4.1 Preparation of the Neutron Beam

**3.4.1.1** We have $v = h/(\lambda m)$ and $E = mv^2/2$, which yields $v = 916\,\mathrm{ms}^{-1}$ and $E = 0.438 \times 10^{-2}$ eV.

**3.4.1.2** The contribution of diffraction to the beam width is $\delta_{\mathrm{diff}} = L \tan\theta \sim L\lambda/a$. With the simple additive prescription (which can be improved, but this would not yield very different results), we obtain $\delta = a + L\lambda/a$ which is minimal for $a = \sqrt{L\lambda} \simeq 21$ μm. The spreading of the beam on the detector is then equal to the Heisenberg minimum $\delta = 2\sqrt{L\lambda} = 42$ μm.

The uncertainty relations forbid $\delta$ to be less than some lower limit. In other words, the spreading of the wave packet, which increases as $a$ decreases competes with the spatial definition of the incoming beam.

**3.4.1.3** For $a = 5$ μm, we have $\delta = 91.5$ μm.

In that case, the effect of diffraction is predominant. The reason for making this choice is that the *shape* of the diffraction peak is known and can be fitted quite nicely. Therefore, this is an advantage in determining the position of the maximum. However, one cannot choose $a$ to be too small, otherwise the neutron flux becomes too small, and the number of events is insufficient.

### 3.4.2 Spin State of the Neutrons

**3.4.2.1** $P_\pm(t) = \int |\psi_\pm(\boldsymbol{r}, t)|^2 \, \mathrm{d}^3 r$
 N.B. The normalization condition (total probability equal to 1) is

$$P_+ + P_- = 1 \Rightarrow \int \left( |\psi_+(\boldsymbol{r}, t)|^2 + |\psi_-(\boldsymbol{r}, t)|^2 \right) \mathrm{d}^3 r = 1. \tag{3.12}$$

The quantity $|\psi_+(\boldsymbol{r}, t)|^2 + |\psi_-(\boldsymbol{r}, t)|^2$ is the probability density of finding the neutron at point $\boldsymbol{r}$.

**3.4.2.2** By definition, the expectation value of $S_x$ is $\langle S_x \rangle = (\hbar/2)\langle \psi | \hat{\sigma}_x | \psi \rangle$ therefore

$$\langle S_x \rangle = \frac{\hbar}{2} \int (\psi_+^*(r, t)\psi_-(r, t) + \psi_-^*(r, t)\psi_+(r, t)) \, \mathrm{d}^3 r. \tag{3.13}$$

**3.4.2.3** Similarly

$$\langle r \rangle = \int r(|\psi_+(r, t)|^2 + |\psi_-(r, t)|^2) \, \mathrm{d}^3 r,$$

$$\langle p \rangle = \frac{\hbar}{i} \int (\psi_+^*(r, t)\nabla \psi_+(r, t) + \psi_-^*(r, t)\nabla \psi_-(r, t)) \, \mathrm{d}^3 r. \tag{3.14}$$

**3.4.2.4** If the variables are factorized, we have the simple results:

$$\langle S_x \rangle = \hbar \, \mathcal{R}e \, (\alpha_+^* \alpha_-) \tag{3.15}$$

and

$$\langle r \rangle = \int r |\psi(r, t)|^2 \mathrm{d}^3 r, \quad \langle p \rangle = \frac{\hbar}{i} \int \psi^*(r, t)\nabla \psi(r, t) \, \mathrm{d}^3 r. \tag{3.16}$$

### 3.4.3 The Stern-Gerlach Experiment

**3.4.3.1** The matrix form of the Hamiltonian is

$$\hat{H} = \frac{\hat{p}^2}{2m} \begin{pmatrix} 1 & 0 \\ 0 & 1 \end{pmatrix} - \mu_0 (B_0 + b'\hat{z}) \begin{pmatrix} 1 & 0 \\ 0 & -1 \end{pmatrix}. \tag{3.17}$$

The Schrödinger equation is

$$i\hbar \frac{\mathrm{d}}{\mathrm{d}t} |\psi(t)\rangle = \hat{H} |\psi(t)\rangle. \tag{3.18}$$

If we write it in terms of the coordinates $\psi_\pm$ we obtain the uncoupled set

$$i\hbar \frac{\partial}{\partial t} \psi_+(r, t) = -\frac{\hbar^2}{2m} \Delta \psi_+ - \mu_0 (B_0 + b'z)\psi_+$$

$$i\hbar \frac{\partial}{\partial t} \psi_-(r, t) = -\frac{\hbar^2}{2m} \Delta \psi_- + \mu_0 (B_0 + b'z)\psi_- \tag{3.19}$$

or, equivalently $i\hbar \frac{d}{dt}|\psi_\pm\rangle = \hat{H}_\pm|\psi_\pm\rangle$, with

$$\hat{H}_\pm = -\frac{\hbar^2}{2m}\Delta \mp \mu_0(B_0 + b'z). \tag{3.20}$$

In other words, we are dealing with two uncoupled Schrödinger equations, where the potentials have opposite values. This is basically what causes the Stern–Gerlach splitting.

**3.4.3.2**  Since both $\psi_+$ and $\psi_-$ satisfy Schrödinger equations, and since $\hat{H}_\pm$ are both hermitian, we have the usual properties of Hamiltonian evolution for $\psi_+$ and $\psi_-$ separately, in particular the conservation of the norm. The probability of finding $\mu_z = \pm\mu_0$, and the expectation value of $\mu_z$ are both time independent.

**3.4.3.3**  By definition, we have

$$\int |\psi_\pm(r, t)|^2 \, d^3r = |\alpha_\pm|^2, \tag{3.21}$$

where $|\alpha_\pm|^2$ is time independent. The quantities $|\psi_+(r, t)|^2/|\alpha_+|^2$ and $|\psi_-(r, t)|^2/|\alpha_-|^2$ are the probability densities for finding a neutron at position $r$ with, respectively, $S_z = +\hbar/2$ and $S_z = -\hbar/2$.

The quantities $\langle A_+\rangle$ and $\langle A_-\rangle$ are the expectation values of the physical quantity $A$ for neutrons which have, respectively, $S_z = +\hbar/2$ and $S_z = -\hbar/2$.

**3.4.3.4**  Applying Ehrenfest's theorem, one has for any observable

$$\frac{d}{dt}\langle A_\pm\rangle = \frac{1}{i\hbar|\alpha_\pm|^2}\int \psi_\pm^*(r, t)[\hat{A}, \hat{H}_\pm]\psi_\pm(r, t) \, d^3r. \tag{3.22}$$

Therefore

$$\frac{d}{dt}\langle r_\pm\rangle = \langle p_\pm\rangle/m \tag{3.23}$$

and

$$\frac{d}{dt}\langle p_{x\pm}\rangle = \frac{d}{dt}\langle p_{y\pm}\rangle = 0 \quad \frac{d}{dt}\langle p_{z\pm}\rangle = \pm\mu_0 b'. \tag{3.24}$$

The solution of these equations is straightforward:

$$\langle p_{x\pm}\rangle = p_0, \quad \langle p_{y\pm}\rangle = 0, \quad \langle p_{z\pm}\rangle = \pm\mu_0 b't$$

$$\langle x_\pm\rangle = \frac{p_0 t}{m} = vt, \quad \langle y_\pm\rangle = 0, \quad \langle z_\pm\rangle = \pm\frac{\mu_0 b' t^2}{2m}.$$

Consequently, the expectation values of the vertical positions of the neutrons which have $\mu_z = +\mu_0$ and $\mu_z = -\mu_0$ diverge as time progresses: there is a separation in space of the support of the two wave functions $\psi_+$ and $\psi_-$. The intensities of the two outgoing beams are proportional to $|\alpha_+|^2$ and $|\alpha_-|^2$.

**3.4.3.5** As the neutrons leave the magnet, one has $\langle x \rangle = L$, therefore $t = L/v$ and $\Delta z = |\mu_0 b'| L^2 / mv^2 = |\mu_0 b'| L^2 / 2E$ where $E$ is the energy of the incident neutrons.

This provides a splitting of $\Delta z = 0.69$ mm. The error on the position of each beam is $\delta z = 5 \mu$m, that is to say a relative error on the splitting of the beams, or equivalently, on the measurement of $\mu_0$

$$\frac{\delta \mu_0}{\mu_0} \simeq \frac{\sqrt{2}\delta z}{\Delta z} \sim 1.5\%, \tag{3.25}$$

which is far from the accuracy of magnetic resonance measurements.

**3.4.3.6** For silver atoms, one has $|\mu_0|/2E = 3.4 \times 10^{-4}$ T$^{-1}$. Hence, in the same configuration, one would obtain, for the same value of the field gradient and the same length $L = 1$ m, a separation $\Delta z = 3.4$ cm, much larger than for neutrons. Actually, Stern and Gerlach, in their first experiment, had a much weaker field gradient and their magnet was 20 cm long.

**3.4.3.7** The condition to be satisfied in order to resolve the two outgoing beams is that the distance $\Delta z$ between the peaks should be larger than the full width of each peak (this is a common criterion in optics; by an appropriate inspection of the line shape, one may lower this limit). We have seen in Sect. 3.1 that the absolute minimum for the total beam extension on the detector is $2\sqrt{L\lambda}$, which amounts to a full width at half-maximum $\sqrt{L\lambda}$. In other words, we must have:

$$\Delta z^2 \geq L\lambda. \tag{3.26}$$

In the previous section, we have obtained the value of $\Delta z$, and, by squaring, we obtain $\Delta z^2 = (\mu_0 b')^2 t^4 / m^2$, where $t$ is the time spent traversing the magnet. On the other hand, the transverse kinetic energy of an outgoing neutron is $E_\perp = p_{z\pm}^2 / (2m) = (\mu_0 b')^2 t^2 / (2m)$.

Putting the two previous relations together, we obtain $\Delta z^2 = 2E_\perp t^2 / m$; inserting this result in the first inequality, we obtain

$$E_\perp t \geq h/2, \tag{3.27}$$

where we have used $L = vt$ and $\lambda = h/mv$. This is nothing but one of the many forms of the time-energy uncertainty relation. The right-hand side is not the standard $\hbar/2$ because we have considered a rectangular shape of the incident beam (and not a Gaussian). This brings in an extra factor of $2\pi$.

Physically, this result is interesting in many respects:

(a) First, it shows that the effort that counts in making the experiment feasible is not to improve individually the magnitude of the field gradient, or the length of the apparatus, etc., but the particular combination of the product of the energy transferred to the system and the interaction time of the system with the measuring apparatus.

(b) Secondly, this is a particular example of the fundamental fact stressed by many authors [see, for instance, L.D. Landau and E.M. Lifshitz, *Quantum Mechanics* (Pergamon Press, Oxford, 1965)] that a measurement is never point-like. It has always a finite extension both in space and in time. The Stern–Gerlach experiment is actually a very good example of a measuring apparatus in quantum mechanics since it transfers quantum information – here the spin state of the neutron – into space–time accessible quantities – here the splitting of the outgoing beams.

(c) This time–energy uncertainty relation is encountered in most, if not all quantum measurements. Here it emerges as a consequence of the spreading of the wave packet. It is a simple and fruitful exercise to demonstrate rigorously the above property by calculating directly the time evolution of the following expectation values:

$$\langle z_\pm \rangle, \quad \Delta z^2 = \langle z_\pm^2 \rangle - \langle z_\pm \rangle^2, \quad \langle E_T \rangle = \left\langle \frac{p_{z\pm}^2}{2m} \right\rangle, \quad \langle z_\pm p_\pm + p_\pm z_\pm \rangle.$$

$$(3.28)$$

# Spectroscopic Measurements on a Neutron Beam

**4**

We present here a very precise method for spectroscopic measurements due to Norman Ramsey. The method, using atomic or molecular beams, can be applied to a very large class of problems. We shall analyse it in the specific case of a neutron beam, where it can be used to determine the neutron magnetic moment with high accuracy, by measuring the Larmor precession frequency in a magnetic field $B_0$.

The beam of neutrons is prepared with velocity $v$ along the $x$ axis. It is placed in a constant uniform magnetic field $B_0$ directed along the $z$ axis. We write $|+\rangle$ and $|-\rangle$ for the eigenstates of the $z$ projection $\hat{S}_z$ of the neutron spin, and $\gamma$ for the gyromagnetic ratio of the neutron: $\hat{\boldsymbol{\mu}} = \gamma \hat{\boldsymbol{S}}$, $\hat{\boldsymbol{\mu}}$ being the neutron magnetic moment operator, and $\hat{\boldsymbol{S}}$ its spin.

The neutrons are initially in the state $|-\rangle$. When they approach the origin, they cross a zone where an oscillating field $B_1(t)$ is applied in the $(x, y)$ plane. The components of $B_1$ are

$$B_{1x} = B_1\, e^{-r/a} \cos \omega(t - z/c)$$

$$B_{1y} = B_1\, e^{-r/a} \sin \omega(t - z/c) \qquad (4.1)$$

$$B_{1z} = 0\,,$$

where $r = \sqrt{x^2 + y^2}$. We assume that $B_1$ is constant (strictly speaking it should vary in order to satisfy $\nabla \cdot \boldsymbol{B} = 0$) and that $B_1 \ll B_0$.

In all parts of the chapter, the neutron motion in space is treated classically as a linear uniform motion. We are only interested in the quantum evolution of the spin state.

© Springer Nature Switzerland AG 2019
J.-L. Basdevant, J. Dalibard, *The Quantum Mechanics Solver*,
https://doi.org/10.1007/978-3-030-13724-3_4

## 4.1    The Ramsey Method of Separated Oscillatory Fields

**4.1.1**  Consider a neutron whose motion in space is $x = vt, y = 0, z = 0$. What is the Hamiltonian $\hat{H}(t)$ describing the coupling of the neutron magnetic moment with the fields $\boldsymbol{B}_0$ and $\boldsymbol{B}_1$? Setting $\omega_0 = -\gamma B_0$ and $\omega_1 = -\gamma B_1$, write the matrix representation of $\hat{H}(t)$ in the basis $\{|+\rangle, |-\rangle\}$.

**4.1.2**  Treating $\boldsymbol{B}_1$ as a perturbation, calculate, in first order time-dependent perturbation theory, the probability of finding the neutron in the state $|+\rangle$ at time $t = +\infty$ (far from the interaction zone) if it was in the state $|-\rangle$ at $t = -\infty$.

One measures the flux of neutrons which have flipped their spins, and are in the state $|+\rangle$ when they leave the field zone. This flux is proportional to the probability $P_{-+}$ that they have undergone the above transition. Show that this probability has a resonant behavior as a function of the applied angular frequency $\omega$. Plot $P_{-+}$ as a function of the distance from the resonance $\omega - \omega_0$. How does the width of the resonance curve vary with $v$ and $a$?

The existence of this width puts a limit on the accuracy of the measurement of $\omega_0$, and therefore of $\gamma$. Is there an explanation of this on general grounds?

**4.1.3**  On the path of the beam, one adds a second zone with an oscillating field $\boldsymbol{B}_1'$. This second zone is identical to the first but is translated along the $x$ axis by a distance $b(b \gg a)$:

$$B_{1x}' = B_1 \, e^{-r'/a} \cos \omega(t - z/c)$$

$$B_{1y}' = B_1 \, e^{-r'/a} \sin \omega(t - z/c) \qquad (4.2)$$

$$B_{1z}' = 0,$$

where $r' = ((x - b)^2 + y^2)^{1/2}$.

Show that the transition probability $P_{-+}$ across the two zones can be expressed in a simple way in terms of the transition probability calculated in the previous question.

Why is it preferable to use a setup with two zones separated by a distance $b$ rather than a single zone, as in question 4.1.2, if one desires a good accuracy in the measurement of the angular frequency $\omega_0$? What is the order of magnitude of the improvement in the accuracy?

**4.1.4**  What would be the probability $P_{-+}$ if one were to use $N$ zones equally spaced by the same distance $b$ from one another? What optical system is this reminiscent of?

**4.1.5** Suppose now that the neutrons, still in the initial spin state $|-\rangle$, propagate along the $z$ axis instead of the $x$ axis. Suppose that the length of the interaction zone is $b$, i.e. that the oscillating field is given by (4.1) for $-b/2 \leq z \leq +b/2$ and is zero for $|z| > b/2$. Calculate the transition probability $P'_{-+}$ in this new configuration.

For what value of $\omega$ is this probability maximum? Explain the difference with the result obtained in question 4.1.2.

**4.1.6** In practice, the neutron beam has some velocity dispersion around the value $v$. Which of the two methods described in questions 4.1.3 and 4.1.5 is preferable?

**4.1.7** The neutrons of the beam have a de Broglie wavelength $\lambda_n = 31$ Å. Calculate their velocity. In order to measure the neutron gyromagnetic ratio $\gamma_n$, one proceeds as in question 4.1.3. One can assume that the accuracy is given by

$$\delta \omega_0 = \frac{\pi}{2} \frac{v}{b}. \tag{4.3}$$

In 1979, Greene, Ramsey and coworkers measured the following value of the neutron gyromagnetic ratio

$$\gamma_n = -1.913\,041\,84\,(88) \times q/M_p \tag{4.4}$$

where $q$ is the unit charge and $M_p$ the proton mass. In a field $B_0 = 1$ T, what must be the length $b$ in order to achieve this accuracy?

---

## 4.2 Solutions

**4.2.1** The magnetic Hamiltonian is

$$\hat{H}(t) = -\hat{\boldsymbol{\mu}}.\boldsymbol{B} = -\gamma \left( B_0 \hat{S}_z + B_{1x}(t) \hat{S}_x + B_{1y}(t) \hat{S}_y \right). \tag{4.5}$$

Since $x = vt$, $y = z = 0$,

$$\hat{H}(t) = -\gamma \left[ B_0 \hat{S}_z + B_1 \, e^{-v|t|/a} \left( \hat{S}_x \cos \omega t + \hat{S}_y \sin \omega t \right) \right] \tag{4.6}$$

whose matrix representation is

$$\hat{H}(t) = \frac{\hbar}{2} \begin{pmatrix} \omega_0 & \omega_1 \exp(-v|t|/a - i\omega t) \\ \omega_1 \exp(-v|t|/a + i\omega t) & -\omega_0 \end{pmatrix}. \tag{4.7}$$

**4.2.2** Let $|\psi(t)\rangle = \alpha(t)|+\rangle + \beta(t)|-\rangle$ be the neutron state at time $t$. The Schrödinger equation gives the evolution of $\alpha$ and $\beta$:

$$i\dot{\alpha} = \frac{\omega_0}{2}\alpha + \frac{\omega_1}{2}e^{-i\omega t - v|t|/a}\beta \tag{4.8}$$

$$i\dot{\beta} = \frac{\omega_1}{2}e^{i\omega t - v|t|/a}\alpha - \frac{\omega_0}{2}\beta. \tag{4.9}$$

We now introduce the variables $\tilde{\alpha}$ and $\tilde{\beta}$:

$$\tilde{\alpha}(t) = \alpha(t)e^{i\omega_0 t/2} \quad \tilde{\beta}(t) = \beta(t)e^{-i\omega_0 t/2}, \tag{4.10}$$

whose evolution is given by

$$i\dot{\tilde{\alpha}} = \frac{\omega_1}{2}e^{i(\omega_0-\omega)t - v|t|/a}\tilde{\beta}$$

$$i\dot{\tilde{\beta}} = \frac{\omega_1}{2}e^{i(\omega-\omega_0)t - v|t|/a}\tilde{\alpha}. \tag{4.11}$$

The equation for $\tilde{\alpha}$ can be formally integrated and it gives

$$\tilde{\alpha}(t) = \frac{\omega_1}{2i}\int_{-\infty}^{t} e^{i(\omega_0-\omega)t' - v|t'|/a}\tilde{\beta}(t')dt', \tag{4.12}$$

where we have used the initial condition $\tilde{\alpha}(-\infty) = \alpha(-\infty) = 0$. Now, since we want the value of $\alpha(t)$ to first order in $B_1$, we can replace $\tilde{\beta}(t')$ by its unperturbed value $\tilde{\beta}(t') = 1$ in the integral. This gives

$$\gamma_{-+} \equiv \tilde{\alpha}(+\infty) = \frac{\omega_1}{2i}\int_{-\infty}^{+\infty} e^{i(\omega_0-\omega)t' - v|t'|/a}dt'$$

$$= \frac{\omega_1 v}{ia}\frac{1}{(\omega-\omega_0)^2 + (v/a)^2}. \tag{4.13}$$

The transition probability is therefore (Fig. 4.1)

$$P_{-+} = \frac{\omega_1^2 v^2}{a^2}\frac{1}{[(\omega_0-\omega)^2 + (v/a)^2]^2}. \tag{4.14}$$

The width of the resonance curve is of the order of $v/a$. This quantity is the inverse of the time $\tau = a/v$ a neutron spends in the oscillating field. From the uncertainty relation $\delta E.\tau \sim \hbar$, when an interaction lasts a finite time $\tau$ the accuracy of the energy measurement $\delta E$ is bounded by $\delta E \geq \hbar/\tau$. Therefore, from first principles, one expects that the resonance curve will have a width of the order of $\hbar/\tau$ in energy, or $1/\tau$ in angular frequency.

Fig. 4.1 Transition
probability $P_{-+}$ in one zone

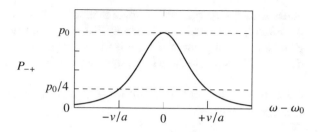

Fig. 4.2 Ramsey fringes in a
two-zone setup

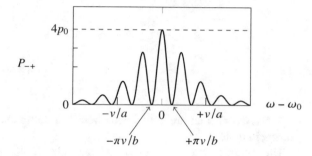

**4.2.3** In the two-zone case, the transition amplitude (in first order perturbation theory) becomes

$$\gamma_{-+} = \frac{\omega_1}{2i} \left( \int_{-\infty}^{+\infty} e^{i(\omega_0-\omega)t-v|t|/a} dt + \int_{-\infty}^{+\infty} e^{i(\omega_0-\omega)t-|vt-b|/a} dt \right). \qquad (4.15)$$

If we make the change of variables $t' = t - b/v$ in the second integral, we obtain

$$\gamma_{-+} = \frac{\omega_1}{2i} \left( 1 + e^{i(\omega_0-\omega)b/v} \right) \int_{-\infty}^{+\infty} e^{i(\omega_0-\omega)t-v|t|/a} dt, \qquad (4.16)$$

which is the same formula as previously but multiplied by $1 + e^{i(\omega_0-\omega)b/v}$. If we square this expression, in order to find the probability, we obtain (Fig. 4.2)

$$P_{-+} = \frac{4\omega_1^2 v^2}{a^2} \frac{1}{[(\omega_0-\omega)^2 + v^2/a^2]^2} \cos^2 \left( \frac{(\omega_0-\omega)b}{2v} \right). \qquad (4.17)$$

The envelope of this curve is, up to a factor of 4, the same as the previous curve. However, owing to the extra oscillating factor, the half-width at half-maximum of the central peak is now of order $\pi v/(2b)$. The parameter which now governs the accuracy is the *total* time $b/v$ that the neutron spends in the apparatus, going from one zone to the other.

In spectroscopic measurements, it is important to locate the exact position of the maximum of the peak. Multiplying the width of the peak by a factor $a/b(\ll 1$ since $a \ll b)$ results in a major improvement of the measurement accuracy. Of course one could in principle build a single interaction zone of large size $\sim b$, but it would be difficult to maintain a well controlled oscillating field over such a large region. From a practical point of view, it is much simpler to use small interaction zones of size $a$ and to separate them by a large distance $b$.

**4.2.4** It is quite straightforward to generalize the previous results to an arbitrary number of zones:

$$\gamma_{-+} = \frac{\omega_1}{2i} \left( 1 + e^{i(\omega_0-\omega)b/v} + \cdots + e^{i(N-1)(\omega_0-\omega)b/v} \right) \tag{4.18}$$

$$\times \int_{-\infty}^{+\infty} e^{i(\omega_0-\omega)t - v|t|/a} \, dt$$

$$P_{-+} = \frac{\omega_1^2 v^2}{a^2} \frac{1}{[(\omega_0-\omega)^2 + v^2/a^2]^2} \frac{\sin^2[N(\omega_0-\omega)b/2v]}{\sin^2[(\omega_0-\omega)b/2v]}. \tag{4.19}$$

As far as amplitudes are concerned, there is a complete analogy with a diffraction grating in optics.

The neutron (more generally, the particle or the atom) has some transition amplitude $t$ for undergoing a spin flip in a given interaction zone. The total amplitude $T$ is the sum

$$T = t + t\, e^{i\phi} + t\, e^{2i\phi} + \dots , \tag{4.20}$$

where $e^{i\phi}$ is the phase shift between two zones.

**4.2.5** We now set $z = vt$, and $x = y = 0$ for the neutron trajectory. This will modify the phase of the field (Doppler effect)

$$\omega(t - z/c) \to \omega(1 - v/c)t = \tilde{\omega}t \quad \text{with} \quad \tilde{\omega} = \omega(1 - v/c) \tag{4.21}$$

and we must integrate the evolution of $\tilde{\alpha}$:

$$i\dot{\tilde{\alpha}} = \frac{\omega_1}{2} e^{i(\omega_0-\tilde{\omega})t} \tilde{\beta} \tag{4.22}$$

(with $\tilde{\beta} \simeq 1$) between $t_i = -b/(2v)$ and $t_f = b/(2v)$. The transition probability is then

$$P'_{-+} = \omega_1^2 \frac{\sin^2[(\omega_0 - \tilde{\omega})b/(2v)]}{(\omega_0 - \tilde{\omega})^2}, \tag{4.23}$$

which has a width of the order of $b/v$ but is centered at

$$\tilde{\omega} = \omega_0 \Rightarrow \omega = \frac{\omega_0}{1 - v/c} \simeq \omega_0(1 + v/c). \tag{4.24}$$

Comparing with question 4.1.2, we find that the resonance frequency is displaced: The neutron moves in the propagation direction of the field, and there is a first order Doppler shift of the resonance frequency.

**4.2.6** If the neutron beam has some velocity dispersion, the experimental result will be the same as calculated above, but smeared over the velocity distribution.

In the method of question 4.1.3, the position of lateral fringes, and the width of the central peak, vary with $v$. A velocity distribution will lead to a broader central peak and lateral fringes of decreasing amplitude. However the *position* of the central peak does not depend on the velocity, and it is therefore not shifted if the neutron beam has some velocity dispersion.

On the contrary, in the method of question 4.1.5, the position of the central peak depends directly on the velocity. A dispersion in $v$ will lead to a corresponding dispersion of the *position* of the peak we want to measure. The first method is highly preferable.

**4.2.7** For $\lambda_n = 31$ Å, one finds $v = h/(M_n \lambda_n) \simeq 128$ m/s. Experimentally, one obtains an accuracy $\delta\omega_0/\omega_0 = \delta\gamma_n/\gamma_n = 4.6 \times 10^{-7}$. For $B = 1$ T, the angular frequency is $\omega_0 = \gamma_n B_0 \simeq 1.8 \times 10^8$ s$^{-1}$, which gives $\delta\omega_0/(2\pi) \simeq 13$ Hz and $b \simeq 2.4$ m.

Actually, one can improve the accuracy considerably by analysing the shape of the peak. In the experiment reported in the reference quoted below, the length $b$ is 2 m and the field is $B_0 = 0.05$ T (i.e. an angular frequency 20 times smaller than above).

**References** The measurement of the neutron magnetic moment by this method is reported by G. L. Greene, N. F. Ramsey, W. Mampe, J. M. Pendlebury, K. F. Smith, W. D. Dress, P. D. Miller, and P. Perrin, *Measurement of the neutron magnetic moment*, Phys. Rev. D **20**, 2139 (1979).

Norman Ramsey was awarded the Nobel Prize in Physics in 1989. More applications of the separated oscillatory field method are described in his Nobel lecture: *Experiments with separated oscillatory fields and hydrogen masers*, Rev. Mod. Phys. **62**, 541 (1990).

# Measuring the Electron Magnetic Moment Anomaly

**5**

In the framework of the Dirac equation, the ratio between the magnetic moment $\mu$ and the spin $S$ of the electron is $q/m$, where $q$ and $m$ are the charge and the mass of the particle. In other words, the gyromagnetic factor $g$ of the electron, defined as $\mu = g\frac{q}{2m}S$, is equal to 2. When one takes into account the interaction of the electron with the quantized electromagnetic field, one predicts a value of $g$ slightly different from 2. The purpose of this chapter is to present an experiment aiming at measuring the quantity $g - 2$.

## 5.1 Spin and Momentum Precession in a Magnetic Field

Consider an electron, of mass $m$ and charge $q$ ($q < 0$), placed in a uniform and static magnetic field $B$ directed along the $z$ axis. The Hamiltonian of the electron is

$$\hat{H} = \frac{1}{2m}(\hat{p} - q\hat{A})^2 - \hat{\mu} \cdot B, \tag{5.1}$$

where $\hat{A}$ is the vector potential $\hat{A} = B \times \hat{r}/2$ and $\hat{\mu}$ is the intrinsic magnetic moment operator of the electron. This magnetic moment is related to the spin operator $\hat{S}$ by $\hat{\mu} = \gamma\hat{S}$, with $\gamma = (1 + a)q/m$. The quantity $a$ is called the magnetic moment "anomaly". In the framework of the Dirac equation, $a = 0$. Using quantum electrodynamics, one predicts at first order in the fine structure constant $a = \alpha/(2\pi)$.

The velocity operator is $\hat{v} = (\hat{p} - q\hat{A})/m$, and we set $\omega = qB/m$.

**5.1.1** Verify the following commutation relations:

$$[\hat{v}_x, \hat{H}] = i\hbar\omega\,\hat{v}_y; \quad [\hat{v}_y, \hat{H}] = -i\hbar\omega\,\hat{v}_x; \quad [\hat{v}_z, \hat{H}] = 0. \tag{5.2}$$

© Springer Nature Switzerland AG 2019
J.-L. Basdevant, J. Dalibard, *The Quantum Mechanics Solver*,
https://doi.org/10.1007/978-3-030-13724-3_5

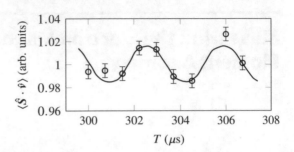

**Fig. 5.1** Variations of the quantity $\langle \hat{S}.\hat{v} \rangle$, as a function of the time $T$

**5.1.2** Consider the three quantities

$$C_1(t) = \langle \hat{S}_z \hat{v}_z \rangle, \quad C_2(t) = \langle \hat{S}_x \hat{v}_x + \hat{S}_y \hat{v}_y \rangle, \quad C_3(t) = \langle \hat{S}_x \hat{v}_y - \hat{S}_y \hat{v}_x \rangle. \quad (5.3)$$

Write the time evolution equations for $C_1, C_2, C_3$. Show that these three equations form a linear differential system with constant coefficients. One will make use of the quantity $\Omega = a\omega$.

**5.1.3** What is the general form for the evolution of $\langle \hat{S} \cdot \hat{v} \rangle$?

**5.1.4** A beam of electrons of velocity $v$ is prepared at time $t = 0$ in a spin state such that one knows the values of $C_1(0)$, $C_2(0)$, and $C_3(0)$. The beam interacts with the magnetic field $B$ during the time interval $[0, T]$. One neglects the interactions between the electrons of the beam. At time $T$, one measures a quantity which is proportional to $\langle \hat{S}.\hat{v} \rangle$.

The result of such a measurement is presented in Fig. 5.1 as a function of the time $T$, for a value of the magnetic field $B = 9.4 \times 10^{-3}$ T (data taken from D.T. Wilkinson and H.R. Crane, Phys. Rev. **130**, 852 (1963)). Deduce from this curve an approximate value for the anomaly $a$.

**5.1.5** Does the experimental value agree with the prediction of quantum electrodynamics?

## 5.2    Solutions

**5.2.1** The electron Hamiltonian is $\hat{H} = m\hat{v}^2/2 - \gamma B \hat{S}_z$. The following commutation relations can be established with no difficulty

$$[\hat{v}_x, \hat{v}_y] = i\hbar q B/m^2 = i\hbar\omega/m, \quad [\hat{v}_x, \hat{v}_z] = [\hat{v}_y, \hat{v}_z] = 0 \quad (5.4)$$

and

$$[\hat{v}_x, \hat{v}_y^2] = [\hat{v}_x, \hat{v}_y]\hat{v}_y + \hat{v}_y[\hat{v}_x, \hat{v}_y] = 2i\hbar\omega \, \hat{v}_y/m. \quad (5.5)$$

Therefore

$$[\hat{v}_x, \hat{H}] = i\hbar\omega\hat{v}_y \; ; \quad [\hat{v}_y, \hat{H}] = -i\hbar\omega\hat{v}_x \; ; \quad [\hat{v}_z, \hat{H}] = 0. \tag{5.6}$$

**5.2.2** We make use of the property $i\hbar(d/dt)\langle\hat{O}\rangle = \langle[\hat{O}, \hat{H}]\rangle$, valid for any observable (Ehrenfest theorem). The time evolution of $C_1$ is trivial:

$$[\hat{S}_z\hat{v}_z, \hat{H}] = 0 \Rightarrow \quad \frac{dC_1}{dt} = 0 \; ; \quad C_1(t) = A_1, \tag{5.7}$$

where $A_1$ is a constant. For $C_2$ and $C_3$, we proceed in the following way:

$$[\hat{S}_x\hat{v}_x, \hat{H}] = [\hat{S}_x\hat{v}_x, m\hat{v}^2/2] - \gamma B[\hat{S}_x\hat{v}_x, \hat{S}_z] = i\hbar\omega(\hat{S}_x\hat{v}_y + (1+a)\hat{S}_y\hat{v}_x). \tag{5.8}$$

Similarly,

$$\left[\hat{S}_y\hat{v}_y, \hat{H}\right] = -i\hbar\omega(\hat{S}_y\hat{v}_x + (1+a)\hat{S}_x\hat{v}_y)$$

$$\left[\hat{S}_x\hat{v}_y, \hat{H}\right] = -i\hbar\omega(\hat{S}_x\hat{v}_x - (1+a)\hat{S}_y\hat{v}_y)$$

$$\left[\hat{S}_y\hat{v}_x, \hat{H}\right] = i\hbar\omega(\hat{S}_y\hat{v}_y - (1+a)\hat{S}_x\hat{v}_x). \tag{5.9}$$

Therefore,

$$\left[\hat{S}_x\hat{v}_x + \hat{S}_y\hat{v}_y, \hat{H}\right] = -i\hbar\omega a(\hat{S}_x\hat{v}_y - \hat{S}_y\hat{v}_x)$$

$$\left[\hat{S}_x\hat{v}_y - \hat{S}_y\hat{v}_x, \hat{H}\right] = i\hbar\omega a(\hat{S}_x\hat{v}_x + \hat{S}_y\hat{v}_y) \tag{5.10}$$

and

$$\frac{dC_2}{dt} = -\Omega C_3, \quad \frac{dC_3}{dt} = \Omega C_2. \tag{5.11}$$

**5.2.3** We therefore obtain $d^2C_2/dt^2 = -\Omega^2 C_2$, whose solution is

$$C_2(t) = A_2 \cos(\Omega t + \varphi), \tag{5.12}$$

where $A_2$ and $\varphi$ are constant. Hence, the general form of the evolution of $\langle \boldsymbol{S} \cdot \boldsymbol{v}\rangle$ is

$$\langle \boldsymbol{S} \cdot \boldsymbol{v}\rangle(t) = C_1(t) + C_2(t) = A_1 + A_2 \cos(\Omega t + \varphi). \tag{5.13}$$

In other words, in the absence of anomaly, the spin and the momentum of the electron would precess with the same angular velocity: the cyclotron frequency (precession of momentum) and the Larmor frequency (precession of magnetic moment) would be equal. Measuring the difference in these two frequencies gives a direct measurement of the anomaly $a$, of fundamental importance in quantum electrodynamics.

**5.2.4** One calculates the anomaly from the relation $a = \Omega/\omega$. The experimental results for $\langle S \cdot v \rangle$ show a periodic behavior in time with a period $\tau \sim 3\ \mu$s, i.e. $\Omega = 2\pi/\tau \sim 2 \times 10^6\,\text{s}^{-1}$. In a field $B = 0.0094$ T, $\omega = 1.65 \times 10^9\,\text{s}^{-1}$, and $a = \Omega/\omega \sim 1.2 \times 10^{-3}$.

**5.2.5** This value is in good agreement with the theoretical prediction $a = \alpha/2\pi = 1.16 \times 10^{-3}$.

**Remarks and references** The value of the anomaly is now known with an impressive accuracy using a single electron trapped in a Penning trap:

$$a^{\text{exp.}} = 0.001\,159\,652\,180\,73\,(28), \tag{5.14}$$

as reported by D. Hanneke, S. Fogwell Hoogerheide, and G. Gabrielse: *Cavity control of a single-electron quantum cyclotron: Measuring the electron magnetic moment*, Phys. Rev. A **83**, 052122 (2011).

The most recent calculations of $a$ using quantum electrodynamics (up to 5th order in $\alpha$) are in agreement with this result if one uses the best independent measurement of $\alpha$. On the other hand, if one trusts the theory of quantum electrodynamics, one can use this measurement to obtain the most precise value of $\alpha$ to date:

$$\alpha^{-1} = 137.035\,999\,174\,(35) \tag{5.15}$$

see T. Aoyama, M. Hayakawa, T. Kinoshita, M. Nio, *Tenth-Order QED Contribution to the Electron g − 2 and an Improved Value of the Fine Structure Constant*, Phys. Rev. Lett. **109**, 111807 (2012).

# Atomic Clocks

**6**

We are interested in the ground state of the external electron of an alkali atom (rubidium, cesium . . .). The atomic nucleus has a spin $s_n$ ($s_n = 3/2$ for $^{87}$Rb, $s_n = 7/2$ for $^{133}$Cs), which carries a magnetic moment $\boldsymbol{\mu}_n$. As in the case of atomic hydrogen, the ground state is split by the hyperfine interaction between the electron magnetic moment and the nuclear magnetic moment $\boldsymbol{\mu}_n$. This splitting of the ground state is used to devise atomic clocks of high accuracy, which have numerous applications such as flight control in aircrafts, the G.P.S. system, the measurement of physical constants etc.

In all the chapter, we shall neglect the effects due to internal core electrons.

## 6.1 The Hyperfine Splitting of the Ground State

**6.1.1** Give the degeneracy of the ground state if one neglects the magnetic interaction between the nucleus and the external electron. We note

$$|m_e; m_n\rangle = |\text{electron}: s_e = 1/2, m_e\rangle \otimes |\text{nucleus}: s_n, m_n\rangle \qquad (6.1)$$

a basis of the total spin states (external electron + nucleus).

**6.1.2** We now take into account the interaction between the electron magnetic moment $\boldsymbol{\mu}_e$ and the nuclear magnetic moment $\boldsymbol{\mu}_n$. As in the hydrogen atom, one can write the corresponding Hamiltonian (restricted to the spin subspace) as:

$$\hat{H} = \frac{A}{\hbar^2} \hat{S}_e \cdot \hat{S}_n, \qquad (6.2)$$

© Springer Nature Switzerland AG 2019
J.-L. Basdevant, J. Dalibard, *The Quantum Mechanics Solver*,
https://doi.org/10.1007/978-3-030-13724-3_6

where $A$ is a characteristic energy, and where $\hat{S}_e$ and $\hat{S}_n$ are the spin operators of the electron and the nucleus, respectively. We want to find the eigenvalues of this Hamiltonian.

We introduce the operators $\hat{S}_{e,\pm} = \hat{S}_{e,x} \pm i\hat{S}_{e,y}$ and $\hat{S}_{n,\pm} = \hat{S}_{n,x} \pm i\hat{S}_{n,y}$.

(a) Show that

$$\hat{H} = \frac{A}{2\hbar^2} \left( \hat{S}_{e,+}\hat{S}_{n,-} + \hat{S}_{e,-}\hat{S}_{n,+} + 2\hat{S}_{e,z}\hat{S}_{n,z} \right). \tag{6.3}$$

(b) Show that the two states

$$|m_e = 1/2; m_n = s_n\rangle \quad \text{and} \quad |m_e = -1/2; m_n = -s_n\rangle \tag{6.4}$$

are eigenstates of $\hat{H}$, and give the corresponding eigenvalues.
(c) What is the action of $\hat{H}$ on the state $|m_e = 1/2; m_n\rangle$ with $m_n \neq s_n$?

What is the action of $\hat{H}$ on the state $|m_e = -1/2; m_n\rangle$ with $m_n \neq -s_n$?
(d) Deduce from these results that the eigenvalues of $\hat{H}$ can be calculated by diagonalizing $2 \times 2$ matrices of the type:

$$\frac{A}{2} \begin{pmatrix} m_n & \sqrt{s_n(s_n+1) - m_n(m_n+1)} \\ \sqrt{s_n(s_n+1) - m_n(m_n+1)} & -(m_n+1) \end{pmatrix}. \tag{6.5}$$

**6.1.3** Show that $\hat{H}$ splits the ground state in two substates of energies $E_1 = E_0 + As_n/2$ and $E_2 = E_0 - A(1 + s_n)/2$. Recover the particular case of the hydrogen atom.

**6.1.4** What are the degeneracies of the two sublevels $E_1$ and $E_2$?

**6.1.5** Show that the states of energies $E_1$ and $E_2$ are eigenstates of the square of the total spin $\hat{S}^2 = \left( \hat{S}_e + \hat{S}_n \right)^2$. Give the corresponding value $s$ of the spin.

## 6.2    The Atomic Fountain

The atoms are initially prepared in the energy state $E_1$, and are sent upwards (Fig. 6.1). When they go up and down they cross a cavity where an electromagnetic wave of frequency $\omega$ is injected. This frequency is close to $\omega_0 = (E_1 - E_2)/\hbar$. At the end of the descent, one detects the number of atoms which have flipped from the $E_1$ level to the $E_2$ level. In all what follows, the motion of the atoms in space (free fall) is treated classically. It is only the evolution of their internal state which is treated quantum-mechanically.

**Fig. 6.1** Sketch of the principle of an atomic clock with an atomic fountain, using laser-cooled atoms

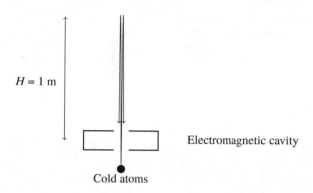

$H = 1$ m

Electromagnetic cavity

Cold atoms

In order to simplify things, we consider only one atom in the sub-level of energy $E_1$. This state (noted $|1\rangle$) is coupled by the electromagnetic wave to only one state (noted $|2\rangle$) of the sublevel of energy $E_2$. By convention, we fix the origin of energies at $(E_1 + E_2)/2$, i.e. $E_1 = \hbar\omega_0/2$, $E_2 = -\hbar\omega_0/2$. We assume that the time $\epsilon$ to cross the cavity is very brief and that this crossing results in an evolution of the state vector of the form:

$$|\psi(t)\rangle = \alpha|1\rangle + \beta|2\rangle \longrightarrow |\psi(t+\epsilon)\rangle = \alpha'|1\rangle + \beta'|2\rangle, \qquad (6.6)$$

with :
$$\begin{pmatrix} \alpha' \\ \beta' \end{pmatrix} = \frac{1}{\sqrt{2}} \begin{pmatrix} 1 & -\mathrm{i}e^{-i\omega t} \\ -\mathrm{i}e^{i\omega t} & 1 \end{pmatrix} \begin{pmatrix} \alpha \\ \beta \end{pmatrix}. \qquad (6.7)$$

**6.2.1** The initial state of the atom is $|\psi(0)\rangle = |1\rangle$. We consider a single round-trip of duration $T$, during which the atom crosses the cavity between $t = 0$ and $t = \epsilon$, then evolves freely during a time $T - 2\epsilon$, and crosses the cavity a second time between $T - \epsilon$ and $T$. Taking the limit $\epsilon \to 0$, show that the state of the atom after this round-trip is given by:

$$|\psi(T)\rangle = \mathrm{i}e^{-i\omega T/2} \, \sin((\omega - \omega_0)T/2) \, |1\rangle - \mathrm{i}e^{i\omega T/2} \, \cos((\omega - \omega_0)T/2) \, |2\rangle \qquad (6.8)$$

**6.2.2** Give the probability $P(\omega)$ to find an atom in the state $|2\rangle$ at time $T$. Determine the half-width $\Delta\omega$ of $P(\omega)$ at the resonance $\omega = \omega_0$. What is the values of $\Delta\omega$ for a 1 m high fountain? We recall the acceleration of gravity $g = 9.81$ ms$^{-2}$.

**6.2.3** We send a pulse of $N$ atoms ($N \gg 1$). After the round-trip, each atom is in the state given by (6.8). We measure separately the numbers of atoms in the states $|1\rangle$ and $|2\rangle$, which we note $N_1$ and $N_2$ (with $N_1 + N_2 = N$). What is the statistical distribution of the random variables $N_1$ and $N_2$? Give their mean values and their r.m.s. deviations $\Delta N_i$. Set $\phi = (\omega - \omega_0)T/2$ and express the results in terms of $\cos\phi$, $\sin\phi$ and $N$.

**Fig. 6.2** Relative accuracy $\Delta\omega/\omega$ of a fountain atomic clock as a function of the number of atoms $N$ sent in each pulse

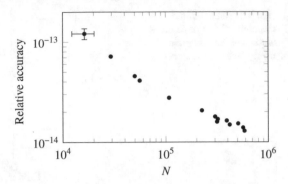

**6.2.4** The departure from resonance $|\omega - \omega_0|$ is characterized by the value of $\cos((\omega - \omega_0)T) = \langle N_2 - N_1 \rangle/N$. Justify this formula. Evaluate the uncertainty $\Delta|\omega - \omega_0|$ introduced by the random nature of the variable $N_2 - N_1$. Show that this uncertainty depends on $N$, but not on $\phi$.

**6.2.5** In Fig. 6.2 we have represented the precision of an atomic clock as a function of the number $N$ of atoms per pulse. Does this variation with $N$ agree with the previous results?

## 6.3 The GPS System

The GPS system uses 24 satellites orbiting around the Earth at 20,000 km. Each of them contains an atomic clock. Each satellite sends, at equal spaced time intervals, an electromagnetic signal composed of a "click" from a clock and the indication of its position. A reception device on Earth, which does not have an atomic clock, detects the signals coming from several satellites. With its own (quartz) clock, it compares the times at which different "clicks" arrive.

**6.3.1** What is the minimum number of satellites that one must see at a given time in order to be able to position oneself in latitude, in longitude, and in altitude on the surface of the Earth?

**6.3.2** We assume that the relative accuracy of each clock is $\Delta\omega/\omega = 10^{-13}$ and that the clocks are synchronized every 24 h. What is the order of magnitude of the accuracy of the positioning just before the clocks undergo a new synchronization?

## 6.4    The Drift of Fundamental Constants

Some cosmological models predict a (small) variation in time of the fine structure constant $\alpha = e^2/(\hbar c) \sim 1/137$. In order to test such an assumption, one can compare two atomic clocks, one using rubidium ($Z = 37$) atoms, the other cesium ($Z = 55$) atoms. In fact, one can show that the hyperfine splitting of an alkali atom varies approximately as:

$$E_1 - E_2 = \hbar\omega_0 \propto \alpha^2 \left(1 + \frac{11}{6}(\alpha Z)^2\right) \quad \text{for } (\alpha Z)^2 \ll 1. \tag{6.9}$$

By comparing a rubidium and a cesium clock at a one year interval, no significant variation of the ratio $R = \omega_0^{(Cs)}/\omega_0^{(Rb)}$ was observed. More precisely, the relative variation $|\delta R|/R$ is smaller than the experimental uncertainty, estimated to be $3 \times 10^{-15}$. What upper bound can one set on the relative variation rate $|\dot{\alpha}/\alpha|$?

## 6.5    Solutions

### 6.5.1  Hyperfine Splitting of the Ground State

**6.5.1.1** The Hilbert space of the ground state is the tensor product of the electron spin space and the nucleus spin space. Its dimension $d$ is therefore the product of their dimensions, i.e. $d = 2 \times (2s_n + 1)$.

**6.5.1.2** Energy levels of the hyperfine Hamiltonian.

**(a)** Making use of

$$\hat{S}_{e,x} = \frac{1}{2}\left(\hat{S}_{e,+} + \hat{S}_{e,-}\right), \quad \hat{S}_{e,y} = \frac{i}{2}\left(\hat{S}_{e,-} - \hat{S}_{e,+}\right), \tag{6.10}$$

and a similar relation for $\hat{S}_{n,x}$ and $\hat{S}_{n,y}$, one obtains the wanted result.

**(b)** The action of $\hat{S}_{e,+}\hat{S}_{n,-}$ and $\hat{S}_{e,-}\hat{S}_{n,+}$ on $|m_e = 1/2; m_n = s_n\rangle$ gives the null vector. The same holds for $|m_e = -1/2, m_n = -s_n\rangle$. Therefore, only the term $\hat{S}_{e,z}\hat{S}_{n,z}$ contributes and one finds:

$$\hat{H}|m_e = 1/2; m_n = s_n\rangle = \frac{As_n}{2}|m_e = 1/2; m_n = s_n\rangle \tag{6.11}$$

$$\hat{H}|m_e = -1/2; m_n = -s_n\rangle = \frac{As_n}{2}|m_e = -1/2; m_n = -s_n\rangle. \tag{6.12}$$

(c) We find:

$$\hat{H}|1/2; m_n\rangle = \frac{A m_n}{2}|1/2; m_n\rangle \tag{6.13}$$

$$+ \frac{A}{2}\sqrt{s_n(s_n+1) - m_n(m_n+1)}|-1/2; m_n+1\rangle \tag{6.14}$$

$$\hat{H}|-1/2; m_n\rangle = -\frac{A m_n}{2}|-1/2; m_n\rangle \tag{6.15}$$

$$+ \frac{A}{2}\sqrt{s_n(s_n+1) - m_n(m_n-1)}|1/2; m_n-1\rangle. \tag{6.16}$$

(d) From the previous question, one concludes that the 2-dimensional subspaces $\mathcal{E}_{m_n}$ generated by $|1/2; m_n\rangle$ and $|-1/2; m_n+1\rangle$ are globally stable under the action of $\hat{H}$. The determination of the eigenvalues of $\hat{H}$ therefore consists in diagonalizing the series of $2 \times 2$ matrices corresponding to its restriction to these subspaces. The matrix corresponding to the restriction of $\hat{H}$ to the subspace $\mathcal{E}_{m_n}$ is the same as given in the text.

**6.5.1.3** The eigenvalues given in the text are actually independent of $m_n$. They are $A s_n/2$ and $-A(1 + s_n)/2$. In the case $s_n = 1/2$ (hydrogen atom), these two eigenvalues are $A/4$ and $-3A/4$.

**6.5.1.4** There are $2s_n$ $2 \times 2$ matrices to be diagonalized, each of which gives a vector associated to $A s_n/2$ and a vector associated to $- A(1 + s_n)/2$. In addition we have found two independent eigenvectors, $|1/2, s_n\rangle$ and $|-1/2, -s_n\rangle$, associated to the eigenvalue $A s_n/2$. We therefore obtain:

$$A s_n/2 \quad \text{degenerated } 2s_n + 2 \text{ times} \tag{6.17}$$

$$-A(1 + s_n/2) \quad \text{degenerated } 2s_n \text{ times} \tag{6.18}$$

We do recover the dimension $2(2s_n + 1)$ of the total spin space of the ground state.

**6.5.1.5** The square of the total spin is:

$$\hat{S}^2 = \hat{S}_e^2 + \hat{S}_n^2 + 2\hat{S}_e \cdot \hat{S}_n = \hat{S}_e^2 + \hat{S}_n^2 + \frac{2\hbar^2}{A}\hat{H}. \tag{6.19}$$

The operators $\hat{S}_e^2$ and $\hat{S}_n^2$ are proportional to the identity and are respectively:

$$\hat{S}_e^2 = \frac{3\hbar^2}{4} \quad \hat{S}_n^2 = \hbar^2 s_n(s_n + 1). \tag{6.20}$$

An eigenstate of $\hat{H}$ is therefore an eigenstate of $\hat{S}^2$. More precisely, an eigenstate of $\hat{H}$ with eigenvalue $As_n/2$ is an eigenstate of $\hat{S}^2$ with eigenvalue $\hbar^2(s_n + 1/2)(s_n + 3/2)$, corresponding to a total spin $s = s_n + 1/2$. An eigenstate of $\hat{H}$ with eigenvalue $-A(1 + s_n)/2$ is an eigenstate of $\hat{S}^2$ with eigenvalue $\hbar^2(s_n - 1/2)(s_n + 1/2)$, i.e. a total spin $s = s_n - 1/2$.

## 6.5.2 The Atomic Fountain

**6.5.2.1** In the limit $\epsilon \to 0$, the final state vector of the atom is simply the matrix product:

$$\begin{pmatrix} \alpha' \\ \beta' \end{pmatrix} = \frac{1}{2} \begin{pmatrix} 1 & -\mathrm{i}e^{-\mathrm{i}\omega T} \\ -\mathrm{i}e^{\mathrm{i}\omega T} & 1 \end{pmatrix} \begin{pmatrix} e^{-\mathrm{i}\omega_0 T/2} & 0 \\ 0 & e^{\mathrm{i}\omega_0 T/2} \end{pmatrix} \begin{pmatrix} 1 & -\mathrm{i} \\ -\mathrm{i} & 1 \end{pmatrix} \begin{pmatrix} 1 \\ 0 \end{pmatrix}, \qquad (6.21)$$

which corresponds to crossing the cavity, at time $t = 0$, then to a free evolution between $t = 0$ and $t = T$, then a second crossing of the cavity at time $t = T$. We therefore obtain the state vector of the text.

**6.5.2.2** One finds $P(\omega) = |\beta'|^2 = \cos^2((\omega - \omega_0)T/2)$. This probability is equal to 1 if one sits exactly at the resonance ($\omega = \omega_0$). It is $1/2$ if $\omega = \omega_0 \pm \pi/(2T)$. For a round-trip free fall motion of height $H = 1$ m, we have $T = 2\sqrt{2H/g}$, i.e. $T = 0.9$ s and $\Delta\omega = 1.7$ s$^{-1}$.

**6.5.2.3** The detection of each atom gives the result $E_1$ with a probability $\sin^2 \phi$ and $E_2$ with a probability $\cos^2 \phi$. Since the atoms are assumed to be independent, the distributions of the random variables $N_1$ and $N_2$ are binomial laws. We therefore have:

$$\langle N_1 \rangle = N \sin^2 \phi \quad \langle N_2 \rangle = N \cos^2 \phi \quad \Delta N_1 = \Delta N_2 = \sqrt{N}|\cos \phi \sin \phi|. \tag{6.22}$$

**6.5.2.4** We do obtain $\langle N_2 - N_1 \rangle/N = \cos 2\phi = \cos((\omega - \omega_0)T)$. The fluctuation on the variable $N_2 - N_1$ induces a fluctuation on the determination of $\omega - \omega_0$. The two fluctuations are related by:

$$\frac{\Delta(N_2 - N_1)}{N} = 2|\sin(2\phi)| \, \Delta\phi. \tag{6.23}$$

Since $\Delta(N_2 - N_1) = 2 \, \Delta N_2 = \sqrt{N}|\sin 2\phi|$, we deduce $\Delta\phi = 1/(2\sqrt{N})$, or equivalently:

$$\Delta|\omega - \omega_0| = \frac{1}{2T\sqrt{N}}. \tag{6.24}$$

The longer the time $T$ and the larger $N$ are, the better the accuracy.

**6.5.2.5** We notice on Fig. 6.2 that the accuracy of the clock improves like $N^{-1/2}$, as $N$ increases. For $N = 10^6$ and $T = 0.9$ s, the above formula gives $5.6 \times 10^{-4}$ s. The hyperfine frequency of cesium is $\omega_0 = 2\pi \times 9.2$ GHz, which corresponds to $\Delta\omega/\omega \sim 10^{-14}$.

### 6.5.3   The GPS System

**6.5.3.1** One must see at least 4 satellites. With two of them, the difference between the two reception times $t_1$ and $t_2$ of the signals localize the observer on a surface (for instance on a plane at equal distances of the two satellites if $t_1 = t_2$); three satellites localize the observer on a line, and the fourth one determines the position of the observer unambiguously (provided of course that one assumes the observer is not deep inside the Earth or on a far lying orbit).

**6.5.3.2** Suppose a satellite sends a signal at time $t_0$. This signal is received by an observer at a distance $D$ at time $t_1 = t_0 + D/c$. If the clock of the satellite has drifted, the signal is not sent at time $t_0$, but at a slightly different time $t_0'$. The observer whom we assume has a correct time reference from another satellite, interprets the time $t_1 - t_0'$ as a distance $D' = c(t_1 - t_0')$, he therefore makes an error $c(t_0' - t_0)$ on his position. For a clock of relative accuracy $10^{-13}$, the typical drift after 24 h ($= 86,000$ s) is $86,000 \times 10^{-13}$ s, i.e. an error on the position of 2.5 m.

   Note that the atomic clocks boarding the GPS satellites are noticeably less accurate than the fountain cold atom clocks in ground laboratories.

### 6.5.4   The Drift of Fundamental Constants

Using the expression given in the text for the dependence on $\alpha$ of the frequencies $\omega_{Cs}$ and $\omega_{Rb}$, we find that a variation of the ratio $R$ would be related to the variation of $\alpha$ by:

$$\frac{1}{R}\frac{dR}{dt} = \frac{1}{\alpha}\frac{d\alpha}{dt}\left[\frac{11\alpha^2}{3}\frac{Z_{Cs}^2 - Z_{Rb}^2}{(1 + 11(\alpha Z_{Rb})^2/6)(1 + 11(\alpha Z_{Cs})^2/6)}\right]. \qquad (6.25)$$

The quantity inside the brackets is 0.22, which leads to an upper bound of $\dot{\alpha}/\alpha$ of 1. $4 \times 10^{-14}$ per year, i.e. $4.3 \times 10^{-22}$ per second. If we extrapolate this variation time to a time of the order of the age of the universe, this corresponds to a variation of $10^{-4}$. Such an effect should be detectable, in principle, by spectroscopic measurements on very far objects.

   Remark: a more precise determination of the $\alpha$ dependence of $\omega_{Cs}$, for which the approximation $Z\alpha \ll 1$ is not very good, gives for the quantity inside the bracket a value of 0.45.

### 6.5.5 References

The experimental data on the stability of a cold atom clock were obtained by the group of Observatoire de Paris: G. Santarelli, P. Laurent, P. Lemonde, A. Clairon, A. G. Mann, S. Chang, A. N. Luiten, and C. Salomon, *Quantum Projection Noise in an Atomic Fountain: A High Stability Cesium Frequency Standard*, Phys. Rev. Lett. **82**, 4619 (1999).

During the last decade the precision of atomic clocks has been increased dramatically, thanks to the use of optical transitions instead of microwave transitions like the one studied in this chapter; see e.g. B. J. Bloom, T. L. Nicholson, J. R. Williams, S. L. Campbell, M. Bishof, X. Zhang, W. Zhang, S. L. Bromley and J. Ye, *An optical lattice clock with accuracy and stability at the $10^{-18}$ level*, Nature **506**, 71 (2014). Recent accurate tests of a possible drift of fundamental constants – either the fine structure constant $\alpha$ or the ratio of the mass of the electron to the mass of the proton $m_e/m_p$ – are presented by N. Huntemann, B. Lipphardt, C. Tamm, V. Gerginov, S. Weyers, and E. Peik, Phys. Rev. Lett. **113**, 210802 (2014).

# The Spectrum of Positronium

**7**

The positron $e^+$ is the antiparticle of the electron. It is a spin-1/2 particle, which has the same mass $m$ as the electron, but an electric charge of opposite sign. In this chapter we consider the system called *positronium* which is an atom consisting of an $e^+e^-$ pair.

## 7.1 Positronium Orbital States

We first consider only the spatial properties of the system, neglecting all spin effects. We only retain the Coulomb interaction between the two particles. No proof is required, an appropriate transcription of the hydrogen atom results suffices.

**7.1.1** Express the reduced mass of the system $\mu$, in terms of the electron mass $m$.

**7.1.2** Write the Hamiltonian of the relative motion of the two particles in terms of their separation $r$ and their relative momentum $p$.

**7.1.3** What are the energy levels of the system, and their degeneracies? How do they compare with those of hydrogen?

**7.1.4** What is the Bohr radius $a_0$ of the system? How do the sizes of hydrogen and positronium compare?

**7.1.5** Give the expression for the normalized ground state wave function $\psi_{100}(r)$. Express $|\psi_{100}(0)|^2$ in terms of the fundamental constants: $m, c, \hbar$, and the fine structure constant $\alpha$.

© Springer Nature Switzerland AG 2019
J.-L. Basdevant, J. Dalibard, *The Quantum Mechanics Solver*,
https://doi.org/10.1007/978-3-030-13724-3_7

## 7.2    Hyperfine Splitting

We now study the hyperfine splitting of the ground state.

**7.2.1** What is the degeneracy of the orbital ground state if one takes into account spin variables (in the absence of a spin–spin interaction)?

**7.2.2** Explain why the (spin) gyromagnetic ratios of the positron and of the electron have opposite signs: $\gamma_1 = -\gamma_2 = \gamma$. Express $\gamma$ in terms of $q$ and $m$.

**7.2.3** One assumes that, as in hydrogen, the spin–spin Hamiltonian in the orbital ground state is:

$$\hat{H}_{SS} = \frac{A}{\hbar^2}\hat{S}_1 \cdot \hat{S}_2, \qquad (7.1)$$

where the constant $A$ has the dimension of an energy.

Recall the eigenstates and eigenvalues of $\hat{H}_{SS}$ in the spin basis $\{|\sigma_1, \sigma_2\rangle\}$, where $\sigma_1 = \pm1, \sigma_2 = \pm1$.

**7.2.4** As in hydrogen, the constant $A$ originates from a contact term:

$$A = -\frac{2}{3}\frac{1}{\epsilon_0 c^2}\,\gamma_1\,\gamma_2\,\hbar^2\,|\psi_{100}(0)|^2. \qquad (7.2)$$

(a) The observed hyperfine line of positronium has a frequency $\nu \simeq 200\,\text{GHz}$, compared to $\nu \simeq 1.4\,\text{GHz}$ for hydrogen. Justify this difference of two orders of magnitude.
(b) Express the constant $A$ in terms of the fine structure constant and the energy $mc^2$. Give the numerical value of $A$ in eV.
(c) What frequency of the hyperfine transition corresponds to this calculated value of $A$?

**7.2.5** Actually, the possibility that the electron and the positron can annihilate, leads to an additional contribution $\hat{H}_A$ in the hyperfine Hamiltonian. One can show that $\hat{H}_A$ does not affect states of total spin equal to zero ($S = 0$), and that it increases systematically the energies of $S = 1$ states by the amount:

$$\hat{H}_A: \quad \delta E^{S=1} = \frac{3A}{4} \quad (\delta E^{S=0} = 0), \qquad (7.3)$$

where $A$ is the same constant as in (7.2).

(a) What are the energies of the $S = 1$ and $S = 0$ states, if one takes into account the above annihilation term?
(b) Calculate the frequency of the corresponding hyperfine transition.

## 7.3    Zeeman Effect in the Ground State

The system is placed in a constant uniform magnetic field $\boldsymbol{B}$ directed along the $z$ axis. The additional Zeeman Hamiltonian has the form

$$\hat{H}_Z = \omega_1 \hat{S}_{1z} + \omega_2 \hat{S}_{2z} , \tag{7.4}$$

where $\omega_1 = -\gamma_1 B$ and $\omega_2 = -\gamma_2 B$.

### 7.3.1  Matrix representation of the Zeeman Hamiltonian

(a) Taking into account the result of question 7.2.2 and setting $\omega = -\gamma B$, write the action of $\hat{H}_Z$ on the basis states $\{|\sigma_1, \sigma_2\rangle\}$.
(b) Write in terms of $A$ and $\hbar\omega$ the matrix representation of

$$\hat{H} = \hat{H}_{SS} + \hat{H}_A + \hat{H}_Z \tag{7.5}$$

in the basis $\{|S, m\rangle\}$ of the *total* spin of the two particles.
(c) Give the numerical value of $\hbar\omega$ in eV for a field $B = 1$ T. Is it easy experimentally to be in a strong field regime, i.e. $\hbar\omega \gg A$?

**7.3.2** Calculate the energy eigenvalues in the presence of the field $B$; express the corresponding eigenstates in the basis $\{|S, m\rangle\}$ of the total spin. The largest eigenvalue will be written $E_+$ and the corresponding eigenstate $|\psi_+\rangle$. For convenience, one can introduce the quantity $x = 8\hbar\omega/(7A)$, and the angle $\theta$ defined by $\sin 2\theta = x/\sqrt{1+x^2}$, $\cos 2\theta = 1/\sqrt{1+x^2}$.

**7.3.3** Draw qualitatively the variations of the energy levels in terms of $B$. Are there any remaining degeneracies?

## 7.4    Decay of Positronium

We recall that when a system $A$ is unstable and decays: $A \to B + \cdots$, the probability for this system to decay during the interval $[t, t + dt]$ if it is prepared at $t = 0$, is $dp = \lambda e^{-\lambda t} dt$, where the decay rate $\lambda$ is related to the lifetime $\tau$ of the system by $\tau = 1/\lambda$. If the decay can proceed via different channels, e.g. $A \to B + \cdots$ and $A \to C + \cdots$, with respective decay rates $\lambda_1$ and $\lambda_2$, the total decay rate is the sum of the partial rates, and the lifetime of $A$ is $\tau = 1/(\lambda_1 + \lambda_2)$.

In all what follows, we place ourselves in the rest frame of the positronium.

**7.4.1** In a two-photon decay, or annihilation, of positronium, what are the energies of the two outgoing photons, and what are their relative directions?

**7.4.2** One can show that the annihilation rate of positronium into photons in an orbital state $|n, l, m\rangle$ is proportional to the probability for the electron and positron to be at the same point, i.e. to $|\psi_{nlm}(0)|^2$. In what orbital states is the annihilation possible?

**7.4.3** In quantum field theory, one can show that, owing to charge conjugation invariance,

- a *singlet* state, $S = 0$, can only decay into an *even* number of photons: $2, 4, \cdots$
- a *triplet* state, $S = 1$, can only decay into an *odd* number of photons: $3, 5, \cdots$

In the orbital ground state $\psi_{100}$, split by spin–spin interactions as calculated in Sect. 7.2, the lifetime of the singlet state is $\tau_2 \sim 1.25 \times 10^{-10}$ s, and the lifetime of either of the three triplet states is $\tau_3 \sim 1.4 \times 10^{-7}$ s. Quantum field theory predicts:

$$\lambda_2 = \frac{1}{\tau_2} = 4\pi\alpha^2 c \left(\frac{\hbar}{mc}\right)^2 |\psi_{100}(0)|^2, \quad \lambda_3 = \frac{1}{\tau_3} = \frac{4}{9\pi}(\pi^2 - 9)\,\alpha\,\lambda_2. \quad (7.6)$$

Compare theory and experiment.

**7.4.4** In order to determine the hyperfine constant $A$ of positronium, it is of interest to study the energy and the lifetime of the level corresponding to the state $|\psi_+\rangle$, defined in question 7.3.2, as a function of the field $B$.

From now on, we assume that the field is weak, i.e. $|x| = |8\hbar\omega/(7A)| \ll 1$, and we shall make the corresponding approximations.

(a) What are, as a function of $x$, the probabilities $p^S$ and $p^T$ of finding the state $|\psi_+\rangle$ in the singlet and triplet states respectively?
(b) Use the result to calculate the decay rates $\lambda_2^+$ and $\lambda_3^+$ of the state $|\psi_+\rangle$ into two and three photons respectively, in terms of the parameter $x$, and of the rates $\lambda_2$ and $\lambda_3$ introduced in question 7.4.4.
(c) What is the lifetime $\tau^+(B)$ of the state $|\psi_+\rangle$? Explain qualitatively its dependence on the applied field $B$, and calculate $\tau^+(B)$ for $B = 0.4\ T$.

**7.4.5** One measures, as a function of $B$, the ratio $R = \tau^+(B)/\tau^+(0)$ of the lifetime of the $|\psi_+\rangle$ state with and without a magnetic field. The dependence on $B$ of $R$ is given in Fig. 7.1, with the corresponding error bars.

(a) What estimate does one obtain for the hyperfine constant, $A$, using the value of the magnetic field for which the ratio $R$ has decreased by a factor two?
(b) How do theory and experiment compare?

**Fig. 7.1** Variation of the ratio $R$ defined in the text as a function of the applied magnetic field $B$

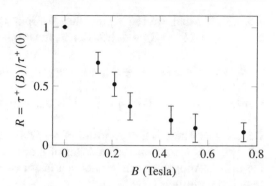

## 7.5 Solutions

### 7.5.1 Positronium Orbital States

In positronium, we have, by scaling:

**7.5.1.1** A reduced mass $\mu = m/2$.

**7.5.1.2** A center of mass Hamiltonian $\hat{H} = \hat{p}^2/2\mu - q^2/4\pi\epsilon_0 r$.

**7.5.1.3** The energy levels $E_n = -(1/2)\mu c^2\alpha^2/n^2 = -(1/4)mc^2\alpha^2/n^2$. The degeneracy is $n^2$ for each level, as in the hydrogen atom; the bound state energies are half of those of hydrogen.

**7.5.1.4** The Bohr radius is $a_0 = \hbar/(\mu c\alpha) = 2\hbar/(mc\alpha) = 2a_0^H \sim 1.06\,\text{Å}$. The diameter of positronium is $\langle r \rangle = 3a_0/2 = 3a_0^H$, and, since the proton is fixed, the diameter of the hydrogen is $2\langle r \rangle_H = 3a_0^H$. Therefore the two systems have the same size.

**7.5.1.5** The ground state wave function is $\psi_{100}(r) = e^{-r/a_0}/\sqrt{\pi a_0^3}$, and we have $|\psi_{100}(0)|^2 = (mc\alpha/(2\hbar))^3/\pi$.

### 7.5.2 Hyperfine Splitting

**7.5.2.1** In the orbital ground state, the degeneracy is 4, corresponding to the number of independent spin states.

**7.5.2.2** Since the masses are equal, but the charges are opposite, we have $\gamma_1 = q/m$, $\gamma_2 = -q/m$, $\gamma = q/m$.

**7.5.2.3** As usual, we can express the spin-spin operator in terms of the total spin $S$ as $S_1 \cdot S_2 = [S^2 - S_1^2 - S_2^2]/2$. Hence, the orbital ground state is split into:

- *the triplet states*: $| + + \rangle$, $(| + - \rangle + | - + \rangle)/\sqrt{2}$, $| - - \rangle$, with the energy shift $E^T = A/4$,
- *the singlet state*: $(| + - \rangle - | - + \rangle)/\sqrt{2}$, with the energy shift $E^S = -3A/4$.

**7.5.2.4** **(a)** There is a mass factor of $\sim 1/2000$, a factor of $\sim 2.8$ for the gyromagnetic ratio of the proton, and a factor of 8 due to the value of the wave function at the origin. Altogether, this results in a factor of $\sim 22/2000 \sim 1\%$ for the ratio of hyperfine splittings $H/(e^+ e^-)$.

**(b)** The numerical value of $A$ is

$$A = \frac{1}{12\pi\epsilon_0} \left( \frac{q\hbar}{mc} \right)^2 \left( \frac{mc\alpha}{\hbar} \right)^3 = \frac{1}{3} mc^2 \alpha^4 \sim 4.84 \times 10^{-4}\,\text{eV}. \tag{7.7}$$

**(c)** This corresponds to a transition frequency of $\nu = A/h \simeq 117\,\text{GHz}$. This prediction is not in agreement with the experimental result ($\sim 200\,\text{GHz}$).

**7.5.2.5** **(a)** Taking into account $\hat{H}_A$, the triplet state energy is $A$ while the singlet state energy is $-3A/4$. The splitting is $\delta E = 7A/4 = 8.47 \times 10^{-4}\,\text{eV}$.

**(b)** The corresponding frequency is $\nu = \delta E/h \sim 205\,\text{GHz}$, in agreement with experiment.

### 7.5.3   Zeeman Effect in the Ground State

**7.5.3.1** Matrix representation of the Zeeman Hamiltonian

**(a)** The Zeeman Hamiltonian is $\hat{H}_Z = \omega(\hat{S}_{1z} - \hat{S}_{2z})$, therefore, we have

$$\hat{H}_Z | + + \rangle = \hat{H}_Z | - - \rangle = 0$$
$$\hat{H}_Z | + - \rangle = \hbar\omega | + - \rangle$$
$$\hat{H}_Z | - + \rangle = -\hbar\omega | - + \rangle. \tag{7.8}$$

In terms of total spin states, this results in

$$\hat{H}_Z |1, 1\rangle = \hat{H}_Z |1, -1\rangle = 0$$
$$\hat{H}_Z |1, 0\rangle = \hbar\omega |0, 0\rangle$$
$$\hat{H}_Z |0, 0\rangle = \hbar\omega |1, 0\rangle. \tag{7.9}$$

**(b)** Hence the matrix representation in the coupled basis:

$$\hat{H}_Z = \begin{pmatrix} 0 & 0 & 0 & 0 \\ 0 & 0 & 0 & 0 \\ 0 & 0 & 0 & \hbar\omega \\ 0 & 0 & \hbar\omega & 0 \end{pmatrix},$$ (7.10)

where the elements are ordered according to: $|1, 1\rangle, |1, -1\rangle, |1, 0\rangle, |0, 0\rangle$.

**7.5.3.2** Similarly, one has the matrix representation of the full spin Hamiltonian:

$$\hat{H} = \begin{pmatrix} A & 0 & 0 & 0 \\ 0 & A & 0 & 0 \\ 0 & 0 & A & \hbar\omega \\ 0 & 0 & \hbar\omega & -3A/4 \end{pmatrix}.$$ (7.11)

In a field of 1 T, $|\hbar\omega| = q\hbar B/m = 2\mu_B B \simeq 1.16 \times 10^{-4}$ eV. The strong field regime corresponds to $|\hbar\omega| \gg A$, i.e. $B \gg 4$ T, which is difficult to reach.

**7.5.3.3** Two eigenstates are obvious: $|1, 1\rangle$ and $|1, -1\rangle$, which correspond to the same degenerate eigenvalue $A$ of the energy. The two others are obtained by diagonalizing a $2 \times 2$ matrix:

$$|\psi_+\rangle = \cos\theta \, |1, 0\rangle + \sin\theta \, |0, 0\rangle,$$
$$|\psi_-\rangle = -\sin\theta \, |1, 0\rangle + \cos\theta \, |0, 0\rangle,$$ (7.12)

corresponding to the energies

$$E\pm = \frac{A}{8} \pm \left[\left(\frac{7A}{8}\right)^2 + (\hbar\omega)^2\right]^{1/2} = \frac{A}{8}\left(1 \pm 7\sqrt{1 + x^2}\right).$$ (7.13)

The triplet states $|++\rangle$ and $|--\rangle$ remain degenerate, as shown in Fig. 7.2.

**Fig. 7.2** Variation of the hyperfine energy levels with applied magnetic field

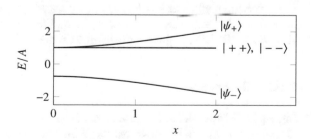

### 7.5.4   Decay of Positronium

**7.5.4.1**  In a two-photon decay, the outgoing photons have opposite momenta, their energies are both $mc^2 = 511$ keV.

**7.5.4.2**  The wave function vanishes at the origin, except for $s$-waves ($|\psi_{nlm}(0)|^2 = 0$ if $l \neq 0$), owing to the centrifugal barrier. Therefore the decay can only occur when the positronium is in an $s$-state.

**7.5.4.3**  The given formulas correspond to $\lambda_2 = mc^2\alpha^5/(2\hbar)$ which yields $\tau_2 = 1.24 \times 10^{-10}$s and $\tau_3 = 1.38 \times 10^{-7}$s, in agreement with experiment.

**7.5.4.4**  **(a)**  For a given value of the applied field, with the positronium prepared in the state $|\psi_+\rangle$, the probabilities of finding the system in the singlet and triplet states are respectively $p^S = \sin^2\theta \sim x^2/4$ and $p^T = \cos^2\theta \sim 1 - x^2/4$.

**(b)**  The rate for $|\psi_+\rangle$ to decay into two photons is the product of the probability of finding $|\psi_+\rangle$ in the singlet state with the singlet state decay rate:

$$\lambda_2^+ = p^S\lambda_2 \sim x^2\lambda_2/4 = x^2/(4\tau_2). \tag{7.14}$$

Similarly, one has

$$\lambda_3^+ = p^T\lambda_3 \sim (1 - x^2/4)\lambda_3 = (1 - x^2/4)/\tau_3. \tag{7.15}$$

**(c)**  The lifetime of the $|\psi_+\rangle$ state is

$$\tau_+ = \frac{1}{\lambda_2^+ + \lambda_3^+} = \frac{\tau_3}{1 - \frac{x^2}{4} + \frac{x^2}{4}\frac{\tau_3}{\tau_2}} \simeq \frac{\tau_3}{1 + \frac{16\hbar^2\omega^2}{49A^2}\frac{\tau_3}{\tau_2}}. \tag{7.16}$$

As the field $B$ increases, the state $|\psi_+\rangle$, which is purely triplet for $B = 0$, acquires a greater and greater singlet component. Therefore its lifetime decreases as $B$ increases. For $B = 0.4$ T, one has $\tau_+ = 0.23\tau_3 = 3.2 \times 10^{-8}$ s.

**(d)**  Experimentally, one has $R \sim 0.5$, i.e. $x^2\tau_3/4\tau_2 \simeq 1$ for $B \sim 0.22$ T. Therefore $x \simeq 6 \times 10^{-2}$ and, since $A = 8\hbar\omega/7x$ and $\hbar\omega = 2.3 \times 10^{-5}$ eV, the result is $A \sim 4.4 \times 10^{-4}$ eV, in good agreement with theoretical expectations.

### 7.5.5   References

See e.g. A.P. Mills and S. Chu, *Precision Measurements in Positronium*, in *Quantum Electrodynamics*, ed. by T. Kinoshita (World Scientific, Singapore 1990) pp. 774–821. Recent references can be found in S. G. Karshenboim, *Precision Study of*

*Positronium: Testing Bound State QED Theory*, Int. Jour. Mod. Phys. A **19**, 3879 (2003).

Positronium is now a common tool for diagnoses in physics and chemistry, see Y. C. Jean, P. E. Mallon, D. M. Schrader (Eds.), *Principles and applications of positron and positronium chemistry*, World Scientific (2003).

# Neutrino Transformations in the Sun

**8**

In weak interactions, the electron is always associated with a neutral particle, the neutrino $\nu_e$. For instance, a neutrino in an accelerator beam can interact with a neutron $(n)$ in a nucleus and give rise to a proton $(p)$ in the reaction $\nu_e + n \to p + e$. There exists another particle, the $\mu$ lepton, or *muon*, whose physical properties are completely analogous to those of the electron, except for its mass $m_\mu \simeq 200\ m_e$. The muon has the same weak interactions as the electron, but it is associated to a *different* neutrino, the $\nu_\mu$, for instance in the reaction $\nu_\mu + n \to p + \mu$. Similarly, the $\tau$ lepton, discovered in 1975 and which is much more massive, $m_\tau \simeq 3500\ m_e$, is associated to its own neutrino $\nu_\tau$. In the 1990s, the experimental measurements at the LEP colliding ring in CERN have shown that these three neutrinos $\nu_e$, $\nu_\mu$, $\nu_\tau$ (and their antiparticles) are *the only ones* of their kinds.

For a long time, physicists believed that neutrinos were massless particles, as is the photon, because experimental limits on these masses were consistent with zero. However, both theoretical and cosmological arguments suggest that this might not be the case, but that their masses (multiplied by $c^2$) are considerably smaller than the energies involved in experiments where they are observed. The proof that neutrino masses are not all zero is a great discovery of the 1990s. For this discovery, the 2002 Nobel prize was awarded to Raymond Davis Jr., a pioneer of this physics, and to Masatoshi Koshiba, the leader of the reactor neutrino KamLAND collaboration in Japan.

A remarkable quantum phenomenon is that the mass differences of neutrinos can be measured by a quantum oscillation effect. The idea is that the "flavor" neutrinos $\nu_e$, $\nu_\mu$ and $\nu_\tau$, which are produced or detected experimentally, do not have definite masses. They are not eigenstates of the *mass*, but rather *linear combinations* of mass eigenstates $\nu_1$, $\nu_2$, $\nu_3$, with masses $m_1$, $m_2$, $m_3$. Therefore, if one prepares a neutrino with a given flavor in vacuum, it will transform spontaneously and periodically into neutrinos with different flavors, i.e., *other* point-like elementary particles. The 2015 Nobel prize for physics was awarded to Takaaki Kajita (Super-Kamiokande) and

© Springer Nature Switzerland AG 2019
J.-L. Basdevant, J. Dalibard, *The Quantum Mechanics Solver*,
https://doi.org/10.1007/978-3-030-13724-3_8

Arthur B. McDonald (SNO) "*for the discovery of neutrino oscillations, which shows that neutrinos have mass*".

In the present study, which we restrict for simplicity to two types of neutrinos, we first explain how the mass differences of neutrinos can be measured by a quantum oscillation effect.

Solar energy is produced by nuclear reactions during which neutrinos are emitted. These neutral particles are very light and cross the Sun to reach the Earth. Recent experiments have shown that the interaction of neutrinos with solar matter allows them not only to oscillate, but to "transform" as they travel through the Sun. This transformation, predicted by Mikheyev, Smirnov and Wolfenstein (MSW), is the subject of the second part of the present problem. We will see that this interaction can affect the rate of electron neutrinos that reach the earth.

We use the units of particle physics: energies in electron-Volt ($1eV=1.6 \times 10^{-19}$ J), linear momentum in eV/$c$, and masses in eV/$c^2$, where $c$ is the speed of light in vacuum. Numerical values of quantities appearing in the text are given in the following table:

| | |
|---|---|
| $m_2^2 - m_1^2 = 8 \times 10^{-5}$ (eV/$c^2$)$^2$ | Earth-Sun distance: $d = 1.5 \times 10^{11}$ m |
| $\sin^2 \theta = 0.3$ | Solar radius: $R = 7 \times 10^8$ m |
| $G = 9.0 \times 10^{-44}$ eV·m$^3$ | Proton mass: $m_p = 1.67 \times 10^{-27}$ kg |
| $\hbar = 1.054 \times 10^{-34}$ J s | $c = 3.0 \times 10^8$ m s$^{-1}$ |

## 8.1    Neutrino Oscillations in Vacuum

The velocity of neutrinos produced in the Sun is close to the speed of light. Their energy $E$ and momentum $p$ are related by $E = \sqrt{p^2c^2 + m^2c^4}$, where $m$ is the neutrino mass. For simplicity, we assume in this problem that there exist only two neutrino species. Their masses are denoted $m_1$ and $m_2$, respectively, with $m_1 < m_2$.

Consider a neutrino arriving on Earth, with momentum $p$ such that $p \gg m_2 c$. We denote by $E_1$ or $E_2$ its energy, depending on whether it is a neutrino of mass $m_1$ or $m_2$.

**8.1.1**  Show that the energy difference $\Delta = E_2 - E_1$ is $\Delta \simeq (m_2^2 - m_1^2)c^3/(2p)$, at leading order in $m_1$ and $m_2$.

**8.1.2**  Calculate $\Delta$ in eV for:

(a) a neutrino of momentum $p = 2 \times 10^5$ eV/$c$,
(b) a neutrino of momentum $p = 8 \times 10^6$ eV/$c$.

These values correspond to two different types of nuclear reactions occurring in the Sun.

**8.1.3** We denote by $|\nu_1\rangle$ and $|\nu_2\rangle$ the state vectors of the two states of masses $m_1$ and $m_2$, respectively, with a given value of the momentum $p$. These state vectors are orthogonal and normalized. Neutrinos produced in solar nuclear reactions are electron neutrinos, $|\nu_e\rangle$. The state $|\nu_e\rangle$ is a linear combination of mass eigenstates $|\nu_1\rangle$ and $|\nu_2\rangle$, fixed by the laws of particle physics. One defines the mixing angle $\theta$ between these states so that $|\nu_e\rangle = \cos\theta\,|\nu_1\rangle + \sin\theta\,|\nu_2\rangle$, where $\theta$ is an angle between 0 and $\pi/2$.

Consider a neutrino produced at time $t = 0$ in the state $|\nu_e\rangle$ with momentum $p$.

(a) Calculate the state of the neutrino at time $t > 0$. Note: the time evolution of an eigenstate $|\psi_E\rangle$ of the Hamiltonian, of energy $E$, is given here by the usual formula $|\psi(t)\rangle = e^{-iEt/\hbar}|\psi_E\rangle$.
(b) What is the probability to detect the neutrino in the state $|\nu_e\rangle$ at time $t$?

**8.1.4** All neutrinos emitted by nuclear reactions in the Sun are initially produced in the state $|\nu_e\rangle$.

(a) Calculate the number of oscillations of a neutrino of momentum $p = 2 \times 10^5$ eV/c as it travels from the Sun to the Earth (take the neutrino velocity equal to the speed of light).
(b) The momentum distribution of solar neutrinos is a smooth curve essentially concentrated between $10^5$ and $4 \times 10^5$ eV/c. What is the fraction of the solar neutrinos that are detected on Earth as electron neutrinos? First give the result as a function of the mixing angle $\theta$, and then calculate its numerical value.

## 8.2    Interaction of Neutrinos with Matter

When electron neutrinos travel through matter, they interact with electrons. The corresponding interaction energy $V$ adds to the mass+kinetic energy $E_j = \sqrt{p^2c^2 + m_j^2c^4}$, $j = 1, 2$. In a homogeneous medium, the corresponding operator can be written as

$$\hat{V} = V|\nu_e\rangle\langle\nu_e|$$

with $V = GN_e$, where $G$ is a positive constant and $N_e$ is the number of electrons per unit volume.

**8.2.1** What does the operator $|\nu_e\rangle\langle\nu_e|$ represent? Give its expression as a matrix in the basis set $\{|\nu_1\rangle, |\nu_2\rangle\}$, as a function of $\cos(2\theta)$ and $\sin(2\theta)$.

**8.2.2** Suppose for simplicity that the Sun is composed only of protons and electrons, in equal numbers per unit volume. The mass density at the center of the Sun is $\rho = 1.5 \times 10^5$ kg·m$^{-3}$. Calculate the value of $V$ in electron-Volt, and compare it to the two values of $\Delta$ calculated in question 8.1.2. Check that $V \ll \Delta$ for one category of neutrinos, and $V > \Delta$ for the other one.

**8.2.3** Calculate in first order perturbation theory in $V$ the shifts $\tilde{E}_1 - E_1$ and $\tilde{E}_2 - E_2$ of the eigenenergies of the Hamiltonian, for a given momentum $p$. Express the results in terms of $E_1$, $E_2$, $V$ and $\theta$.

**8.2.4** We now want to diagonalize the Hamiltonian exactly.

(a) Write the Hamiltonian $\hat{H}_0$ in the absence of interaction and the interaction potential $\hat{V}$, in the basis $\{|\nu_1\rangle, |\nu_2\rangle\}$. Show that the total Hamiltonian $\hat{H} = \hat{H}_0 + \hat{V}$ can be written as:

$$\hat{H} = \left( E_1 + E_2 + \frac{V}{2} \right) \hat{1} - \frac{1}{2} \sqrt{\Delta^2 + V^2 - 2V\Delta \cos 2\theta} \begin{pmatrix} \cos 2\tilde{\theta} & \sin 2\tilde{\theta} \\ \sin 2\tilde{\theta} & -\cos 2\tilde{\theta} \end{pmatrix}$$

where $\hat{1}$ is the identity matrix and $\Delta = E_2 - E_1$. Give the expression of $\cos 2\tilde{\theta}$ and $\sin 2\tilde{\theta}$ in terms of $\theta$, $V$ and $\Delta$.

(b) Show that an eigenbasis of $\hat{H}$ is given by

$$|\tilde{\nu}_1\rangle = \cos \tilde{\theta} \, |\nu_1\rangle + \sin \tilde{\theta} \, |\nu_2\rangle \qquad\qquad |\tilde{\nu}_2\rangle = -\sin \tilde{\theta} \, |\nu_1\rangle + \cos \tilde{\theta} \, |\nu_2\rangle$$

and calculate the eigenvalues $\tilde{E}_1$ and $\tilde{E}_2$ associated to these two vectors.

(c) In the limit $V \ll \Delta$, recover the expressions obtained for $\tilde{E}_1$ and $\tilde{E}_2$ using perturbation theory.

(d) Give an approximate value of $\tilde{\theta}$ (possibly as a function of $\theta$) in the two limiting cases $V \ll \Delta$ and $V \gg \Delta$.

(e) Plot $\tilde{E}_1$ and $\tilde{E}_2$ as a function of $V$.

(f) Give the value of $V$ for which the difference $\tilde{E}_2 - \tilde{E}_1$ is minimum. What is then the minimum value of $\tilde{E}_2 - \tilde{E}_1$?

**8.2.5 The MSW effect.** Consider a neutrino produced at the center of the Sun in the state $|\nu_e\rangle$. As it travels to the boundary of the Sun, the neutrino sees a decreasing electronic density $N_e$. Therefore the Hamiltonian $\hat{H}$ in the frame of the neutrino depends on time. We shall use the following result (adiabatic approximation): if the characteristic time $T$ over which the Hamiltonian varies, i.e., the time needed for the neutrino to get out of the Sun, is much larger than $\hbar/(\tilde{E}_2 - \tilde{E}_1)$, then the probability to remain in a given eigenstate of $\hat{H}$ is practically a constant as a function of time.

(a) Check that the condition $T \gg \hbar/(\tilde{E}_2 - \tilde{E}_1)$ is indeed satisfied for neutrinos of momentum $p = 8 \times 10^6$ eV/c.

(b) Show that a neutrino produced in the center of the Sun in the state $|\nu_e\rangle$ arrives on Earth in the state

$$|\psi\rangle = e^{i\phi_1}\cos(\theta - \tilde{\theta})\,|\nu_1\rangle + e^{i\phi_2}\sin(\theta - \tilde{\theta})\,|\nu_2\rangle,$$

where $\phi_1$ and $\phi_2$ are phases that we shall not determine explicitly.
(c) What is the probability to detect the neutrino on Earth in the state $|\nu_e\rangle$ ?
(d) Explain why it is legitimate to average this probability over $\phi_1$ and $\phi_2$. Give this average value in the limit $V \gg \Delta$ (use the relation found in 2.4d between $\tilde{\theta}$ and $\theta$).
(e) Compare this probability to the one found in the first section of the problem.
(f) The neutrinos fluxes are detected on Earth with an accuracy of 10%. Is it necessary to take into account the MSW effect, when comparing theory and experiment for neutrinos of momentum $p = 8 \times 10^6$ eV/c?

## 8.3 Solutions

### 8.3.1 Neutrino Oscillations in the Vacuum

**8.3.1.1** $E \simeq pc + m^2c^3/(2p)$ therefore $\Delta \simeq (m_2^2 - m_1^2)c^3/(2p)$.

**8.3.1.2** For $p = 2 \times 10^5$ eV/c, $\Delta \simeq 2 \times 10^{-10}$ eV ; for $p = 8 \times 10^6$ eV/c, $\Delta \simeq 5 \times 10^{-12}$ eV.

**8.3.1.3** The basis vectors $|\nu_1\rangle$ and $|\nu_2\rangle$ are defined up to a phase factor that can be chosen so that the coefficients of the expansion of $|\nu_e\rangle$ on this basis are real and positive. The normalization condition $\langle \nu_e | \nu_e \rangle = 1$ then justifies the form $|\nu_e\rangle = \cos\theta\,|\nu_1\rangle + \sin\theta\,|\nu_2\rangle$.

**(a)** For $t > 0$ the state is $|\psi(t)\rangle = e^{-iE_1t/\hbar}\cos\theta\,|\nu_1\rangle + e^{-iE_2t/\hbar}\sin\theta\,|\nu_2\rangle$.

**(b)** The probability $P(t)$ to detect the neutrino with the electron flavour is given by

$$P(t) = |\langle \nu_e | \psi(t)\rangle|^2 = \cos^4\theta + \sin^4\theta + 2\sin^2\theta\cos^2\theta\cos(\Delta t/\hbar).$$

**8.3.1.4 (a)** The neutrinos travel nearly at the speed of light, so that their travel time from the Sun to the Earth is $t \approx d/c = 500$ s. With the value of $\Delta$ obtained at question 8.1.2 for $p = 2 \times 10^5$ eV/c, one finds $t\Delta/\hbar \simeq 1.5 \times 10^8$, i.e., $2.4 \times 10^7$ periods of oscillation.

**(b)** When $p$ varies in the interval from $10^5$ to $4 \times 10^5$ eV/c, the term $\cos(\Delta t/\hbar)$ in the result of 1.4b averages to zero, and we obtain

$$P = \cos^4\theta + \sin^4\theta = 1 - 2\sin^2\theta\cos^2\theta = 0.58.$$

Fifty eight percentage of the solar neutrinos that reach us should be electron neutrinos.

## 8.3.2    Interaction of Neutrinos with Matter

**8.3.2.1** The operator $|\nu_e\rangle\langle\nu_e|$ is the projector on the state $|\nu_e\rangle$. In the basis $\{|\nu_1\rangle, |\nu_2\rangle\}$, its expression is

$$|\nu_e\rangle\langle\nu_e| = \begin{pmatrix} \cos\theta \\ \sin\theta \end{pmatrix} (\cos\theta, \ \sin\theta) = \begin{pmatrix} \cos^2\theta & \cos\theta \ \sin\theta \\ \cos\theta \ \sin\theta & \sin^2\theta \end{pmatrix}$$

$$= \frac{1}{2} \begin{pmatrix} 1 + \cos 2\theta & \sin 2\theta \\ \sin 2\theta & 1 - \cos 2\theta \end{pmatrix}. \tag{8.1}$$

The form of $\hat{V}$ is therefore chosen so that it only acts on $|\nu_e\rangle$, and not on the state orthogonal to $|\nu_e\rangle$.

NB. In practice, all varieties of neutrinos interact with matter through neutral currents with the same intensity, and that interaction does not produce an energy difference. However, it is only the electron neutrinos that interact with electrons through charged currents because of energy and lepton number conservation (the muon that must be produced in $\nu_\mu + e \rightarrow \mu + \nu_e$ has a mass $m_\mu \sim 200 m_e$, which is much too large compared to nuclear energies.)

**8.3.2.2** Neglecting the electron mass compared to the proton mass, $N_e \simeq \rho/m_p$, therefore $V = G\rho/m_p = 8.1 \times 10^{-12}$ eV. $V$ is much smaller than $\Delta$ for neutrinos of momentum $2 \times 10^5$ eV/c, but larger than $\Delta$ for neutrinos of momentum $8 \times 10^6$ eV/c.

**8.3.2.3** In first order perturbation theory, $\tilde{E}_1 - E_1 = \langle\nu_1|\hat{V}|\nu_1\rangle = V\cos^2\theta$. Similarly, $\tilde{E}_2 - E_2 = V\sin^2\theta$.

**8.3.2.4** **(a)** In the basis $\{|\nu_1\rangle, |\nu_2\rangle\}$, the hamiltonian $\hat{H} = \hat{H}_0 + \hat{V}$ is

$$\hat{H} = \begin{pmatrix} E_1 & 0 \\ 0 & E_2 \end{pmatrix} + \frac{V}{2} \begin{pmatrix} 1 + \cos 2\theta & \sin 2\theta \\ \sin 2\theta & 1 - \cos 2\theta \end{pmatrix}$$

$$= \frac{E_1 + E_2 + V}{2} - \frac{1}{2} \begin{pmatrix} \Delta - V\cos 2\theta & -V\sin 2\theta \\ -V\sin 2\theta & -\Delta + V\cos 2\theta \end{pmatrix}$$

One can therefore choose $\tilde{\theta}$ such that

$$\cos 2\tilde{\theta} = \frac{\Delta - V\cos 2\theta}{\sqrt{\Delta^2 + V^2 - 2V\Delta\cos 2\theta}} \qquad \sin 2\tilde{\theta} = \frac{-V\sin 2\theta}{\sqrt{\Delta^2 + V^2 - 2V\Delta\cos 2\theta}}$$

in order to find the result.

**(b)** One checks immediately that the vectors $|\tilde{v}_i\rangle$, $i = 1, 2$ are indeed eigenvectors of $\hat{H}$. The corresponding energy eigenvalues are

$$\tilde{E}_1 = \frac{1}{2}\left(E_1 + E_2 + V - \sqrt{\Delta^2 + V^2 - 2V\Delta\cos 2\theta}\right)$$

$$\tilde{E}_2 = \frac{1}{2}\left(E_1 + E_2 + V + \sqrt{\Delta^2 + V^2 - 2V\Delta\cos 2\theta}\right)$$

**(c)** By expanding to first order in $V$, one recovers the result of perturbation theory:

$$\tilde{E}_1 \simeq \frac{1}{2}\left[E_1 + E_2 + V - \Delta\left(1 - \frac{V}{\Delta}\cos 2\theta\right)\right] = E_1 + \frac{V}{2}(1 + \cos 2\theta)$$

$$\tilde{E}_2 \simeq \frac{1}{2}\left[E_1 + E_2 + V + \Delta\left(1 - \frac{V}{\Delta}\cos 2\theta\right)\right] = E_2 + \frac{V}{2}(1 - \cos 2\theta)$$

**(d)** If $V \ll \Delta$, one finds $\cos 2\tilde{\theta} \simeq 1$ and $\sin 2\tilde{\theta} \simeq -V/\Delta \ll 1$, i.e., $\tilde{\theta} \simeq 0$. If $V \gg \Delta$, one finds $\cos 2\tilde{\theta} \simeq -\cos 2\theta$ and $\sin 2\tilde{\theta} \simeq -\sin 2\theta$, i.e., $2\tilde{\theta} \simeq 2\theta \pm \pi$ or $\tilde{\theta} \simeq \theta \pm (\pi/2)$.

**(e)** The variation of $\tilde{E}_1$ and $\tilde{E}_2$ as a function of $V$ is given in Fig. 8.1.

**(f)** The energy difference between the levels is $\sqrt{\Delta^2 + V^2 - 2V\Delta\cos 2\theta}$. Its minimum occurs for $V = \Delta\cos 2\theta$. For that value of $V$, the difference is $\tilde{E}_2 - \tilde{E}_1 = \Delta\sin 2\theta$.

**8.3.2.5 (a)** The minimum value of $\tilde{E}_2 - \tilde{E}_1$ is $\Delta\sin 2\theta \sim 4 \times 10^{-12}$ eV. One therefore obtains $\hbar/(\tilde{E}_2 - \tilde{E}_1) < 2 \times 10^{-4}$ s, which is much smaller than the time to cross the Sun: roughly 2 s.

**(b)** When a neutrino is produced in the center of the Sun, it is in the electron neutrino state $|v_e\rangle$, which can be decomposed on the eigenbasis of the hamiltonian at that point $\{|\tilde{v}_i\rangle\}$, $i = 1, 2$:

$$|v_e\rangle = \langle\tilde{v}_1|v_e\rangle|\tilde{v}_1\rangle + \langle\tilde{v}_2|v_e\rangle|\tilde{v}_2\rangle = \cos(\theta - \tilde{\theta})|\tilde{v}_1\rangle + \sin(\theta - \tilde{\theta})|\tilde{v}_2\rangle .$$

**Fig. 8.1** Energy curves $\tilde{E}_i$ as a function of $V$. The constant energy $(E_1 + E_2)/2$ has been subtracted

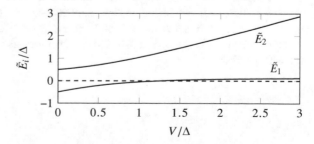

We assume that the adiabatic approximation is valid. The hamiltonian eigenbasis evolves progressively from $\{|\tilde{\nu}_i\rangle\}$ to $\{|\nu_i\rangle\}$ as the neutrino leaves the Sun. In the adiabatic approximation, the absolute values of the coefficients of the expansion of $|\psi\rangle$ in that basis will not change, and one therefore has, for the neutrino leaving the Sun:

$$|\psi\rangle = e^{i\phi_1} \cos(\theta - \tilde{\theta}) |\nu_1\rangle + e^{i\phi_2} \sin(\theta - \tilde{\theta}) |\nu_2\rangle,$$

where the phases $\phi_1$ and $\phi_2$ are unknown a priori.

**(c)** The probability to detect the neutrino on the Earth in the state $|\nu_e\rangle$ is

$$P = |\langle \nu_e | \psi \rangle|^2 = \left| e^{i\phi_1} \cos\theta \cos(\theta - \tilde{\theta}) + e^{i\phi_2} \sin\theta \sin(\theta - \tilde{\theta}) \right|^2$$

or, by the expanding,

$$P = \cos^2\theta \cos^2(\theta - \tilde{\theta}) + \sin^2\theta \sin^2(\theta - \tilde{\theta}) + 2\cos(\phi_1 - \phi_2) \cos\theta \cos(\theta - \tilde{\theta}) \sin\theta \sin(\theta - \tilde{\theta}).$$

**(d)** We have seen in the first part that the phases $\phi_1$ and $\phi_2$ are very large and that they have a rapid variation with the neutrino momentum. It is therefore legitimate to average the previous result over $\phi_1$ and $\phi_2$. The last term cancels and one obtains

$$P = \cos^2(\theta - \tilde{\theta}) \cos^2\theta + \sin^2(\theta - \tilde{\theta}) \sin^2\theta.$$

In the limit $V \gg \Delta$, we have seen that $\tilde{\theta} = \theta \pm \pi/2$. In that limit, the probability is $\sin^2\theta$.

**(e)** The probability to detect the neutrino in the electron form is reduced to $\sin^2\theta = 0.3$ by the MSW effect, whereas we had found a probability $\cos^4\theta + \sin^4\theta \simeq 0.58$ in the first part, in the absence of the MSW effect.

**(f)** With an accuracy of the order of 10% on the measurement of the neutrino flux, one can certainly distinguish between the two predictions, i.e., with or without the MSW effect.

The experimental data confirm the existence of the effect on energetic neutrinos $p \sim 8 \times 10^6 \text{eV}/c$. For less energetic neutrinos, $p \sim 2 \times 10^5 \text{eV}/c$, the effect is negligible ; in fact in that case, one has $V \ll \Delta$, i.e., $\tilde{\theta} \sim 0$. The probability to detect a neutrino of that energy on Earth in the electron form is practically not modified by the MSW effect.

**Further reading:**

- Original publications: Lincoln Wolfenstein, *Neutrino oscillations in matter*, Phys. Rev. D **17**, 2369 (1978); S. P. Mikheyev and A. Yu. Smirnov, *Resonance amplification of oscillations in matter and spectroscopy of solar neutrinos*, Soviet Journal of Nuclear Physics **42**, 913 (1985).
- Experimental results: SNO Collaboration (2005), Phys. Rev. C **72**, 055502 (2005); BOREXINO collaboration, *Final results of Borexino Phase-I on low energy solar neutrino spectroscopy*, Phys. Rev. D **89**, 112007 (2014).
- For a very clear theoretic explanation of the subject: H. A. Bethe, *Possible explanation of the solar-neutrino puzzle*, Phys. Rev. Lett. **56**, 1305 (1986). For further developments, see e.g. S. J. Parke, *Nonadiabatic level crossing in resonant neutrino oscillations*, Phys. Rev. Lett. **57**, 1275 (1986); M. Fukugita, T. Yanagida, *Physics of Neutrinos*, (Springer, 2003).

# The Hydrogen Atom in Crossed Fields

**9**

In this chapter we study the modification of the energy spectrum of a hydrogen atom placed in crossed static electric and magnetic fields in perturbation theory. We thus recover a result first derived by Wolfgang Pauli.

In his famous 1925–1926 work on the hydrogen atom, Pauli made use of the particular symmetry of the Coulomb problem. In addition to the hydrogen spectrum, he was able to calculate the splitting of the levels in an electric field (Stark effect) or in a magnetic field (Zeeman effect). Pauli also noticed that he could obtain a simple and compact formula for the level splitting in a superposition of a magnetic field $B_0$ and an electric field $E_0$ both static and uniform, and perpendicular to each other. In this case, he found that a level with principal quantum number $n$ is split into $2n - 1$ sublevels $E_n + \delta E_n^{(k)}$ with

$$\delta E_n^{(k)} = \hbar k \, (\omega_0^2 + \omega_e^2)^{1/2}, \tag{9.1}$$

where $k$ is an integer ranging from $-(n - 1)$ to $n - 1$, $\omega_0$ and $\omega_e$ are respectively proportional to $B_0$ and $E_0$, and $\omega_e$ can be written as

$$\omega_e = \frac{3}{2} \, \Omega_e f(n) \quad \text{with} \quad \Omega_e = \frac{4\pi \epsilon_0 \hbar}{M q_e} E_0, \tag{9.2}$$

where $M$ and $q_e$ are the mass and charge of the electron, and where $f(n)$ depends on $n$ only.

Pauli's result was verified experimentally only 60 years later. Our purpose, here, is to prove (9.1) in the special case $n = 2$, to calculate $\omega_0$ and $\omega_e$ in that case, and, by examining the experimental result for $n = 34$, to guess what was the very simple formula found by Pauli for $f(n)$.

© Springer Nature Switzerland AG 2019
J.-L. Basdevant, J. Dalibard, *The Quantum Mechanics Solver*,
https://doi.org/10.1007/978-3-030-13724-3_9

## 9.1    The Hydrogen Atom in Crossed Electric and Magnetic Fields

We consider the $n = 2$ level of the hydrogen atom. We neglect all spin effects. We assume that $\boldsymbol{B_0}$ is along the $z$ axis and $\boldsymbol{E_0}$ along the $x$ axis. We use first order perturbation theory.

**9.1.1** What are the energy levels and the corresponding eigenstates in the presence of $\boldsymbol{B_0}$ only? Check that (9.1) is valid in this case and give the value of $\omega_0$?

**9.1.2** In the presence of $\boldsymbol{E_0}$ only, the perturbing Hamiltonian is the electric dipole term $\hat{H}_E = -\hat{\boldsymbol{D}}.\boldsymbol{E_0} = -q_e\hat{\boldsymbol{r}}.\boldsymbol{E_0}$. Write the matrix representing $\hat{H}_E$ in the $n = 2$ subspace under consideration.
  We recall that:

(a) $\int_0^\infty r^3 R_{2s}(r)R_{2p}(r)\,dr = 3\sqrt{3}a_1$ where $R_{2s}$ and $R_{2p}$ are the radial wave functions for the level $n = 2, l = 0$ and $n = 2, l = 1$ respectively, and where $a_1 = \hbar^2/(Me^2)$ is the Bohr radius ($e^2 = q_e^2/4\pi\epsilon_0$).
(b) In spherical coordinates ($\theta$ polar angle and $\phi$ azimuthal angle), the $l = 0$ and $l = 1$ spherical harmonics are

$$Y_0^0(\theta, \phi) = \frac{1}{\sqrt{4\pi}}, \quad Y_1^{\pm 1}(\theta, \phi) = \mp\sqrt{\frac{3}{8\pi}}\,\sin\theta\,e^{\pm i\phi},$$

$$Y_1^0(\theta, \phi) = \sqrt{\frac{3}{4\pi}}\cos\theta. \tag{9.3}$$

**9.1.3** Calculate the energies of the levels originating from the $n = 2$ level in the presence of the crossed fields $\boldsymbol{E_0}$ and $\boldsymbol{B_0}$. Show that one recovers (9.1) with $\omega_e = (3/2)f(2)\Omega_e$, and give the value of $f(2)$.

## 9.2    Pauli's Result

The first experimental verification of Pauli's result was performed in 1983. In Fig. 9.1, the points correspond to a sub-level with a given value of $k$ arising from the $n = 34$ level of an hydrogen-like atom. All points correspond to the *same energy* of this level, but to different values of the static fields $E_0$ and $B_0$.
  Knowing that $\omega_e$ is a function of the principal quantum number $n$ of the form: $\omega_e = (3/2)f(n)\Omega_e$, and that $\omega_0$ and $\Omega_e$ are the constants introduced above, answer the following questions:

**9.2.1** Does the experimental data agree with (9.1)?

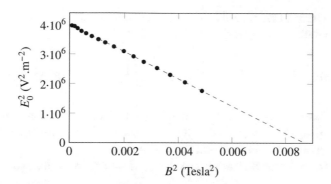

**Fig. 9.1** Values of the electric and magnetic fields giving rise to the same sub-level energy of the $n = 34$ level of a hydrogen-like atom. The dashed line is a guide to the eye

**9.2.2** Write the quantity $\omega_0^2 + \omega_e^2$ in the form $\lambda \left( \gamma B_0^2 + f^2(n) E_0^2 \right)$, give the value of the constant $\gamma$, and calculate $f(34)$.

**9.2.3** Guess Pauli's result concerning $f(n)$.

## 9.3 Solutions

### 9.3.1 The Hydrogen Atom in Crossed Electric and Magnetic Fields

**9.3.1.1** Consider a state $|n, l, m\rangle$. The orbital magnetic moment of the electron is $\hat{\boldsymbol{\mu}}_{\text{orb}} = \gamma_0 \hat{\boldsymbol{L}}$, with $\gamma_0 = q_e/(2M)$. The magnetic Hamiltonian is $\hat{H} = -\hat{\boldsymbol{\mu}}_{\text{orb}} \cdot \boldsymbol{B} = -(q_e/2M)\hat{L}_z B_0$.

At first order perturbation theory, the energy levels originating from the $n = 2$ subspace (angular momentum $l = 0$ or $l = 1$) are $m\hbar\omega_0$ with $m = -1, 0, +1$, and $\omega_0 = -q_e B_0/(2M)$ $(\omega_0 > 0$ for $B_0 > 0)$. The corresponding states are

$$
\begin{array}{lll}
|2s\rangle \text{ and } |2p, m = 0\rangle & \delta E = 0 & \\
|2p, m = -1\rangle & \delta E = -\hbar\omega_0 & \quad (9.4) \\
|2p, m = +1\rangle & \delta E = +\hbar\omega_0. &
\end{array}
$$

**9.3.1.2** The Hamiltonian is $\hat{H}_E = -q_e \hat{x} E_0$. We have to calculate the 16 matrix elements $\langle 2, l', m'|\hat{x}|2, l, m\rangle$. The integral to be evaluated is

$$
\langle 2, l', m'|\hat{x}|2, l, m\rangle = \int\int \left( Y_{l'}^{m'}(\theta, \phi) \right)^* \sin\theta \, \cos\phi \, Y_l^m(\theta, \phi) \, d^2\Omega
$$

$$
\times \int_0^\infty r^3 (R_{2,l'}(r))^* R_{2,l}(r) \, dr. \quad (9.5)
$$

The angular integral vanishes if $l = l'$. We need only consider the terms $l' = 0, l = 1$ (and the hermitian conjugate $l' = 1, l = 0$), i.e.

$$3\sqrt{3}\,a_1 \int\int \frac{1}{\sqrt{4\pi}}\sqrt{\frac{2\pi}{3}}(-Y_1^1(\theta,\phi) + Y_1^{-1}(\theta,\phi))\,Y_1^m(\theta,\phi)\,\mathrm{d}^2\Omega \qquad (9.6)$$

where we have incorporated the radial integral given in the text. One therefore obtains $3\,a_1(\delta_{m,-1} - \delta_{m,1})/\sqrt{2}$. The only non-vanishing matrix elements are $\langle 2s|\hat{H}|2p, m = \pm 1\rangle$ and their hermitian conjugates.

Setting $\Omega_e = 4\pi\epsilon_0\hbar E_0/(Mq_e) = q_e E_0 a_1/\hbar$, we obtain the matrix

$$\hat{H}_E = \frac{3\hbar\Omega_e}{\sqrt{2}}\begin{pmatrix} 0 & 0 & 0 & 1 \\ 0 & 0 & 0 & 0 \\ 0 & 0 & 0 & -1 \\ 1 & 0 & -1 & 0 \end{pmatrix}. \qquad (9.7)$$

where the rows (columns) are ordered as $2p, m = 1, 0, -1; 2s$.

**9.3.1.3** We want to find the eigenvalues of the matrix

$$\hbar\begin{pmatrix} \omega_0 & 0 & 0 & 3\Omega_e/\sqrt{2} \\ 0 & 0 & 0 & 0 \\ 0 & 0 & -\omega_0 & -3\Omega_e/\sqrt{2} \\ 3\Omega_e/\sqrt{2} & 0 & -3\Omega_e/\sqrt{2} & 0 \end{pmatrix}. \qquad (9.8)$$

There is an obvious eigenvalue $\lambda = 0$ since the $|2p, m = 0\rangle$ and $|2s\rangle$ states do not mix in the presence the electric field. The three other eigenvalues are easily obtained as the solutions of:

$$\lambda(\hbar^2\omega_0^2 - \lambda^2) + 9\hbar^2\Omega_e^2\lambda = 0, \qquad (9.9)$$

i.e. $\lambda = 0$ and $\lambda = \pm\hbar\sqrt{\omega_0^2 + 9\,\Omega_e^2}$.

The shifts of the energy levels are therefore: $\delta E = 0$ twice degenerate, and $\delta E = \pm\hbar\sqrt{\omega_0^2 + 9\Omega_e^2}$. If we adopt the prescription given in the text, we obtain

$$\omega_e = 3\Omega_e \Rightarrow f(2) = 2. \qquad (9.10)$$

### 9.3.2  Pauli's Result

**9.3.2.1** We remark that the experimental points are aligned on a straight line $aB_0^2 + bE_0^2 = constant$ which is in agreement with (9.1), i.e. a constant value of $\omega_0^2 + \omega_e^2$ corresponds to a constant value of each energy level.

**9.3.2.2** Given the definitions of $\omega_0$ and $\Omega_e$, one has

$$\omega_0^2 + \omega_e^2 = \frac{9}{4} \left( \frac{4\pi\epsilon_0\hbar}{Mq_e} \right)^2 \left[ \left( \frac{\alpha c}{3} \right)^2 B_0^2 + f^2(n) E_0^2 \right], \tag{9.11}$$

where $\alpha$ is the fine structure constant and $c$ the velocity of light. The experimental line

$$(\alpha c/3)^2 B_0^2 + f^2(34) E_0^2 \tag{9.12}$$

goes through the points $E_0^2 = 0$, $B_0^2 \simeq 8.7 \times 10^{-3}$ T$^2$ and $B_0^2 = 0$, $E_0^2 \simeq 4 \times 10^6$ V$^2$m$^{-2}$. This gives $f(34) \simeq 34$.

**9.3.2.3** Indeed, the very simple result found by Pauli was $f(n) = n$.

### 9.3.3 References

W. Pauli published his work on the spectrum of the Hydrogen atom in *Über das wasserstoffspektrum vom standpunkt der neuen quantenmechanik*, Zeitschrift für Physik **36**, 336 (1926).

The data shown in Fig. 9.1 are extracted form the work by F. Biraben, D. Delande, J.-C. Gay, and F. Penent, *Rydberg atoms in crossed electric and magnetic fields. Experimental evidence for Pauli quantization*, Optics Comm. **49**, 184 (1984). Note that the data shown here have been slightly corrected with respect to the original publication, in order to account for the role of second order Stark effect and diamagnetism in the measurement procedure. See also J.-C. Gay, in *Atoms in unusual situations*, J.-P. Briand ed., p. 107 (Plenum, New York, 1986).

# Energy Loss of Ions in Matter

**10**

When a charged particle travels through condensed matter, it loses its kinetic energy gradually by transferring it to the electrons of the medium. In this chapter we evaluate the energy loss of the particle as a function of its mass and its charge, by studying the modifications that the state of an atom undergoes when a charged particle passes in its vicinity. We show how this process can be used to identify the products of a nuclear reaction.

The electric potential created by the moving particle appears as a time-dependent perturbation in the atom's Hamiltonian. In order to simplify the problem, we shall consider the case of an atom with a single external electron. The nucleus and the internal electrons will be treated globally as a core of charge $+q$, infinitely massive and, therefore, fixed in space. We also assume that the incident particle of charge $Z_1 q$ is heavy and non-relativistic, and that its kinetic energy is large enough so that in good approximation its motion can be considered linear and uniform, of constant velocity $v$, when it interacts with an atom.

Here $q$ denotes the unit charge and we set $e^2 = q^2/(4\pi \varepsilon_0)$. We consider the $x, y$ plane defined by the trajectory of the particle and the center of gravity of the atom, which is chosen to be the origin, as shown on Fig. 10.1.

Let $R(t)$ be the position of the particle at time $t$ and $r = (x, y, z)$ the coordinates of the electron of the atom. The impact parameter is $b$ and the notation is specified in Fig. 10.1. The time at which the particle passes nearest to the atom, i.e. $x = b, y = 0$ is denoted $t = 0$. We write $E_n$ and $|n\rangle$ for the energy levels and corresponding eigenstates of the atom in the absence of an external perturbation.

© Springer Nature Switzerland AG 2019
J.-L. Basdevant, J. Dalibard, *The Quantum Mechanics Solver*,
https://doi.org/10.1007/978-3-030-13724-3_10

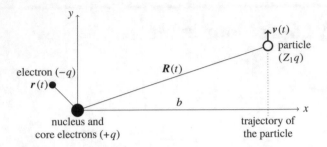

**Fig. 10.1** Definition of the coordinates

## 10.1   Energy Absorbed by One Atom

**10.1.1** Write the expression for the time-dependent perturbing potential $\hat{V}(t)$ due to the presence of the charged particle.

**10.1.2** We assume that the impact parameter $b$ is much larger than the typical atomic size, i.e. $b \gg \langle r \rangle$, so that $|R(t)| \gg |r|$ for all $t$. Replace $\hat{V}(t)$ by its first order expansion in $|r|/|R|$ and express the result in terms of the coordinates $x$ and $y$ of the electron, and of $b$, $v$ and $t$.

**10.1.3** Initially, at time $t = -\infty$, the atom is in a state $|i\rangle$ of energy $E_i$. Using first order time-dependent perturbation theory, write the probability amplitude $\gamma_{if}$ to find the atom in the final state $|f\rangle$ of energy $E_f$ after the charged particle has passed ($t = +\infty$). We set $\omega_{fi} = (E_f - E_i)/\hbar$ and we only consider the case $E_f \neq E_i$.

**10.1.4** The calculation of $\gamma_{if}$ involves the Bessel function $K_0(z)$. One has

$$\int_0^\infty \frac{\cos \omega t}{(\beta^2 + t^2)^{1/2}} dt = K_0(\omega\beta) \qquad \int_0^\infty t \frac{\sin \omega t}{(\beta^2 + t^2)^{3/2}} dt = \omega K_0(\omega\beta). \qquad (10.1)$$

Express $\gamma_{if}$ in terms of $K_0$ and its derivative.

The asymptotic behavior of $K_0$ is $K_0(z) \simeq -\ln z$ for $z \ll 1$, and $K_0(z) \simeq \sqrt{2\pi/z}\, e^{-z}$ for $z \gg 1$. Under what condition on the parameters $\omega_{fi}$, $b$ and $v$ is the transition probability $P_{if} = |\gamma_{if}|^2$ large?

Show that, in that case, one obtains

$$P_{if} \simeq \left(\frac{2Z_1 e^2}{\hbar v}\right)^2 |\langle f|\hat{x}|i\rangle|^2. \qquad (10.2)$$

**10.1.5** Give the physical interpretation of the condition derived above. Show that, given the parameters of the atom, the crucial parameter is the *effective interaction time*, and give a simple explanation of this effect.

## 10.2 Energy Loss in Matter

We assume in the following that the Hamiltonian of the atom is of the form

$$\hat{H}_0 = \frac{\hat{p}^2}{2m} + V(\hat{r}). \tag{10.3}$$

### 10.2.1 Thomas–Reiche–Kuhn sum rule

(a) Calculate the commutator $[\hat{x}, \hat{H}_0]$.
(b) Deduce from this commutator a relation between the matrix elements $\langle i|\hat{x}|f \rangle$ and $\langle i|\hat{p}|f \rangle$, where $|i\rangle$ and $|f\rangle$ are eigenstates of $\hat{H}_0$.
(c) Applying a closure relation to $[\hat{x}, \hat{p}] = i\hbar$, show that:

$$\frac{2m}{\hbar^2} \sum_f (E_f - E_i)|\langle f|\hat{x}|i \rangle|^2 = 1 \tag{10.4}$$

for all eigenstates $|i\rangle$ of $H_0$.

**10.2.2** Using the Thomas–Reiche–Kuhn sum rule, calculate the expectation value $\delta E$ of the energy loss of the incident particle when it interacts with the atom.

Let $E$ be the energy of the particle before the interaction. Which parameters does the product $E \, \delta E$ depend on?

**10.2.3** Experimental Application. We are now interested in incident particles which are fully ionized atoms ($Z_1 = Z$, where $Z$ is the atomic number), whose masses are, to a good approximation, proportional to the mass number $A = Z + N$ (where $N$ is the number of neutrons of the isotope). When these ions traverse condensed matter, they interact with many atoms of the medium, and their energy loss implies some averaging over the random impact parameter $b$. The previous result then takes the form

$$E \, \delta E = kZ^2 A, \tag{10.5}$$

where the constant $k$ depends on the nature of the medium.

Semiconductor detectors used for the identification of the nuclei in nuclear reactions are based on this result. In the following example, the ions to be identified are the final state products of a reaction induced by 113 MeV nitrogen ions impinging on a target of silver atoms.

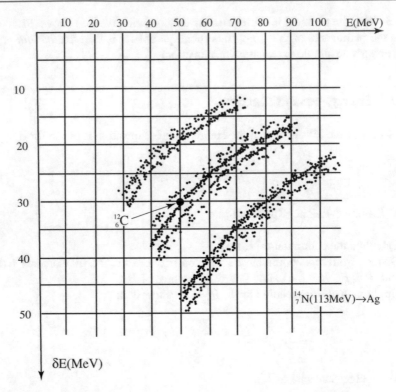

**Fig. 10.2** Energy loss $\delta E$ versus energy $E$ through a silicon detector, of the final products of a reaction corresponding to 113 MeV nitrogen ions impinging on a target of silver atoms

In Fig. 10.2 each point represents an event, i.e. the energy $E$ and energy loss $\delta E$ of an ion when it crosses a silicon detector. The reference point corresponds to the isotope $A = 12$ of carbon $^{12}_{6}C$ (we use the notation $^{A}_{Z}N$ for a nucleus charge $Z$ and mass number $A$) which loses $\delta E = 30$ MeV at an energy $E = 50$ MeV.

(a) Calculate the constant $k$ and the theoretical prediction for the energy loss at 60 and 70 MeV. Put the corresponding points on the figure.
(b) Assuming the reaction could produce the following isotopes:
    – boron, $Z = 5$, $A = 10, 11, 12$
    – carbon, $Z = 6$, $A = 11, 12, 13, 14$
    – nitrogen $Z = 7$, $A = 13, 14, 15, 16$,

    what are the nuclei effectively produced in the reaction? Justify the answers by inserting the points corresponding to $E = 50$ MeV and $E = 70$ MeV on the figure.

## 10.3 Solutions

### 10.3.1 Energy Absorbed by One Atom

**10.3.1.1** The interaction potential between the particle and the atom is the sum of the Coulomb interactions between the particle and the core, and those between the particle and the outer electron:

$$\hat{V}(t) = \frac{Z_1 e^2}{R(t)} - \frac{Z_1 e^2}{|R(t) - \hat{r}|}. \tag{10.6}$$

**10.3.1.2** For $|R| \gg \langle |r| \rangle$, we have

$$\frac{1}{|R - r|} = (R^2 - 2R \cdot r + r^2)^{-1/2} \simeq \frac{1}{R} + \frac{r \cdot R}{R^3}. \tag{10.7}$$

Therefore

$$\hat{V}(t) \simeq -\frac{Z_1 e^2}{R^3(t)} \hat{r} \cdot R(t). \tag{10.8}$$

Since $R(t) = (b, vt, 0)$, we obtain

$$\hat{V}(t) \simeq -\frac{Z_1 e^2}{(b^2 + v^2 t^2)^{3/2}} (\hat{x}b + \hat{y}vt). \tag{10.9}$$

**10.3.1.3** To first order in $\hat{V}$, the probability amplitude is

$$\gamma_{if} = \frac{1}{i\hbar} \int_{-\infty}^{+\infty} e^{i\omega_{fi}t} \langle f|\hat{V}(t)|i\rangle \mathrm{d}t. \tag{10.10}$$

Inserting the value found above for $\hat{V}(t)$, we find

$$\gamma_{if} = -\frac{1}{i\hbar} \int_{-\infty}^{+\infty} \frac{Z_1 e^2 e^{i\omega_{fi}t}}{(b^2 + v^2 t^2)^{3/2}} (b\langle f|\hat{x}|i\rangle + vt\langle f|\hat{y}|i\rangle) \, \mathrm{d}t. \tag{10.11}$$

**10.3.1.4** One has

$$\int_0^\infty \frac{\cos \omega t \, \mathrm{d}t}{(\beta^2 + t^2)^{3/2}} = -\frac{1}{\beta} \frac{d}{d\beta} K_0(\omega\beta) = -\frac{\omega}{\beta} K_0'(\omega\beta). \tag{10.12}$$

Setting $\beta = b/v$, the amplitude $\gamma_{if}$ is

$$\gamma_{if} = i\frac{2Z_1 e^2 \omega_{fi}}{\hbar v^2} (K_0(\omega_{fi}b/v)\langle f|\hat{y}|i\rangle - K_0'(\omega_{fi}b/v)\langle f|\hat{x}|i\rangle). \tag{10.13}$$

The probability $P_{if} = |\gamma_{if}|^2$ is large if $K_0$ or $K_0'$ are also large. This happens for $\omega_{fi}b/v \ll 1$. In this limit, $K_0(z) \sim -\ln z$ and $K_0'(z) \sim -1/z$, and we obtain

$$\gamma_{if} = i\frac{2Z_1e^2}{\hbar v b} \left( \langle f|\hat{x}|i\rangle - \langle f|\hat{y}|i\rangle \frac{\omega_{fi}b}{v} \ln \frac{\omega_{fi}b}{v} \right). \tag{10.14}$$

Since $|\langle f|\hat{x}|i\rangle| \simeq |\langle f|\hat{y}|i\rangle|$, one can neglect the second term ($|x \ln x| \ll 1$ for $x \ll 1$) and we obtain, for $\omega_{fi}b/v \ll 1$,

$$P_{if} = |\gamma_{if}|^2 \simeq \left(\frac{2Z_1e^2}{\hbar b v}\right)^2 |\langle f|\hat{x}|i\rangle|^2. \tag{10.15}$$

**10.3.1.5** The time $\tau = b/v$ is the characteristic time during which the interaction is important, as we can see on the above formulas. For $t \gg \tau$, the interaction is negligible.

The condition $\omega_{fi}\tau \ll 1$ means that the interaction time $\tau$ must be much smaller than the Bohr period $\sim 1/\omega_{fi}$ of the atom. The perturbation $\hat{V}(t)$ must have a large Fourier component at $\omega = \omega_{fi}$ if we want the probability $P_{if}$ to be significant (the shorter in time the perturbation, the larger the spread of its Fourier transform in frequency). In the opposite limiting case, where the perturbation is infinitely slow, the atom is not excited.

This observation provides an alternative way to evaluate the integrals of question 10.3.1.3. The only values of $t$ which contribute significantly are those for which $t$ is not too large, compared to $\tau$ (say $|t| \ll 10\,\tau$). If $\omega_{fi}\tau \ll 1$, one can replace $e^{i\omega_{fi}t}$ by 1 in these integrals; the second integral is then zero for symmetry reasons and the first one is easily evaluated, and gives the desired result.

### 10.3.2   Energy Loss in Matter

**10.3.2.1** Thomas–Reiche–Kuhn sum rule.

**(a)** We find $[\hat{x}, \hat{H}_0] = i\hbar\hat{p}/m$.

**(b)** Taking the matrix element of this commutator between two eigenstates $|i\rangle$ and $|f\rangle$ of $\hat{H}_0$, we obtain:

$$\frac{i\hbar}{m}\langle f\,|\hat{p}\,|\,i\rangle = \langle f\,|\,[\hat{x}, \hat{H}_0]\,|\,i\rangle = (E_i - E_f)\langle f\,|\,\hat{x}\,|\,i\rangle. \tag{10.16}$$

(c) We now take the matrix element of $[\hat{x}, \hat{p}] = i\hbar$ between $\langle i|$ and $|i\rangle$ and we use the closure relation:

$$i\hbar = \sum_f \langle i \mid \hat{x} \mid f\rangle \langle f \mid \hat{p} \mid i\rangle - \sum_f \langle i \mid \hat{p} \mid f\rangle \langle f \mid \hat{x} \mid i\rangle$$

$$= \frac{m}{i\hbar} \sum_f (E_i - E_f) \mid \langle f \mid \hat{x} \mid i\rangle \mid^2 - \frac{m}{i\hbar} \sum_f (E_f - E_i) \mid \langle i \mid \hat{x} \mid f\rangle \mid^2$$

$$= \frac{2m}{i\hbar} \sum_f (E_i - E_f) \mid \langle f \mid \hat{x} \mid i\rangle \mid^2 \qquad (10.17)$$

This proves the Thomas–Reiche–Kuhn sum rule.

**10.3.2.2** The expectation value $\delta E$ of the energy transferred to the atom is

$$\delta E = \sum_f (E_f - E_i) P_{if} = \left(\frac{2Z_1 e^2}{\hbar v}\right)^2 \sum_f (E_f - E_i) \mid \langle f \mid \hat{x} \mid i\rangle \mid^2. \qquad (10.18)$$

Making use of the Thomas–Reiche–Kuhn sum rule, we obtain

$$\delta E = \frac{2Z_1^2 e^4}{mb^2 v^2}, \qquad (10.19)$$

where $m$ is the electron mass. If the ion has mass $M$, its kinetic energy is $E = Mv^2/2$, and we therefore obtain a very simple expression:

$$E \, \delta E = \frac{M}{m} \left(\frac{Z_1 e^2}{b}\right)^2, \qquad (10.20)$$

where we see that the product $E \, \delta E$ does not depend on the energy of the incident particle, but is proportional to its mass and to the square of its charge.

**10.3.2.3** With the $^{12}_{6}$C point, one obtains $k = 3.47$. We have put the calculated points of the various isotopes on Fig. 10.3.

We make the following observations:

(a) For boron, the isotopes $^{10}$B and $^{11}$B are produced, but not $^{12}$B.

(b) For carbon, $^{12}$C is produced more abundantly than $^{13}$C, $^{14}$C and $^{11}$C.

(c) For nitrogen, there is an abundant production of $^{14}$N, a smaller production of $^{15}$N, but practically no $^{13}$N or $^{16}$N.

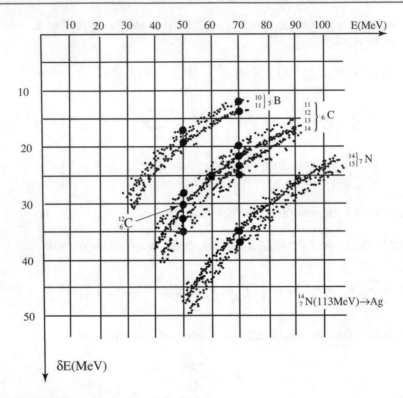

**Fig. 10.3** Interpretation of the data of Fig. 10.2

### 10.3.3 Comments

Ionization of matter has numerous applications, for instance in developing detectors
for particle and nuclear physics, or in defining protection regulations against
radioactivity. In order to calculate the energy loss of an ion in matter, one must
integrate the above results over the impact parameter. In practice, taking everything
into account, one ends up with the following formula, due to Hans Bethe and Felix
Bloch, for the rate of energy loss per unit length:

$$-\frac{dE}{dx} = \frac{4\pi K^2 Z^2 e^4 N}{m_e c^2 \beta^2} \left[ \ln \left( \frac{2 m_e c^2 \beta^2}{I(1 - \beta^2)} \right) - \beta^2 \right] \tag{10.21}$$

where $\beta = v/c$, $K$ is a constant, $N$ is the number density of atoms in the medium
and $I$ is the mean excitation energy of the medium ($I \sim 11.5\,\text{eV}$).

The cases of protons or heavy ions is of great interest and, in comparatively recent
years, it has allowed a major improvement in the medical treatment of tumors in
the eyes (proton therapy) and in the brain (ion therapy). Owing to the factor $1/\beta^2$,

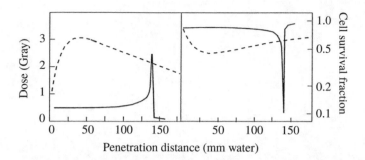

**Fig. 10.4** Energy loss of ions (*left*) and survival rate of cells (*right*) as a function of the penetration depth. The *dashed* curve corresponds to the same quantities for photons. We can see the considerable medical advantage to use heavy ion beams. Document from the data of *Heavy ion therapy at GSI*, Darmstadt, http://www.gsi.de. (Courtesy James Rich)

or equivalently $1/v^2$, in (10.21), practically *all* the energy is deposited in a very localized region near the stopping point.

Figure 10.4 shows the comparison between the effect of ion beams and photons. One can see the enormous advantage, from the medical point of view, of heavy ion beams. These permit to attack and destroy tumors in a very accurate and localized manner, as opposed to $\gamma$ rays which produce damages all around the point of interest. Pioneering work on brain tumor therapy has been developed in Darmstadt at the Heavy ion accelerator facility. This promising sector of medical applications in rapidly developing at present.

**Additional references** A. V. Solov'yov et al., *Physics of ion beam cancer therapy: A multiscale approach* Phys. Rev. E **79**, 011909 (2009); E. Surdutovich, D. C. Gallagher and A. V. Solov'yov, *Calculation of complex DNA damage induced by ions*, Phys. Rev.E **84**, 051918 (2011); W. D. Newhauser and R. Zhang, *The physics of proton therapy*, Phys. Med. Biol. **60**, R155 (2015).

# Part II

# Quantum Entanglement and Measurement

# The EPR Problem and Bell's Inequality

# 11

When a quantum system possesses more than one degree of freedom, the associated Hilbert space is a tensor product of the spaces associated to each degree of freedom. This structure leads to specific properties of quantum mechanics, whose paradoxical character has been pointed out by Einstein, Podolsky and Rosen. Here we study an example of such a situation, by considering *entangled states* for the spins of two particles.

The model system under consideration is a hydrogen atom which is dissociated into an electron and a proton. We consider the spin states of these two particles when they have left the dissociation region and are located in geometrically distinct regions, e.g. a few meters from one another. They are then free particles whose spin states do not evolve.

## 11.1 The Electron Spin

Consider a unit vector $\boldsymbol{u}_\varphi$ in the $xz$ plane: $\boldsymbol{u}_\varphi = \cos\varphi\,\boldsymbol{u}_z + \sin\varphi\,\boldsymbol{u}_x$, where $\boldsymbol{u}_z$ and $\boldsymbol{u}_x$ are unit vectors along the $z$ and $x$ axes. We note $\hat{S}_{e\varphi} = \hat{S}_e.\boldsymbol{u}_\varphi$ the component of the electron spin along the $\boldsymbol{u}_\varphi$ axis.

**11.1.1** What are the eigenvalues of $\hat{S}_{e\varphi}$?

**11.1.2** We denote the eigenvectors of $\hat{S}_{e\varphi}$ by $|e : +\varphi\rangle$ and $|e, -\varphi\rangle$ which, in the limit $\varphi = 0$, reduce respectively to the eigenvectors $|e : +\rangle$ and $|e : -\rangle$ of $\hat{S}_{ez}$. Express $|e : +\varphi\rangle$ and $|e : -\varphi\rangle$ in terms of $|e : +\rangle$ and $|e : -\rangle$.

**11.1.3** Assume the electron is emitted in the state $|e : +\varphi\rangle$. One measures the component $\hat{S}_{e\alpha}$ of the spin along the direction $\boldsymbol{u}_\alpha = \cos\alpha\,\boldsymbol{u}_z + \sin\alpha\,\boldsymbol{u}_x$. What is the probability $P_+(\alpha)$ of finding the electron in the state $|e : +\alpha\rangle$? What is the expectation value $\langle \hat{S}_{e\alpha} \rangle$ in the spin state $|e : +\varphi\rangle$?

© Springer Nature Switzerland AG 2019
J.-L. Basdevant, J. Dalibard, *The Quantum Mechanics Solver*,
https://doi.org/10.1007/978-3-030-13724-3_11

## 11.2   Correlations Between the Two Spins

We first assume that, after the dissociation, the electron–proton system is in the factorized spin state $|e : +\varphi\rangle \otimes |p : -\varphi\rangle$.

We recall that if $|u_1\rangle$ and $|u_2\rangle$ are vectors of the Hilbert space $E$, and $|v_1\rangle$ and $|v_2\rangle$ of the Hilbert space $F$, if $|u\rangle \otimes |v\rangle$ belongs to the tensor product $G = E \otimes F$, and if $\hat{A}$ and $\hat{B}$ act respectively in $E$ and $F$, $\hat{C} = \hat{A} \otimes \hat{B}$ acting in $G$, one has:

$$\langle u_2| \otimes \langle v_2| \hat{C} |u_1\rangle \otimes |v_1\rangle = \langle u_2|\hat{A}|u_1\rangle \, \langle v_2|\hat{B}|v_1\rangle. \qquad (11.1)$$

**11.2.1** What is the probability $P_+(\alpha)$ of finding $+\hbar/2$ when measuring the component $\hat{S}_{e\alpha}$ of the electron spin in this state?

(a) Having found this value, what is the state of the system after the measurement?
(b) Is the proton spin state affected by the measurement of the electron spin?

**11.2.2** Calculate the expectation values $\langle \hat{S}_{e\alpha}\rangle$ and $\langle \hat{S}_{p\beta}\rangle$ of the components of the electron and the proton spins along axes defined respectively by $\boldsymbol{u}_\alpha$ and $\boldsymbol{u}_\beta = \cos \beta \, \boldsymbol{u}_z + \sin \beta \, \boldsymbol{u}_x$.

**11.2.3** The correlation coefficient between the two spins $E(\alpha, \beta)$ is defined as

$$E(\alpha, \beta) = \frac{\langle \hat{S}_{e\alpha} \otimes \hat{S}_{p\beta}\rangle - \langle \hat{S}_{e\alpha}\rangle \, \langle \hat{S}_{p\beta}\rangle}{\left(\langle \hat{S}_{e\alpha}^2\rangle \, \langle \hat{S}_{p\beta}^2\rangle\right)^{1/2}}. \qquad (11.2)$$

Calculate $E(\alpha, \beta)$ in the state under consideration.

## 11.3   Correlations in the Singlet State

We now assume that, after the dissociation, the two particles are in the singlet spin state:

$$|\Psi_s\rangle = \frac{1}{\sqrt{2}} \left(|e : +\rangle \otimes |p : -\rangle - |e : -\rangle \otimes |p : +\rangle\right). \qquad (11.3)$$

**11.3.1** One measures the component $\hat{S}_{e\alpha}$ of the electron spin along the direction $\boldsymbol{u}_\alpha$. Give the possible results and their probabilities.

**11.3.2** Suppose the result of this measurement is $+\hbar/2$. Later on, one measures the component $\hat{S}_{p\beta}$ of the proton spin along the direction $\boldsymbol{u}_\beta$. Here again give the possible results and their probabilities.

**11.3.3**  Would one have the same probabilities if the proton spin had been measured *before* the electron spin?

Why was this result shocking for Einstein who claimed that "the real states of two spatially separated objects must be independent of one another"?

**11.3.4**  Calculate the expectation values $\langle \hat{S}_{e\alpha} \rangle$ and $\langle \hat{S}_{p\beta} \rangle$ of the electron and the proton spin components if the system is in the singlet state (11.3).

**11.3.5**  Calculate $E(\alpha, \beta)$ in the singlet state.

## 11.4   A Simple Hidden Variable Model

For Einstein and several other physicists, the solution to the "paradox" uncovered in the previous section could come from the fact that the states of quantum mechanics, in particular the singlet state (11.3), provide an incomplete description of reality. A "complete" theory (for predicting spin measurements, in the present case) should incorporate additional variables or parameters, whose knowledge would render measurements independent for two spatially separated objects. However, present experiments cannot determine the values of these parameters, which are therefore called "hidden variables". The experimental result should then consist in some averaging over these unknown parameters.

In the case of interest, a very simplified example of such a theory is the following. We assume that, after each dissociation, the system is in a factorized state $|e : +\varphi\rangle \otimes |p : -\varphi\rangle$, but that the direction $\varphi$ varies from one event to the other. In this case, $\varphi$ is the hidden variable. We assume that all directions $\varphi$ are equally probable, i.e. the probability density that the decay occurs with direction $\varphi$ is uniform and equal to $1/2\pi$.

Owing to this ignorance of the value of $\varphi$, the expectation value of an observable $\hat{A}$ is now *defined* to be:

$$\langle \hat{A} \rangle = \frac{1}{2\pi} \int_0^{2\pi} \langle e : +\varphi| \otimes \langle p : -\varphi| \hat{A} |e : +\varphi\rangle \otimes |p : -\varphi\rangle \, d\varphi. \tag{11.4}$$

**11.4.1**  Using the definition (11.2) for $E(\alpha, \beta)$ and the new definition (11.4) for expectation values, calculate $F(\alpha, \beta)$ in this new theory. Compare the result with the one found using "orthodox" quantum mechanics in Sect. 11.3.5.

**11.4.2**  The first precise experimental tests of hidden variable descriptions vs. quantum mechanics have been performed on correlated pairs of photons emitted in an atomic cascade. Although one is not dealing with spin-1/2 particles in this case, the physical content is basically the same as here. As an example, Fig. 11.1 presents experimental results obtained by A. Aspect and his collaborators in 1982.

**Fig. 11.1** Measured
variation of $E(\alpha, \beta)$ as a
function of $\alpha - \beta$

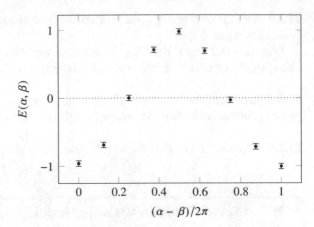

It gives the variation of $E(\alpha, \beta)$ as a function of the difference $\alpha - \beta$, which is found
to be the only experimentally relevant quantity.

Which theory, quantum mechanics or the simple hidden variable model developed above, gives a good account of the experimental data?

## 11.5   Bell's Theorem and Experimental Results

As proved by Bell in 1965, the disagreement between the predictions of quantum
mechanics and those of hidden variable theories is actually very general when one
considers correlation measurements on entangled states. We now show that the
correlation results for hidden variable theories are constrained by what is known as
Bell's inequality, which, however, can be violated by quantum mechanics in specific
experimental configurations.

Consider a hidden variable theory, whose result consists in two functions
$A(\lambda, \boldsymbol{u}_\alpha)$ and $B(\lambda, \boldsymbol{u}_\beta)$ giving respectively the results of the electron and proton
spin measurements. Each of these two functions takes only the two values $\hbar/2$ and
$-\hbar/2$. It depends on the value of the hidden variable $\lambda$ for the considered electron–
proton pair. The nature of this hidden variable does not need to be further specified
for the proof of Bell's theorem. The result $A$ of course depends on the axis $\boldsymbol{u}_\alpha$
chosen for the measurement of the electron spin, but it *does not* depend on the axis
$\boldsymbol{u}_\beta$. Similarly $B$ does not depend on $\boldsymbol{u}_\alpha$. This *locality* hypothesis is essential for the
following discussion.

**11.5.1**   Give the correlation coefficient $E(\alpha, \beta)$ for a hidden variable theory in terms
of the functions $A$ and $B$ and the (unknown) distribution law $P(\lambda)$ for the hidden
variable $\lambda$.

**11.5.2** Show that for any set $u_\alpha, u'_\alpha, u_\beta, u'_\beta$, one has

$$A(\lambda, u_\alpha)\, B(\lambda, u_\beta) + A(\lambda, u_\alpha)\, B(\lambda, u'_\beta)$$

$$+ A(\lambda, u'_\alpha)\, B(\lambda, u'_\beta) - A(\lambda, u'_\alpha)\, B(\lambda, u_\beta) = \pm\frac{\hbar^2}{2}. \tag{11.5}$$

**11.5.3** We define the quantity $S$ as

$$S = E(\alpha, \beta) + E(\alpha, \beta') + E(\alpha', \beta') - E(\alpha', \beta). \tag{11.6}$$

Derive Bell's inequality

$$|S| \leq 2. \tag{11.7}$$

**11.5.4** Consider the particular case $\alpha - \beta = \beta' - \alpha = \alpha' - \beta' = \pi/4$, and compare the predictions of quantum mechanics with the constraint imposed by Bell's inequality.

**11.5.5** The experimental results obtained by A. Aspect et al. are

$$E(\alpha, \beta) = -0.66\ (\pm 0.04) \quad \text{for } \alpha - \beta = \pi/4 \tag{11.8}$$

$$E(\alpha, \beta) = +0.68\ (\pm 0.03) \quad \text{for } \alpha - \beta = 3\pi/4 \tag{11.9}$$

Is a description of these experimental results by a local hidden variable theory possible? Are these results compatible with quantum mechanics?

## 11.6   Solutions

### 11.6.1 The Electron Spin

**11.6.1.1** In the eigenbasis $|e : \pm\rangle$ of $\hat{S}_{ez}$, the matrix of $\hat{S}_{e\varphi}$ is

$$\frac{\hbar}{2}\begin{pmatrix} \cos\varphi & \sin\varphi \\ \sin\varphi & -\cos\varphi \end{pmatrix}. \tag{11.10}$$

The eigenvalues of this operator are $+\hbar/2$ and $-\hbar/2$.

**11.6.1.2** The corresponding eigenvectors are

$$|e : +\varphi\rangle = \cos\frac{\varphi}{2}|e : +\rangle + \sin\frac{\varphi}{2}|e : -\rangle \tag{11.11}$$

$$|e : -\varphi\rangle = -\sin\frac{\varphi}{2}|e : +\rangle + \cos\frac{\varphi}{2}|e : -\rangle. \tag{11.12}$$

**11.6.1.3** The probability amplitude is $\langle e : +\alpha | e : +\varphi \rangle = \cos((\varphi - \alpha)/2)$ and the probability $P_+(\alpha) = \cos^2((\varphi - \alpha)/2)$. Similarly $P_-(\alpha) = \sin^2((\varphi - \alpha)/2)$, and the expectation value is, finally,

$$\langle \hat{S}_{e\alpha} \rangle = \frac{\hbar}{2} \cos(\varphi - \alpha). \tag{11.13}$$

## 11.6.2 Correlations Between the Two Spins

**11.6.2.1** The projector on the eigenstate $|e : +\alpha\rangle$, corresponding to the measured value, is $|e : +\alpha\rangle\langle e : +\alpha| \otimes \hat{I}_p$, where $\hat{I}_p$ is the identity operator on the proton states. Therefore

$$P_+(\alpha) = |\langle e : +\alpha | e : +\varphi \rangle|^2 = \cos^2 \frac{\varphi - \alpha}{2}, \tag{11.14}$$

and the state after measurement is $|e : +\alpha\rangle \otimes |p : -\varphi\rangle$. The proton spin is not affected, because the initial state is factorized (all probability laws are factorized).

**11.6.2.2** One has $\langle \hat{S}_{e\alpha} \rangle = \frac{\hbar}{2} \cos(\varphi - \alpha)$ and $\langle \hat{S}_{p\beta} \rangle = -\frac{\hbar}{2} \cos(\varphi - \beta)$.

**11.6.2.3** By definition, one has:

$$\hat{S}_{e\alpha}^2 = \frac{\hbar^2}{4} \hat{I}_e \quad \text{and} \quad \hat{S}_{p\beta}^2 = \frac{\hbar^2}{4} \hat{I}_p \tag{11.15}$$

and

$$\langle \hat{S}_{e\alpha} \otimes \hat{S}_{p\beta} \rangle = \langle e : +\varphi | \hat{S}_{e\alpha} | e : +\varphi \rangle \langle p : -\varphi | \hat{S}_{p\beta} | p : -\varphi \rangle$$

$$= -\frac{\hbar^2}{4} \cos(\varphi - \alpha) \cos(\varphi - \beta). \tag{11.16}$$

Therefore $E(\alpha, \beta) = 0$. This just reflects the fact that in a factorized state, the two spin variables are independent.

## 11.6.3 Correlations in the Singlet State

**11.6.3.1** There are two possible values:
   $\hbar/2$, corresponding to the projector $|e : +\alpha\rangle\langle e : +\alpha| \otimes \hat{I}_p$, and $-\hbar/2$, corresponding to the projector $|e : -\alpha\rangle\langle e : -\alpha| \otimes \hat{I}_p$.
   Therefore, the probabilities are

$$P_+(\alpha) = \frac{1}{2}\left( |\langle e : +\alpha | e : +\rangle|^2 + |\langle e : +\alpha | e : -\rangle|^2 \right) = 1/2 \tag{11.17}$$

and similarly $P_-(\alpha) = 1/2$. This result is a consequence of the rotational invariance of the singlet state.

**11.6.3.2** The state after the measurement of the electron spin, yielding the result $+\hbar/2$, is

$$\cos\frac{\alpha}{2}|e : +\alpha\rangle \otimes |p : -\rangle - \sin\frac{\alpha}{2}|e : +\alpha\rangle \otimes |p : +\rangle = |e : +\alpha\rangle \otimes |p : -\alpha\rangle.$$

(11.18)

This simple result is also a consequence of the rotational invariance of the singlet state, which can be written as

$$|\Psi_s\rangle = \frac{1}{\sqrt{2}}(|e : +\alpha\rangle \otimes |p : -\alpha\rangle - |e : -\alpha\rangle \otimes |p : +\alpha\rangle).$$

(11.19)

Now the two possible results for the measurement of the proton spin $\pm\hbar/2$ have probabilities

$$P_+(\beta) = \sin^2\frac{\alpha - \beta}{2} \qquad P_-(\beta) = \cos^2\frac{\alpha - \beta}{2}.$$

(11.20)

**11.6.3.3** If one had measured $\hat{S}_{p\beta}$ first, one would have found $P_+(\beta) = P_-(\beta) = 1/2$.

The fact that a measurement on the electron affects the probabilities for the results of a measurement on the proton, although the two particles are spatially separated, is in contradiction with Einstein's assertion, or belief. This is the starting point of the Einstein–Podolsky–Rosen paradox. Quantum mechanics is *not a local theory* as far as measurement is concerned.

Note, however, that this non-locality does not allow the instantaneous transmission of information. From a single measurement of the proton spin, one cannot determine whether the electron spin has been previously measured. It is only when, for a series of experiments, the results of the measurements on the electron and the proton are later compared, that one can find this non-local character of quantum mechanics.

**11.6.3.4** Individually, the expectation values vanish, since one does not worry about the other variable:

$$\langle \hat{S}_{e\alpha}\rangle = \langle \hat{S}_{p\beta}\rangle = 0.$$

(11.21)

**11.6.3.5** However, the spins are correlated and we have

$$\langle \hat{S}_{e\alpha} \otimes \hat{S}_{p\beta}\rangle = \frac{\hbar^2}{4}\left(\sin^2\frac{\alpha - \beta}{2} - \cos^2\frac{\alpha - \beta}{2}\right)$$

(11.22)

and therefore $E(\alpha, \beta) = -\cos(\alpha - \beta)$.

## 11.6.4  A Simple Hidden Variable Model

**11.6.4.1**  Using the results of Sect. 11.2, we have:

$$\langle \hat{S}_{e\alpha} \rangle = \frac{\hbar}{2} \int \cos(\varphi - \alpha) \frac{d\varphi}{2\pi} = 0 \tag{11.23}$$

and similarly $\langle \hat{S}_{p\beta} \rangle = 0$. We also obtain

$$\langle \hat{S}_{e\alpha} \otimes \hat{S}_{p\beta} \rangle = -\frac{\hbar^2}{4} \int \cos(\varphi - \alpha) \cos(\varphi - \beta) \frac{d\varphi}{2\pi}$$

$$= -\frac{\hbar^2}{8} \cos(\alpha - \beta). \tag{11.24}$$

Therefore, in this simple hidden variable model,

$$E(\alpha, \beta) = -\frac{1}{2} \cos(\alpha - \beta). \tag{11.25}$$

In such a model, one finds a non-vanishing correlation coefficient, which is an interesting observation. Even more interesting is that this correlation is smaller than the prediction of quantum mechanics by a factor 2.

**11.6.4.2**  The experimental points agree with the predictions of quantum mechanics, and undoubtedly disagree with the results of the particular hidden variable model we have considered.

## 11.6.5  Bell's Theorem and Experimental Results

**11.6.5.1**  In the framework of a hidden variable theory, the correlation coefficient is

$$E(\alpha, \beta) = \frac{4}{\hbar^2} \int P(\lambda) \, A(\lambda, \boldsymbol{u}_\alpha) \, B(\lambda, \boldsymbol{u}_\beta) \, d\lambda, \tag{11.26}$$

where $P(\lambda)$ is the (unknown) distribution law for the variable $\lambda$, with

$$P(\lambda) > 0 \quad \forall \lambda \quad \text{and} \quad \int P(\lambda) \, d\lambda = 1. \tag{11.27}$$

Note that we assume here that the hidden variable theory reproduces the one-operator averages found for the singlet state:

$$\langle S_{e\alpha} \rangle = \int P(\lambda) \, A(\lambda, \boldsymbol{u}_\alpha) \, d\lambda = 0 \quad \langle S_{p\beta} \rangle = \int P(\lambda) \, B(\lambda, \boldsymbol{u}_\beta) \, d\lambda = 0. \tag{11.28}$$

If this was not the case, such a hidden variable theory should clearly be rejected since it would not reproduce a well established experimental result.

**11.6.5.2** The quantity of interest can be written:

$$A(\lambda, \boldsymbol{u}_\alpha) \, (B(\lambda, \boldsymbol{u}_\beta) + B(\lambda, \boldsymbol{u}'_\beta)) + A(\lambda, \boldsymbol{u}'_\alpha) \, (B(\lambda, \boldsymbol{u}'_\beta) - B(\lambda, \boldsymbol{u}_\beta)). \qquad (11.29)$$

The two quantities $B(\lambda, \boldsymbol{u}_\beta)$ and $B(\lambda, \boldsymbol{u}'_\beta)$ can take only the two values $\pm\hbar/2$. Therefore one has either

$$B(\lambda, \boldsymbol{u}_\beta) + B(\lambda, \boldsymbol{u}'_\beta) = \pm\hbar \quad B(\lambda, \boldsymbol{u}_\beta) - B(\lambda, \boldsymbol{u}'_\beta) = 0 \qquad (11.30)$$

or

$$B(\lambda, \boldsymbol{u}_\beta) + B(\lambda, \boldsymbol{u}'_\beta) = 0 \quad B(\lambda, \boldsymbol{u}_\beta) - B(\lambda, \boldsymbol{u}'_\beta) = \pm\hbar, \qquad (11.31)$$

hence the result, since $|A(\lambda, \boldsymbol{u}_\alpha)| = |A(\lambda, \boldsymbol{u}'_\beta)| = \hbar/2$.

**11.6.5.3** We multiply the result (11.5) by $P(\lambda)$ and integrate over $\lambda$. Bell's inequality follows immediately.

**11.6.5.4** The quantum mechanical result for $S$ is

$$S_Q = -\cos(\alpha - \beta) - \cos(\alpha - \beta') - \cos(\alpha' - \beta') + \cos(\alpha' - \beta). \qquad (11.32)$$

In general, if we set $\theta_1 = \alpha - \beta, \theta_2 = \beta' - \alpha, \theta_3 = \alpha' - \beta'$, we can look for the extrema of

$$f(\theta_1, \theta_2, \theta_3) = \cos(\theta_1 + \theta_2 + \theta_3) - (\cos\theta_1 + \cos\theta_2 + \cos\theta_3). \qquad (11.33)$$

The extrema correspond to $\theta_1 = \theta_2 = \theta_3$ and $\sin\theta_1 = \sin 3\theta_1$, whose solutions between 0 and $\pi$ are $\theta_1 = 0, \pi/4, 3\pi/4, \pi$. Defining the function $g(\theta_1) = -3\cos\theta_1 + \cos 3\theta_1$ we have: $g(0) = -2, \ g(\pi/4) = -2\sqrt{2}, g(3\pi/4) = 2\sqrt{2}, g(\pi) = 2$.

We have represented the variation of $g(\theta)$ in Fig. 11.2. The gray areas correspond to results which *cannot* be explained by hidden variable theories. In particular, for $\alpha - \beta = \beta' - \alpha = \alpha' - \beta' = \pi/4$, we get $S_Q = -2\sqrt{2}$, which clearly violates Bell's inequality. This system constitutes therefore a test of the predictions of quantum mechanics vs. *any* local hidden variable theory.

**11.6.5.5** The numbers given in the text lead to $|3E(\pi/4) - E(3\pi/4)| = 2.66 \ (\pm0.15)$ in excellent agreement with quantum mechanics $(2\sqrt{2})$ but incompatible with hidden variable theories.

As in the previous question, the actual measurements were in fact $E(\pi/4) = -0.62 \ (\pm0.04)$, $E(3\pi/4) = 0.60 \ (\pm0.03)$, therefore $|3E(\pi/4) - E(3\pi/4)| =$

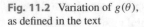

**Fig. 11.2** Variation of $g(\theta)$, as defined in the text

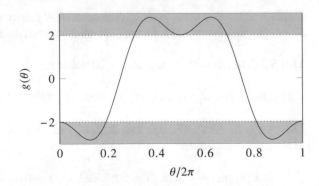

2.46 ($\pm 0.15$) which violates unquestionably Bell's inequality, and is consistent with the quantum mechanical prediction.

It is therefore not possible to find a local hidden variable theory which gives a good account of experiment.

### 11.6.6 References

The famous EPR paradox was presented by A. Einstein, B. Podolsky, and N. Rosen in *Can quantum-mechanical description of physical reality be considered complete?* Phys. Rev. **47**, 777 (1935), to which N. Bohr replied shortly after in Phys. Rev. 48, 696 (1935).

Bell's theorem was first shown in J. Bell, Physics **1**, 195 (1964); see also J. Bell, *Speakable and unspeakable in quantum mechanics*, Cambridge University Press, Cambridge (1993).

The data used in this chapter are adapted from the experimental results of A. Aspect, P. Grangier, and G. Roger, Phys. Rev. Lett. **49**, 91 (1982), which are shown in Fig. 11.3. Note that this experiment was performed with pairs of correlated photons rather than with electron-proton pairs. The statistical errors on the experimental results were low enough to conclude to a violation of Bell's inequality by more than 40 standard deviations. Once experimental imperfections are taken into account (mostly the acceptance of the detectors), the agreement between these results and the prediction of quantum mechanics (dotted line in Fig. 11.3) is excellent.

The precision of this type of tests has now been improved even more, in particular with the use of photon pairs produced by nonlinear splitting of ultraviolet photons. For a recent review, see e.g. A. Aspect, *Closing the Door on Einstein and Bohr's Quantum Debate*, Physics **8**, 123 (2015).

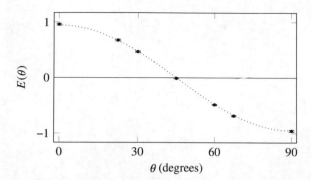

**Fig. 11.3** Measured correlation function $E(\theta)$ for Bell pairs of photons emitted in an atomic cascade between atomic levels with angular momenta $J = 0 \to J = 1 \to J = 0$. Here $\theta$ is the angle between the polarizers in front of the photon detectors. The indicated errors are $\pm 2$ standard deviations. The dotted line is the quantum mechanical prediction, taking into account experimental imperfections

# Quantum Correlations in a Multi-particle System

When looking for a possible conflict between the predictions of quantum theory with those of a hidden variable model, it is interesting to consider correlated states involving more than two particles. The violation of Bell's inequality that is known for a two-particle system (see Chap. 11) is then replaced by an even more spectacular contradiction, as we show in this chapter.

*In what follows, we are interested only in the spin degree of freedom and we will thus omit the orbital degrees of freedom when describing the state of a particle.*

## 12.1 Measurements on a Single Spin

Consider a spin 1/2 particle with $|+\rangle$, $|-\rangle$ denoting the eigenstates of the operator $\hat{S}_z$, projection on the $z$ axis of the spin operator.

**12.1.1** Recall the expression of the matrices $\hat{S}_j$ ($j = x, y, z$) in the basis $|\pm\rangle$.

**12.1.2** One measures the spin component $S_n = \boldsymbol{n} \cdot \boldsymbol{S}$ along the direction of unit vector $\boldsymbol{n}$, characterized by the angles $(\theta, \varphi)$ in spherical coordinates

$$\boldsymbol{n} = \sin\theta \cos\varphi \, \boldsymbol{u}_x + \sin\theta \sin\varphi \, \boldsymbol{u}_y + \cos\theta \, \boldsymbol{u}_z. \tag{12.1}$$

To simplify the writing, we will divide the result of each spin measurement by $\hbar/2$, so that the operator associated with the measurement is $\hat{\sigma}_n = \frac{2}{\hbar} \boldsymbol{n} \cdot \hat{\boldsymbol{S}} = \boldsymbol{n} \cdot \hat{\boldsymbol{\sigma}}$.

Show that the matrix of the operator $\hat{\sigma}_n$ in the basis $|\pm\rangle$ reads

$$\begin{pmatrix} \cos\theta & e^{-i\varphi} \sin\theta \\ e^{i\varphi} \sin\theta & -\cos\theta \end{pmatrix}. \tag{12.2}$$

© Springer Nature Switzerland AG 2019
J.-L. Basdevant, J. Dalibard, *The Quantum Mechanics Solver*,
https://doi.org/10.1007/978-3-030-13724-3_12

**12.1.3** What are the possible results of a measurement of $\sigma_n$?

**12.1.4** One repeats the measurement of $\sigma_n$ on a large number of particles, each being prepared in the state $|+\rangle$. Give the mean value of the result in terms of $\theta$ et $\varphi$. Give also the result when each particle is prepared in the state $|-\rangle$.

---

## 12.2   Measurements on a Two-Spin System

Consider a system of two spins 1/2 that we label $a$ and $b$ and that are discernable (no a priori requirement to symmetrize or antisymmetrize the state vector). This two-spin system is prepared in the state $|\Psi\rangle$. To simplify the writing we shall denote $|\eta_a \eta_b\rangle \equiv |a : \eta_a\rangle \otimes |b : \eta_b\rangle$, i.e., the eigenstate of $\hat{S}_{az}$ and $\hat{S}_{bz}$ with the eigenvalues $\hbar\eta_a/2$ and $\hbar\eta_b/2$ ($\eta_a = \pm$, $\eta_b = \pm$).

**12.2.1** One measures the spin component of $a$ along the direction $\boldsymbol{n}_a$ and divides the result by $\hbar/2$ as indicated above. What is the operator associated with this measurement and what are the possible results? Write the average value $E(\boldsymbol{n}_a)$ of the results in terms of $|\Psi\rangle$, $\boldsymbol{n}_a$ and $\hat{\boldsymbol{\sigma}}_a$.

**12.2.2** Calculate $E(\boldsymbol{n}_a)$ when the two-spin system is prepared in the singlet state $|\Psi_s\rangle = (|+-\rangle - |-+\rangle)/\sqrt{2}$. One can introduce the angles $(\theta_a, \varphi_a)$ characterizing the direction of $\boldsymbol{n}_a$ and use the result of question 12.1.4. Justify the statement that spin $a$ appears in this case as *unpolarized*.

**12.2.3** One prepares a large number of spin pairs $ab$ in the same state $|\Psi\rangle$. For each pair, one measures the component $\sigma_{a,n_a}$ of spin $a$ along $\boldsymbol{n}_a$ and the component $\sigma_{b,n_b}$ of spin $b$ along $\boldsymbol{n}_b$. For each pair, one then multiplies the results of the measurements of $\sigma_{a,n_a}$ and $\sigma_{b,n_b}$. Consider the mean value $E(\boldsymbol{n}_a, \boldsymbol{n}_b)$ of this product and indicate the correct statement among the following ones:

- The average is $E(\boldsymbol{n}_a, \boldsymbol{n}_b) = \langle\Psi| (\boldsymbol{n}_a \cdot \hat{\boldsymbol{\sigma}}_a) \otimes (\boldsymbol{n}_b \cdot \hat{\boldsymbol{\sigma}}_b) |\Psi\rangle$.
- The average is $E(\boldsymbol{n}_a, \boldsymbol{n}_b) = \langle\Psi|\boldsymbol{n}_a \cdot \hat{\boldsymbol{\sigma}}_a|\Psi\rangle \langle\Psi|\boldsymbol{n}_b \cdot \hat{\boldsymbol{\sigma}}_b|\Psi\rangle$.
- Both of the above since these two forms are equal for all $|\Psi\rangle$.
- None of the above. Indicate in this case the correct result.

**12.2.4** The vectors $\boldsymbol{n}_a$ and $\boldsymbol{n}_b$ are characterized by the angles $(\theta_a, \varphi_a)$ and $(\theta_b, \varphi_b)$, and the two-spin system is supposed to be prepared in the singlet state. Show that

$$E(\boldsymbol{n}_a, \boldsymbol{n}_b) = -(\cos\theta_a \cos\theta_b + \cos(\varphi_a - \varphi_b)\sin\theta_a \sin\theta_b) = -\boldsymbol{n}_a \cdot \boldsymbol{n}_b. \quad (12.3)$$

**12.2.5** Give the value of $E(\boldsymbol{n}_a, \boldsymbol{n}_b)$ when $\boldsymbol{n}_a = \boldsymbol{n}_b$ and comment the result.

## 12.3    Measurements on a Four-Spin System

We consider now a system composed of four particles with spin $1/2$, which we denote $a, b, c, d$. As above the particles are supposed to discernable.

**12.3.1** Recall the dimension of the Hilbert space associated with the spin state of these four particles.

**12.3.2** The 4-particle system is prepared in the state

$$|\Psi\rangle = \frac{1}{\sqrt{2}} (|++--\rangle - |--++\rangle) , \qquad (12.4)$$

where the state vectors $|\eta_a \; \eta_b \; \eta_c \; \eta_d\rangle$ (with $\eta_j = \pm$) are eigenstates of $\hat{S}_{az}, \hat{S}_{bz}, \hat{S}_{cz}, \hat{S}_{dz}$ with eigenvalues $\hbar\eta_a/2, \hbar\eta_b/2, \hbar\eta_c/2, \hbar\eta_d/2$.

One measures the component $\sigma_{j,n_j}$ of the spin of each particle $j$ along $n_j$ $(j = a, b, c, d)$, with all vectors $n_j$ in the $xy$ plane: $\theta_j = \pi/2$. One then multiplies the four results and calculates the mean value $E(n_a, n_b, n_c, n_d)$ of this product. Following a procedure similar to that of Sect. 12.2, show that

$$E(n_a, n_b, n_c, n_d) = -\cos(\varphi_a + \varphi_b - \varphi_c - \varphi_d) . \qquad (12.5)$$

**12.3.3** Discuss the expected correlations when the angles are such that $\varphi_a + \varphi_b = \varphi_c + \varphi_d$.

**12.3.4** Let us now try to reproduce the prediction (12.5) with a local hidden variable model. This model should provide four functions $A(\lambda, \varphi_a)$, $B(\lambda, \varphi_b)$, $C(\lambda, \varphi_c)$, $D(\lambda, \varphi_d)$, giving the measurement results for each particle. Here we introduced the *hidden variable* $\lambda$ which characterizes a given quadruplet, and which is not accessible within the quantum mechanics framework. One thus has $A(\lambda, \varphi_a) = \pm 1$, $B(\lambda, \varphi_b) = \pm 1, \ldots$

Show that for all angles $\varphi$, the functions $A, B, C, D$ must be such that

$$A(\lambda, 0) \; B(\lambda, 0) \; C(\lambda, 0) \; D(\lambda, 0) = -1 , \qquad (12.6)$$

$$A(\lambda, \varphi) \; B(\lambda, 0) \; C(\lambda, \varphi) \; D(\lambda, 0) = -1 , \qquad (12.7)$$

$$A(\lambda, \varphi) \; B(\lambda, 0) \; C(\lambda, 0) \; D(\lambda, \varphi) = -1 , \qquad (12.8)$$

$$A(\lambda, 2\varphi) \; B(\lambda, 0) \; C(\lambda, \varphi) \; D(\lambda, \varphi) = -1 . \qquad (12.9)$$

**12.3.5** Show using (12.6), (12.7), (12.8), and (12.9) that the function $A$ cannot depend on $\varphi$ (Hint: Consider the product of these four equations).

**12.3.6** Explain why no local hidden variable theory can reproduce the result (12.5).

## 12.4   Solutions

### 12.4.1 Measurements on a Single Spin

**12.4.1.1** The matrices $\hat{S}_j$ read $\hat{S}_j = \frac{\hbar}{2}\hat{\sigma}_j$ where the $\hat{\sigma}_j$ are the Pauli matrices

$$\hat{\sigma}_x = \begin{pmatrix} 0 & 1 \\ 1 & 0 \end{pmatrix} \qquad \hat{\sigma}_y = \begin{pmatrix} 0 & -i \\ i & 0 \end{pmatrix} \qquad \hat{\sigma}_z = \begin{pmatrix} 1 & 0 \\ 0 & -1 \end{pmatrix}. \tag{12.10}$$

**12.4.1.2** The matrix of the operator $\hat{\sigma}_n$ is

$$\hat{\sigma}_n = \boldsymbol{n} \cdot \hat{\boldsymbol{\sigma}} = \sin\theta \cos\varphi\, \hat{\sigma}_x + \sin\theta \sin\varphi\, \hat{\sigma}_y + \cos\theta\, \hat{\sigma}_z = \begin{pmatrix} \cos\theta & e^{-i\varphi}\sin\theta \\ e^{+i\varphi}\sin\theta & -\cos\theta \end{pmatrix}. \tag{12.11}$$

**12.4.1.3** In a measurement of $\sigma_n$, the possible results are the eigenvalues of the matrix (12.11). This $2 \times 2$ hermitian matrix has a trace equal to 0 and a determinant equal to $-1$, which entails that its two eigenvalues are $+1$ and $-1$. Coming back to $\hat{S}_n$, we find the eigenvalues $\pm\hbar/2$, as expected for the measurement of the component of a spin 1/2 particle along any axis.

**12.4.1.4** For a spin prepared in a given state $|\psi\rangle$, the mean value of the measurement results of $\sigma_n$ is $\langle\psi|\hat{\sigma}_n|\psi\rangle$. If one takes $|\psi\rangle = |+\rangle$ (resp. $|\psi\rangle = |-\rangle$), the mean value is $+\cos\theta$ (resp. $-\cos\theta$).

### 12.4.2 Measurements on a Two-Spin System

**12.4.2.1** The operator associated with the measurement of the projection of $\sigma_a$ along $\boldsymbol{n}_a$ is $\hat{\sigma}_{a,\boldsymbol{n}_a} \otimes \hat{1}_b = (\boldsymbol{n}_a \cdot \hat{\boldsymbol{\sigma}}_a) \otimes \hat{1}_b$. The possible results are the eigenvalues of $\hat{\sigma}_{a,\boldsymbol{n}_a}$, i.e. $\pm 1$. The mean value of the results is

$$E(\boldsymbol{n}_a) = \langle\Psi|\boldsymbol{n}_a \cdot \hat{\boldsymbol{\sigma}}_a|\Psi\rangle. \tag{12.12}$$

**12.4.2.2** One inserts the expression of the singlet state into (12.12):

$$E(\boldsymbol{n}_a) = \frac{1}{2}\langle + - |\boldsymbol{n}_a \cdot \hat{\boldsymbol{\sigma}}_a| + -\rangle - \frac{1}{2}\langle + - |\boldsymbol{n}_a \cdot \hat{\boldsymbol{\sigma}}_a| - +\rangle$$

$$- \frac{1}{2}\langle - + |\boldsymbol{n}_a \cdot \hat{\boldsymbol{\sigma}}_a| + -\rangle + \frac{1}{2}\langle - + |\boldsymbol{n}_a \cdot \hat{\boldsymbol{\sigma}}_a| - +\rangle. \tag{12.13}$$

The operator $\boldsymbol{n}_a \cdot \hat{\boldsymbol{\sigma}}_a$ acts only on spin $a$. In each matrix element of the sum (12.13), the spin state of $b$ must thus be the same on the left and right hand sides, in order to have a nonzero contribution. This occurs only in the first and last terms of the sum:

$$E(\boldsymbol{n}_a) = \frac{1}{2}\langle +|\boldsymbol{n}_a \cdot \hat{\boldsymbol{\sigma}}_a|+\rangle + \frac{1}{2}\langle -|\boldsymbol{n}_a \cdot \hat{\boldsymbol{\sigma}}_a|-\rangle$$

$$= \frac{1}{2}(+\cos\theta_a) + \frac{1}{2}(-\cos\theta_a) = 0 \qquad (12.14)$$

One finds a zero mean value of the spin projection along any direction $\boldsymbol{n}_a$, which matches with the definition of a non-polarized spin state.

**12.4.2.3** The correct answer is the first one:

$$E(\boldsymbol{n}_a, \boldsymbol{n}_b) = \langle\Psi|\left(\boldsymbol{n}_a \cdot \hat{\boldsymbol{\sigma}}_a\right) \otimes \left(\boldsymbol{n}_b \cdot \hat{\boldsymbol{\sigma}}_b\right)|\Psi\rangle . \qquad (12.15)$$

Each measurement result is indeed the product of the result for $\sigma_{a,\boldsymbol{n}_a}$ by the result for $\sigma_{b,\boldsymbol{n}_b}$. The operator corresponding to the pair measurement is thus $\hat{\sigma}_{a,\boldsymbol{n}_a} \otimes \hat{\sigma}_{b,\boldsymbol{n}_b}$. The result of the second proposed answer would correspond to the following procedure. One would first average all measurement results obtained with spin $a$, leading to $\langle\Psi|\boldsymbol{n}_a \cdot \hat{\boldsymbol{\sigma}}_a|\Psi\rangle$. Similarly the average of all measurements performed on spin $b$ would give $\langle\Psi|\boldsymbol{n}_b \cdot \hat{\boldsymbol{\sigma}}_b|\Psi\rangle$. Finally one would multiply these average values, which would give the other proposed result:

$$E(\boldsymbol{n}_a, \boldsymbol{n}_b) = \langle\Psi|\boldsymbol{n}_a \cdot \hat{\boldsymbol{\sigma}}_a|\Psi\rangle \langle\Psi|\boldsymbol{n}_b \cdot \hat{\boldsymbol{\sigma}}_b|\Psi\rangle. \qquad (12.16)$$

The results (12.15) and (12.16) are generally different: The mean value of the product is not equal to the product of the mean values! They coincide only when the state $|\Psi\rangle$ does not exhibit any correlation between particles, i.e. , it can be written $|\Psi\rangle = |\psi_a\rangle \otimes |\psi_b\rangle$.

**12.4.2.4** We develop (12.15) using the singlet state and we get

$$E(\boldsymbol{n}_a) = \frac{1}{2}\langle +-|\left(\boldsymbol{n}_a \cdot \hat{\boldsymbol{\sigma}}_a\right)\left(\boldsymbol{n}_b \cdot \hat{\boldsymbol{\sigma}}_b\right)|+-\rangle - \frac{1}{2}\langle +-|\left(\boldsymbol{n}_a \cdot \hat{\boldsymbol{\sigma}}_a\right)\left(\boldsymbol{n}_b \cdot \hat{\boldsymbol{\sigma}}_b\right)|-+\rangle$$

$$- \frac{1}{2}\langle -+|\left(\boldsymbol{n}_a \cdot \hat{\boldsymbol{\sigma}}_a\right)\left(\boldsymbol{n}_b \cdot \hat{\boldsymbol{\sigma}}_b\right)|+-\rangle + \frac{1}{2}\langle -+|\left(\boldsymbol{n}_a \cdot \hat{\boldsymbol{\sigma}}_a\right)\left(\boldsymbol{n}_b \cdot \hat{\boldsymbol{\sigma}}_b\right)|-+\rangle$$

$$= \frac{1}{2}(+\cos\theta_a)(-\cos\theta_b) - \frac{1}{2}(\sin\theta_a e^{-i\varphi_a})(\sin\theta_b e^{+i\varphi_b})$$

$$- \frac{1}{2}(\sin\theta_a e^{+i\varphi_a})(\sin\theta_b e^{-i\varphi_b}) + \frac{1}{2}(-\cos\theta_a)(+\cos\theta_b)$$

$$= -\cos\theta_a \cos\theta_b - \cos(\varphi_a - \varphi_b)\sin\theta_a \sin\theta_b = -\boldsymbol{n}_a \cdot \boldsymbol{n}_b .$$

**12.4.2.5** If $n_a = n_b$ we find $E(n_a, n_b) = -1$: The two results are fully correlated and any measurement can only yield one of the two possibilities $(+-)$ or $(-+)$; The two other ones $(++)$ and $(--)$ never occur for the singlet state.

### 12.4.3  Measurements on a Four-Spin System

**12.4.3.1** The dimension of the Hilbert space is $2^4 = 16$.

**12.4.3.2** The desired mean value is

$$E(n_a, n_b, n_c, n_d) = \langle \Psi | \hat{\sigma}_{a,n_a} \hat{\sigma}_{b,n_b} \hat{\sigma}_{c,n_c} \hat{\sigma}_{d,n_d} | \Psi \rangle. \tag{12.17}$$

Inserting the expression of $|\Psi\rangle$ leads to the sum of four terms:

$$
\begin{aligned}
E(n_a, n_b, n_c, n_d) = {} & \frac{1}{2} \langle + + - - | \hat{\sigma}_{a,n_a} \hat{\sigma}_{b,n_b} \hat{\sigma}_{c,n_c} \hat{\sigma}_{d,n_d} | + + - - \rangle \\
& - \frac{1}{2} \langle + + - - | \hat{\sigma}_{a,n_a} \hat{\sigma}_{b,n_b} \hat{\sigma}_{c,n_c} \hat{\sigma}_{d,n_d} | - - + + \rangle \\
& - \frac{1}{2} \langle - - + + | \hat{\sigma}_{a,n_a} \hat{\sigma}_{b,n_b} \hat{\sigma}_{c,n_c} \hat{\sigma}_{d,n_d} | + + - - \rangle \\
& + \frac{1}{2} \langle - - + + | \hat{\sigma}_{a,n_a} \hat{\sigma}_{b,n_b} \hat{\sigma}_{c,n_c} \hat{\sigma}_{d,n_d} | - - + + \rangle . \tag{12.18}
\end{aligned}
$$

Each term can be easily calculated. For example the first term is:

$$
\begin{aligned}
\langle + + - - | \hat{\sigma}_{a,n_a} \hat{\sigma}_{b,n_b} \hat{\sigma}_{c,n_c} \hat{\sigma}_{d,n_d} | + + - - \rangle = {} \\
\langle + | \hat{\sigma}_{a,n_a} | + \rangle \langle + | \hat{\sigma}_{b,n_b} | + \rangle \langle - | \hat{\sigma}_{c,n_c} | - \rangle \langle - | \hat{\sigma}_{d,n_d} | - \rangle . \tag{12.19}
\end{aligned}
$$

Since we choose $\theta_j = \pi/2$, we have $\langle + | \hat{\sigma}_{j,n_j} | + \rangle = \langle - | \hat{\sigma}_{j,n_j} | - \rangle = 0$ and the first and last term of (12.18) are equal to 0. The second term can be written $(-1/2) e^{i(\varphi_c + \varphi_d - \varphi_a - \varphi_b)}$ and the third term is the complex conjugate of the second one. We thus obtain

$$E(n_a, n_b, n_c, n_d) = -\cos(\varphi_c + \varphi_d - \varphi_a - \varphi_b) . \tag{12.20}$$

**12.4.3.3** If the angles are chosen such that $\varphi_c + \varphi_d = \varphi_a + \varphi_b$, then $E(n_a, n_b, n_c, n_d) = -1$. This means that there is a strong correlation between the four measurement results. Either three of them give $+$ and the fourth one gives $-$, or the reverse. Note that $\langle \Psi | \hat{\sigma}_{a,n_a} | \Psi \rangle = 0$ (and similarly for $b, c, d$), which means that the measurements on a given spin lead in average to the same numbers of $+$ and $-$. Each spin state thus appears as unpolarized. It is only when one combines the measurement results on the four spins of a given realization of the experiment that their correlations are revealed.

**12.4.3.4** The hidden variable theory should reproduce the strong correlation that we found at the previous question. Therefore for all quadruplet $\varphi_a$, $\varphi_b$, $\varphi_c$, $\varphi_d$ such that $\varphi_c + \varphi_d = \varphi_a + \varphi_b$, one should have:

$$A(\lambda, \varphi_a)\, B(\lambda, \varphi_b)\, C(\lambda, \varphi_c)\, D(\lambda, \varphi_d) = -1 \,, \tag{12.21}$$

which expresses that the product of the four measurement results should always be equal to $-1$. The equations (12.6), (12.7), (12.8), and (12.9) correspond to the four following choices, all satisfying $\varphi_c + \varphi_d = \varphi_a + \varphi_b$:

1. $\varphi_a = \varphi_b = \varphi_c = \varphi_d = 0$,
2. $\varphi_a = \varphi_c = \varphi$ and $\varphi_b = \varphi_d = 0$,
3. $\varphi_a = \varphi_d = \varphi$ and $\varphi_b = \varphi_c = 0$,
4. $\varphi_a = 2\varphi$, $\varphi_c = \varphi_d = \varphi$ and $\varphi_b = 0$.

**12.4.3.5** The functions $A$, $B$, $C$, $D$ are equal to $\pm 1$. One thus has $A^2(\lambda, \varphi_a) = 1$ for all $\lambda$ and $\varphi_a$, and similarly for $B$, $C$, $D$. When multiplying the four equations (12.6), (12.7), (12.8), and (12.9), one thus gets

$$A(\lambda, 0) A(\lambda, 2\varphi) = 1 \,, \quad \forall \lambda, \varphi. \tag{12.22}$$

Since $A^2(\lambda, \varphi_a) = 1$, it can also be written:

$$A(\lambda, 0) = A(\lambda, 2\varphi) \,, \quad \forall \lambda, \varphi \,. \tag{12.23}$$

One then arrives to the conclusion that the function $A(\lambda, \varphi_a)$ does not depend on the angle $\varphi_a$ and is only a function of $\lambda$. The same conclusion holds for $B$, $C$, $D$ which should not depend on the angles $\varphi_b$, $\varphi_c$, $\varphi_d$.

**12.4.3.6** If $A$, $B$, $C$, $D$ do not depend on the angles $\varphi_a$, $\varphi_b$, $\varphi_c$, $\varphi_d$, then for any local hidden variable model, the correlation function

$$E(n_a, n_b, n_c, n_d) = A(\lambda, \varphi_a)\, B(\lambda, \varphi_b)\, C(\lambda, \varphi_c)\, D(\lambda, \varphi_d) \tag{12.24}$$

will not depend on the orientations $n_a, n_b, n_c, n_d$. However, this result cannot reproduce the predictions of quantum mechanics which are such that $E = -1$ for $\varphi_a = \varphi_b = \varphi_c = \varphi_d = 0$ and $E = 1$ for $\varphi_a = \varphi_b = 0$, $\varphi_c = \varphi_d = \pi/2$.

To summarize, one cannot construct a local hidden variable model that would reproduce all results predicted by quantum mechanics for this four-spin system.

### 12.4.4 References

Such correlated states involving more than two particles are often called GHZ states, from Greenberger, Horne and Zeilinger who were the first to suggest their usefulness for testing quantum mechanics. See for example D. M. Greenberger, M. A. Horne, A. Shimony, and A. Zeilinger, *Bell's theorem without inequalities*, American Journal of Physics **58**, 1131 (1990).

The practical implementation of a four-particle (photon) state was first performed by J. W. Pan, M. Daniell, S. Gasparoni, G. Weihs, and A. Zeilinger, *Experimental demonstration of four-photon entanglement and high-fidelity teleportation*, Phys. Rev. Lett. **86**, 4435 (2001). For a review, see e.g. Jian-Wei Pan, Zeng-Bing Chen, Chao-Yang Lu, Harald Weinfurter, Anton Zeilinger, Marek Zukowski, *Multiphoton entanglement and interferometry*, Rev. Mod. Phys. **84**, 777 (2012).

# A Non-destructive Bomb Detector

13

We suppose that there exist two kinds of bombs, real ones and factitious ones (Fig. 13.1). The distinction between a real and a factitious bomb can be made by sending a neutron: a real bomb will explode when hit by the neutron, whereas a factitious bomb will transmit the neutron without exploding and without changing the state of the neutron. The purpose of this chapter is to propose a quantum method allowing one to determine if a given bomb is real or factitious without exploding it.

## 13.1 A Neutron Beam Splitter

**13.1.1** Consider a quantum system described by the Hamiltonian $\hat{H}(t)$, and two arbitrary solutions $|\psi(t)\rangle$ and $|\psi'(t)\rangle$ of the corresponding Schrödinger equation. Show that the scalar product $\langle \psi(t)|\psi'(t)\rangle$ does not vary with $t$.

**13.1.2** One sends a neutron on a slab of matter (Fig. 13.2). This slab creates a potential barrier on which the neutron may be transmitted or reflected. We recall that the motion of the neutron can be studied in two equivalent ways:

- (i) One describes the state of the neutron using wavepackets that evolve with time.
- (ii) One supposes that the neutron is in a state of well defined energy $E$ and one describes its states as a superposition of a few plane waves.

In this section as well as in Sects. 13.2 et 13.3, we shall use the point of view (i). In Sect. 13.4, we shall use the point of view (ii).

We suppose that at time $t_0$, before reaching the slab, the state of the neutron is

$$|\psi(t_0)\rangle = |\phi_1\rangle \tag{13.1}$$

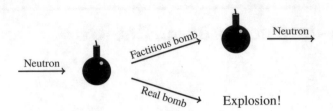

**Fig. 13.1** A factitious bomb transmits a neutron without exploding. A real bomb explodes when hit by a neutron

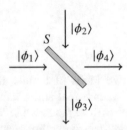

**Fig. 13.2** Partially reflecting slab for neutrons

where $|\phi_1\rangle$ represents the incident wavepacket sketched on Fig. 13.2. The interaction of the neutron with the slab is described by a Hamiltonian that we will not try to write explicitly. At the subsequent time $t_1$, the neutron state is a superposition of the state $|\phi_3\rangle$ (the neutron was reflected by the slab) and the state $|\phi_4\rangle$ (the neutron was transmitted by the slab):

$$|\psi(t_1)\rangle = \rho|\phi_3\rangle + \tau|\phi_4\rangle \tag{13.2}$$

where $\langle\phi_i|\phi_j\rangle = \delta_{i,j}$ $(i, j = 1, 3, 4)$. The coefficients $\rho$ and $\tau$ are chosen positive or zero.
Explain why $\rho^2 + \tau^2 = 1$.

**13.1.3** We suppose now that the neutron is sent on the second entrance port of the beam splitter in the state that is symmetric of $|\phi_1\rangle$ with respect to the slab:

$$|\psi'(t_0)\rangle = |\phi_2\rangle \tag{13.3}$$

such that $\langle\phi_j|\phi_2\rangle = \delta_{j,2}$, $(j = 1, \ldots, 4)$. The state at time $t_1$ then reads

$$|\psi'(t_1)\rangle = \tau'|\phi_3\rangle + \rho'|\phi_4\rangle \ . \tag{13.4}$$

Using the result of question 1, explain why $\tau'$ et $\rho'$ satisfy the relations

$$|\rho'|^2 + |\tau'|^2 = 1 \ , \quad \tau\rho' + \rho\tau' = 0 \ . \tag{13.5}$$

We will take in the following $\tau' = \tau$, $\rho' = -\rho$. We will thus describe the action of the beam splitter by

$$|\phi_1\rangle \quad \text{evolves to} \quad \rho|\phi_3\rangle + \tau|\phi_4\rangle \, , \tag{13.6}$$

$$|\phi_2\rangle \quad \text{evolves to} \quad \tau|\phi_3\rangle - \rho|\phi_4\rangle \, . \tag{13.7}$$

## 13.2 A Mach–Zehnder Interferometer for Neutrons

We consider the setup sketched in Fig. 13.3. It is composed of two beam splitters $S$ et $S'$ and two ideal mirrors $M_a$ and $M_b$. We define the times $t_0, \ldots, t_3$ such that:

- $t_0$: the neutron is in the state $|\psi(t_0)\rangle = |\phi_1\rangle$, before hitting the first beam splitter $S$.
- $t_1$: after interacting with $S$, the neutron is in the state (13.6), i.e. a superposition of $|\phi_3\rangle$ and $|\phi_4\rangle$.
- $t_2$: the neutron propagates in the interferometer. During this propagation,

$$|\phi_3\rangle \quad \text{evolves to} \quad e^{i\Delta_a}|\phi_1'\rangle \, , \tag{13.8}$$

$$|\phi_4\rangle \quad \text{evolves to} \quad e^{i\Delta_b}|\phi_2'\rangle \, , \tag{13.9}$$

where $|\phi_1'\rangle$ and $|\phi_2'\rangle$ correspond to the entrance ports of the second beam splitter $S'$, and where the phases $\Delta_a$ and $\Delta_b$ can be adjusted by the experimentalist.
- $t_3$: the neutron interacts with the second beam splitter $S'$ and its state can now be written

$$|\psi(t_3)\rangle = \alpha|\phi_3'\rangle + \beta|\phi_4'\rangle \, . \tag{13.10}$$

The evolution of the states $|\phi_{1,2}'\rangle$ to the states $|\phi_{3,4}'\rangle$ during the interaction with $S'$ is described in a way similar to (13.6)–(13.7).

**Fig. 13.3** Mach-Zehnder interferometer for neutrons

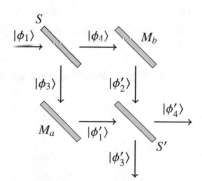

**13.2.1** Show that the probability $P_B$ to detect the neutron in the outport $B$ of the interferometer, i.e. in the state $|\phi_4'\rangle$, is equal to

$$P_B = 2\rho^2\tau^2(1 - \cos\Delta) , \qquad (13.11)$$

where we set $\Delta = \Delta_a - \Delta_b$.

**13.2.2** Calculate the probability $P_A$ to detect the neutron in the outport $A$.

**13.2.3** Calculate $P_A + P_B$ and comment the result.

**13.2.4** Show that there exist phase choices for $\Delta$ such that we are sure that the neutron will exit in the $A$ port, irrespective of the values of $\rho$ and $\tau$. We will choose this value of $\Delta$ in the following.

## 13.3 A First Step Towards a Non Destructive Detection

We have a bomb with an unknown status, real or factitious, and we place it in the interferometer arm $SM_bS'$ (see Fig. 13.3).

**13.3.1** If the bomb is factitious, what are the probabilities to detect the neutron in the outports $A$ and $B$. We remind that we assume that the phase $\Delta$ has been chosen according to the result of question 13.2.4.

**13.3.2** If the bomb is real, what is the probability that it explodes?

**13.3.3** Suppose that the bomb is real and that it did not explode when the neutron crossed the interferometer. What are the probabilities to detect the neutron in the outports $A$ and $B$?

**13.3.4** In which case can we be sure that the bomb is real without having it explode?

**13.3.5** We define the merit factor $F$ in the following way: $F$ is the probability to detect a real bomb without any explosion. What is the merit factor $F(\rho, \tau)$ of the interferometer of Fig. 13.3?

**13.3.6** What is the set of parameters $(\rho, \tau)$ that maximizes the merit factor?

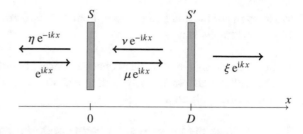

Fig. 13.4 Fabry–Perot interferometer allowing one to assess efficiently the nature of a bomb without explosion

---

## 13.4  Towards an Efficient Non-destructive Detection

Consider now the linear Fabry–Perot interferometer shown in Fig. 13.4. The beam splitters $S$ and $S'$ are the same as above, but they are now placed at normal incidence and used as semi-transparent mirrors. The distance between them is denoted $D$.

We describe the state of the neutron by a superposition of plane waves $e^{\pm ikx}$, where $k$ is the wavevector of the neutron. The amplitudes $\eta$, $\mu$, $\nu$, $\xi$ of the waves depend on the region of space and are indicated in Fig. 13.4. By convention the amplitude of the incident wave coming from $x = -\infty$ is set equal to 1. The amplitudes of the plane waves $e^{\pm ikx}$ on each side of a beam splitter ($S$ or $S'$) are related by the same coefficients $\tau$ and $\rho$ as in the previous sections.

**13.4.1 Beam splitter S.** Show that $\mu = \tau - \rho \nu$ and $\eta = \rho + \tau \nu$.

**13.4.2 Beam splitter S'.** Show that $\xi \, e^{ikD} = \tau \, \mu \, e^{ikD}$ and $\nu \, e^{-ikD} = \rho \, \mu \, e^{ikD}$.

**13.4.3** Show that for any choice of $(\rho, \tau)$, there exist values of $D$ such that one is certain that an incident neutron will always be transmitted by this interferometer (i.e. $|\xi| = 1$ and $\eta = 0$).

**13.4.4** Suppose that a bomb, either real of factitious, is placed between $S$ and $S'$. Show that one can reach merit factors arbitrarily close to 1 for the detection of a real bomb.

---

## 13.5  Solutions

### 13.5.1  A Neutron Beam Splitter

**13.5.1.1** $|\psi(t)\rangle$ and $\psi'(t)\rangle$ are both supposed to be solutions of $i\hbar|\dot{\psi}\rangle = \hat{H}(t)|\psi\rangle$. Therefore:

$$\frac{d}{dt}\langle\psi(t)|\psi'(t)\rangle = \left(\frac{i}{\hbar}\langle\psi(t)|\hat{H}^{\dagger}(t)\right)|\psi'(t)\rangle + \langle\psi(t)|\left(\frac{-i}{\hbar}\hat{H}(t)|\psi'(t)\rangle\right) = 0 ,$$

$$(13.12)$$

where we used the fact that the Hamiltonian $\hat{H}$ is Hermitian: $\hat{H}^\dagger = \hat{H}$. In particular this result entails that

- the norm of a state remains unchanged during the evolution,
- two states that are initially orthogonal to each other will remain orthogonal during the evolution.

**13.5.1.2** Since the initial state is normalized, the final state is also normalized. Since the states $|\phi_3\rangle$ and $|\phi_4\rangle$ are supposed to be orthogonal to each other, we have $\rho^2 + \tau^2 = 1$. In other words the sum of the probabilities for reflection ($\rho^2$) and transmission ($\tau^2$) is equal to 1.

**13.5.1.3** The initial state is again normalized as well as the final state, hence $|\rho'|^2 + |\tau'|^2 = 1$. In addition the initial state $|\phi_2\rangle$ is orthogonal to the initial state $|\phi_1\rangle$. The corresponding final states $\rho\,|\phi_3\rangle + \tau\,|\phi_4\rangle$ and $\tau'\,|\phi_3\rangle + \rho'\,|\phi_4\rangle$ must also be orthogonal to each other, hence $\rho\tau' + \rho'\tau = 0$. This condition is verified for the proposed choice $\rho' = -\rho$, $\tau' = \tau$.

## 13.5.2 A Mach-Zehnder Interferometer for Neutrons

**13.5.2.1** At time $t_3$, just before crossing the beam splitter $S'$, the state of the neutron is:

$$|\psi(t_3)\rangle = \rho\,e^{i\Delta_a}|\phi_1'\rangle + \tau\,e^{i\Delta_b}|\phi_2'\rangle . \tag{13.13}$$

Just after crossing $S'$, the state is now

$$
\begin{aligned}
|\psi(t_4)\rangle &= \rho\,e^{i\Delta_a}\left(\rho|\phi_3'\rangle + \tau|\phi_4'\rangle\right) + \tau\,e^{i\Delta_b}\left(\tau|\phi_3'\rangle - \rho|\phi_4'\rangle\right) \\
&= \left(e^{i\Delta_a}\rho^2 + e^{i\Delta_b}\tau^2\right)|\phi_3'\rangle + \rho\tau\left(e^{i\Delta_a} - e^{i\Delta_b}\right)|\phi_4'\rangle .
\end{aligned} \tag{13.14}
$$

The probability $P_B$ to find the neutron in the channel $B$ corresponding to the state $|\phi_4'\rangle$ is thus:

$$P_B = |\rho\tau\left(e^{i\Delta_a} - e^{i\Delta_b}\right)|^2 = 2\rho^2\tau^2(1 - \cos\Delta) . \tag{13.15}$$

**13.5.2.2** The probability $P_A$ is

$$P_A = |e^{i\Delta_a}\rho^2 + e^{i\Delta_b}\tau^2|^2 = \rho^4 + \tau^4 + 2\rho^2\tau^2\cos\Delta . \tag{13.16}$$

**13.5.2.3** We find $P_A + P_B = (\rho^2 + \tau^2)^2 = 1$, which was of course expected. In the absence of absorption by the beam splitters or by the mirrors, the neutron will exit in either of the two ports $A$ and $B$.

**13.5.2.4** The choice $\Delta = 0$ (modulo $2\pi$) leads to $P_B = 0$ for any value of $(\rho, \tau)$: the two paths "reflection on $S$ – transmission by $S'$" and "transmission by $S$ – reflection of $S'$" interfere destructively in the output channel $B$.

### 13.5.3 A First Step Towards a Non Destructive Detection

**13.5.3.1** A factitious bomb has no influence on the neutron state. One is certain to find the neutron in the output port $A$: $P_A = 1$, $P_B = 0$.

**13.5.3.2** The neutron state after $S$ is $\rho|\phi_3\rangle + \tau|\phi_4\rangle$. Detecting if a real bomb located around the mirror $M_b$ does explode or not amounts to measuring whether the neutron has been reflected or transmitted by the beam splitter $S$. The probability that the bomb explodes is thus the modulus square of the amplitude of $|\phi_4\rangle$, i.e. $\tau^2$.

**13.5.3.3** If the bomb is real and does not explode, it means that the neutron has followed the path $LM_aL'$. In this case, it reached the beam splitter $S'$ in the state $|\phi_1'\rangle$. After crossing $S'$, its state is $\rho|\phi_3'\rangle + \tau|\phi_4'\rangle$. The probabilities to detect this neutron in the output channels $A$ and $B$ are thus $\rho^2$ and $\tau^2$, respectively.

**13.5.3.4** Once the neutron has crossed the interferometer, if the bomb did not explode and if the neutron is detected in the output channel $B$, one is certain that the bomb was real. In the other case, when the neutron is detected in channel $A$, one cannot conclude regarding the nature of the bomb.

**13.5.3.5** The merit factor is the product of two probabilities: (i) In order to achieve a non destructive detection of a real bomb, one requires that the bomb does not explode (probability $\rho^2$ according to question 13.5.3.2) and (ii) one needs the neutron to be detected in channel $B$ (probability $\tau^2$ according to question 13.5.3.3). One has thus $F = \rho^2\tau^2$.

**13.5.3.6** We have to maximize the product $\rho^2\tau^2$ while keeping the sum $\rho^2 + \tau^2 = 1$. The maximum is reached for $\rho^2 = \tau^2 = 1/2$, corresponding to the factor of merit $F = 1/4$. In this case, half of the real bombs will explode on the average. Among the remaining half, only 50% lead to a detection of the neutron in the channel $B$ and are thus detected in a non destructive way. That's better than nothing, but not really efficient...

### 13.5.4 Towards an Efficient Non-destructive Detection

**13.5.4.1** There are two incident waves at $x = 0$, one coming from the left ($e^{ikx}$) and one coming from the right ($ve^{-ikx}$). These two waves give rise to two waves emerging from $S$, one towards the right ($\mu e^{ikx}$) and the other one towards the left

$(\eta e^{-ikx})$. Using (13.6) and (13.7), we find that the amplitudes of these waves are related by

$$\mu = \tau - \rho v \qquad \eta = \rho + \tau v. \qquad (13.17)$$

**13.5.4.2** At $x = D$, there is only one wave arriving on $S'$ with the complex amplitude $\mu e^{ikD}$. This wave can be either reflected or transmitted, hence:

$$\xi e^{ikD} = \tau \mu e^{ikD} \qquad v e^{-ikD} = \rho \mu e^{ikD}. \qquad (13.18)$$

**13.5.4.3** We write $v$ in terms of $\mu$ using (13.18), $v = \rho \mu e^{2ikD}$, and we insert this result into (13.17), which gives:

$$\mu = \frac{\tau}{1 + \rho^2 e^{2ikD}}. \qquad (13.19)$$

We obtain in this way the transmitted ($\xi$) and reflected ($\eta$) amplitudes for this Fabry–Perot interferometer:

$$\xi = \frac{\tau^2}{1 + \rho^2 e^{2ikD}} \qquad \eta = \rho \frac{1 + e^{2ikD}}{1 + \rho^2 e^{2ikD}}. \qquad (13.20)$$

We check that $|\xi|^2 + |\eta|^2 = 1$. The choice $2kD = \pi$ modulo $2\pi$ leads to $\xi = 1$, $\eta = 0$. Even if the reflection coefficients of the beam splitters are very close to unity, one can reach a full transmission of the setup when the distance $D$ between them is properly chosen.

**13.5.4.4** Consider the situation where $2kD = \pi$ modulo $2\pi$, with a neutron arriving on the first beam splitter. For a factitious bomb, we are certain that the neutron will be transmitted by the setup. When the bomb is real, it will explode if the neutron is transmitted by the first beam splitter (probability $\tau^2$) and it will not explode if the neutron is reflected by this beam splitter (probability $\rho^2$).

To obtain a high factor of merit, we choose $\rho$ close to 1 (and thus $\tau$ close to 0). Then we send a neutron from the left and detect if it was transmitted or reflected. If the bomb is factitious, the neutron is transmitted by the setup with probability 1. If the bomb is real, the neutron is reflected with probability $\rho^2$ close to 1, or there may be an explosion with a small probability $\tau^2$. Therefore the detection of a reflected neutron means that one has detected the presence of a real bomb without seeing it explode. The factor of merit is $\rho^2 \sim 1$. A real bomb is present if the neutron does *not* have the behavior predicted in the case of a factitious bomb, in other words if being "real" means being "non-factitious".

### 13.5.5 References

The initial proposal for this scheme is due to A. C. Elitzur and L. Vaidman, *Quantum mechanical interaction-free measurements*, Foundations of Physics **23**, 987 (1993). It was implemented experimentally by P. Kwiat, H. Weinfurter, T. Herzog, A. Zeilinger, and M. Kasevich, *Experimental Realization of Interaction-free Measurements*, Annals of the New York Academy of Sciences, **755**, 383 (1995). For a recent discussion of the link between this problem and tests of quantum mechanics, see C. Robens, W. Alt, C. Emary, D. Meschede, and A. Alberti, *Atomic "bomb testing": the Elitzur-Vaidman experiment violates the Leggett-Garg inequality*, Applied Physics B **123**, 12 (2017).

# Direct Observation of Field Quantization

**14**

We consider here a two-level atom interacting with a single mode of the electromagnetic field. When this mode is treated quantum mechanically, specific features occur in the atomic dynamics, such as damping and revivals of the Rabi oscillations.

## 14.1  Quantization of a Mode of the Electromagnetic Field

We recall that in *classical mechanics*, a harmonic oscillator of mass $m$ and frequency $\omega/2\pi$ obeys the equations of motion $dx/dt = p/m$ and $dp/dt = -m\omega^2 x$ where $x$ is the position and $p$ the momentum of the oscillator. Defining the *reduced* variables $X(t) = x(t)\sqrt{m\omega/\hbar}$ and $P(t) = p(t)/\sqrt{\hbar m\omega}$, the equations of motion of the oscillator are

$$\frac{dX}{dt} = \omega P \qquad \frac{dP}{dt} = -\omega X, \tag{14.1}$$

and the total energy $U(t)$ is given by

$$U(t) = \frac{\hbar\omega}{2}(X^2(t) + P^2(t)). \tag{14.2}$$

**14.1.1** Consider a cavity for electromagnetic waves, of volume $V$ Throughout this chapter, we consider a single mode of the electromagnetic field, of the form

$$\boldsymbol{E}(\boldsymbol{r}, t) = \boldsymbol{u}_x\, e(t)\sin kz \quad \boldsymbol{B}(\boldsymbol{r}, t) = \boldsymbol{u}_y\, b(t)\cos kz, \tag{14.3}$$

© Springer Nature Switzerland AG 2019
J.-L. Basdevant, J. Dalibard, *The Quantum Mechanics Solver*,
https://doi.org/10.1007/978-3-030-13724-3_14

where $u_x$, $u_y$ and $u_z$ are an orthonormal basis. We recall Maxwell's equations in vacuum:

$$\nabla \cdot E(r,t) = 0 \qquad \nabla \wedge E(r,t) = -\frac{\partial B(r,t)}{\partial t} \tag{14.4}$$

$$\nabla \cdot B(r,t) = 0 \qquad \nabla \wedge B(r,t) = \frac{1}{c^2}\frac{\partial E(r,t)}{\partial t} \tag{14.5}$$

and the total energy $U(t)$ of the field in the cavity:

$$U(t) = \int_V \left( \frac{\epsilon_0}{2} E^2(r,t) + \frac{1}{2\mu_0} B^2(r,t) \right) d^3r \quad \text{with } \epsilon_0\mu_0 c^2 = 1. \tag{14.6}$$

(a) Express $de/dt$ and $db/dt$ in terms of $k, c, e(t), b(t)$.
(b) Express $U(t)$ in terms of $V, e(t), b(t), \epsilon_0, \mu_0$. One can take

$$\int_V \sin^2 kz \, d^3r = \int_V \cos^2 kz \, d^3r = \frac{V}{2}. \tag{14.7}$$

(c) Setting $\omega = ck$ and introducing the reduced variables

$$\chi(t) = \sqrt{\frac{\epsilon_0 V}{2\hbar\omega}}\, e(t) \quad \Pi(t) = \sqrt{\frac{V}{2\mu_0\hbar\omega}}\, b(t) \tag{14.8}$$

show that the equations for $d\chi/dt, d\Pi/dt$ and $U(t)$ in terms of $\chi$, $\Pi$ and $\omega$ are formally identical to equations (14.1) and (14.2).

**14.1.2** The quantization of the mode of the electromagnetic field under consideration is performed in the same way as that of an ordinary harmonic oscillator. One associates to the physical quantities $\chi$ and $\Pi$, Hermitian operators $\hat{\chi}$ and $\hat{\Pi}$ which satisfy the commutation relation

$$[\hat{\chi}, \hat{\Pi}] = i. \tag{14.9}$$

The Hamiltonian of the field in the cavity is

$$\hat{H}_C = \frac{\hbar\omega}{2}\left( \hat{\chi}^2 + \hat{\Pi}^2 \right). \tag{14.10}$$

The energy of the field is quantized: $E_n = (n+1/2)\hbar\omega$ ($n$ is a non-negative integer); one denotes by $|n\rangle$ the eigenstate of $\hat{H}_C$ with eigenvalue $E_n$.

The *quantum states* of the field in the cavity are linear combinations of the set $\{|n\rangle\}$. The state $|0\rangle$, of energy $E_0 = \hbar\omega/2$, is called the "vacuum", and the state $|n\rangle$ of energy $E_n = E_0 + n\hbar\omega$ is called the "$n$ photon state". A "photon" corresponds to an elementary excitation of the field, of energy $\hbar\omega$.

One introduces the "creation" and "annihilation" operators of a photon as $\hat{a}^\dagger = (\hat{\chi} - i\hat{\Pi})/\sqrt{2}$ and $\hat{a} = (\hat{\chi} + i\hat{\Pi})/\sqrt{2}$ respectively. These operators satisfy the usual relations:

$$\hat{a}^\dagger|n\rangle = \sqrt{n+1}\,|n+1\rangle \tag{14.11}$$

$$\hat{a}|n\rangle = \sqrt{n}\,|n-1\rangle \quad \text{if} \quad n \neq 0 \qquad \text{and} \quad \hat{a}|0\rangle = 0. \tag{14.12}$$

(a) Express $\hat{H}_C$ in terms of $\hat{a}^\dagger$ and $\hat{a}$. The observable $\hat{N} = \hat{a}^\dagger\hat{a}$ is called the "number of photons".

The *observables* corresponding to the electric and magnetic fields at a point $r$ are defined as:

$$\hat{E}(r) = u_x\sqrt{\frac{\hbar\omega}{\epsilon_0 V}}\,(\hat{a} + \hat{a}^\dagger)\sin kz \tag{14.13}$$

$$\hat{B}(r) = iu_y\sqrt{\frac{\mu_0\hbar\omega}{V}}(\hat{a}^\dagger - \hat{a})\cos kz. \tag{14.14}$$

The interpretation of the theory in terms of states and observables is the same as in ordinary quantum mechanics.

(b) Calculate the expectation values $\langle E(r)\rangle$, $\langle B(r)\rangle$, and $\langle n|\hat{H}_C|n\rangle$ in an $n$-photon state.

**14.1.3** The following superposition:

$$|\alpha\rangle = e^{-|\alpha|^2/2}\sum_{n=0}^{\infty}\frac{\alpha^n}{\sqrt{n!}}|n\rangle, \tag{14.15}$$

where $\alpha$ is any complex number, is called a "quasi-classical" state of the field.

(a) Show that $|\alpha\rangle$ is a normalized eigenvector of the annihilation operator $\hat{a}$ and give the corresponding eigenvalue. Calculate the expectation value $\langle n\rangle$ of the number of photons in that state.

(b) Show that if, at time $t = 0$, the state of the field is $|\psi(0)\rangle = |\alpha\rangle$, then, at time $t$, $|\psi(t)\rangle = e^{-i\omega t/2}|(\alpha e^{-i\omega t})\rangle$.

(c) Calculate the expectation values $\langle E(r)\rangle_t$ and $\langle B(r)\rangle_t$ at time $t$ in a quasi-classical state for which $\alpha$ is real.

(d) Check that $\langle E(r)\rangle_t$ and $\langle B(r)\rangle_t$ satisfy Maxwell's equations.

(e) Calculate the energy of a *classical* field such that $E_{cl}(r, t) = \langle E(r) \rangle_t$ and $B_{cl}(r, t) = \langle \hat{B}(r) \rangle_t$. Compare the result with the expectation value of $\hat{H}_C$ in the same quasi-classical state.

(f) Why do these results justify the name "quasi-classical" state for $|\alpha\rangle$ if $|\alpha| \gg 1$?

## 14.2    The Coupling of the Field with an Atom

Consider an atom at point $r_0$ in the cavity. The motion of the center of mass of the atom in space is treated classically. Hereafter we restrict ourselves to the *two-dimensional* subspace of internal atomic states generated by the ground state $|f\rangle$ and an excited state $|e\rangle$. The origin of atomic energies is chosen in such a way that the energies of $|f\rangle$ and $|e\rangle$ are respectively $-\hbar\omega_A/2$ and $+\hbar\omega_A/2$ ($\omega_A > 0$). In the basis $\{|f\rangle, |e\rangle\}$, one can introduce the operators:

$$\hat{\sigma}_z = \begin{pmatrix} 1 & 0 \\ 0 & -1 \end{pmatrix} \qquad \hat{\sigma}_+ = \begin{pmatrix} 0 & 0 \\ 1 & 0 \end{pmatrix} \qquad \hat{\sigma}_- = \begin{pmatrix} 0 & 1 \\ 0 & 0 \end{pmatrix}, \tag{14.16}$$

that is to say $\hat{\sigma}_+|f\rangle = |e\rangle$ and $\hat{\sigma}_-|e\rangle = |f\rangle$, and the atomic Hamiltonian can be written as: $\hat{H}_A = -\frac{\hbar\omega_A}{2}\hat{\sigma}_z$.

The set of orthonormal states $\{|f, n\rangle, |e, n\rangle, n \geq 0\}$ where $|f, n\rangle \equiv |f\rangle \otimes |n\rangle$ and $|e, n\rangle \equiv |e\rangle \otimes |n\rangle$ forms a basis of the Hilbert space of the {atom + photons} states.

**14.2.1**  Check that it is an eigenbasis of $\hat{H}_0 = \hat{H}_A + \hat{H}_C$, and give the corresponding eigenvalues.

**14.2.2**  In *the remaining parts of the problem* we assume that the frequency of the cavity is exactly tuned to the Bohr frequency of the atom, i.e. $\omega = \omega_A$. Draw schematically the positions of the first 5 energy levels of $\hat{H}_0$. Show that, except for the ground state, the eigenstates of $\hat{H}_0$ are grouped in degenerate pairs.

**14.2.3**  The Hamiltonian of the electric dipole coupling between the atom and the field can be written as:

$$\hat{W} = \gamma(\hat{a}\hat{\sigma}_+ + \hat{a}^\dagger\hat{\sigma}_-), \tag{14.17}$$

where $\gamma = -d\sqrt{\hbar\omega/\epsilon_0 V} \sin kz_0$, and where the electric dipole moment $d$ is determined experimentally.

(a) Write the action of $\hat{W}$ on the states $|f, n\rangle$ and $|e, n\rangle$.
(b) To which physical processes do $\hat{a}\hat{\sigma}_+$ and $\hat{a}^\dagger\hat{\sigma}_-$ correspond?

**14.2.4** Determine the eigenstates of $\hat{H} = \hat{H}_0 + \hat{W}$ and the corresponding energies. Show that the problem reduces to the diagonalization of a set of $2 \times 2$ matrices. One hereafter sets:

$$|\phi_n^\pm\rangle = \frac{1}{\sqrt{2}}(|f, n + 1\rangle \pm |e, n\rangle) \qquad (14.18)$$

$$\frac{\hbar\Omega_0}{2} = \gamma = -d\sqrt{\frac{\hbar\omega}{\epsilon_0 V}} \sin kz_0 \quad \Omega_n = \Omega_0\sqrt{n + 1}. \qquad (14.19)$$

The energies corresponding to the eigenstates $|\phi_n^\pm\rangle$ are denoted $E_n^\pm$.

## 14.3   Interaction of the Atom with an "Empty" Cavity

*In the following, one assumes that the atom crosses the cavity along a line where* $\sin kz_0 = 1$.

An atom in the excited state $|e\rangle$ is sent into the cavity prepared in the vacuum state $|0\rangle$. At time $t = 0$, when the atom enters the cavity, the state of the system is $|e, n = 0\rangle$.

**14.3.1** What is the state of the system at a later time $t$?

**14.3.2** What is the probability $P_f(T)$ to find the atom in the state $f$ at time $T$ when the atom leaves the cavity? Show that $P_f(T)$ is a periodic function of $T$ ($T$ is varied by changing the velocity of the atom).

**14.3.3** The experiment has been performed on rubidium atoms for a couple of states $(f, e)$ such that $d = 1.1 \times 10^{-26}$ C.m and $\omega/2\pi = 5.0 \times 10^{10}$ Hz. The volume of the cavity is $1.87 \times 10^{-6}$ m$^3$ (we recall that $\epsilon_0 = 1/(36\pi\,10^9)$ S.I.).

The curve $P_f(T)$, together with the real part of its Fourier transform $J(\nu) = \int_0^\infty \cos(2\pi\nu T)\, P_f(T)\, dT$, are shown in Fig. 14.1. One observes a damped oscillation, the damping being due to imperfections of the experimental setup.

How do theory and experiment compare?

(We recall that the Fourier transform of a damped sinusoid in time exhibits a peak at the frequency of this sinusoid, whose width is proportional to the inverse of the characteristic damping time.)

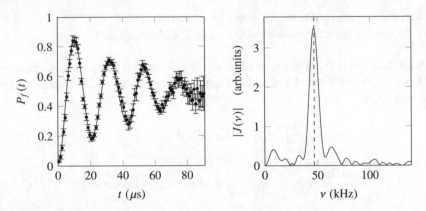

**Fig. 14.1** Left: Probability $P_f(T)$ of detecting the atom in the ground state after it crosses a cavity containing zero photons. Right: Fourier transform of this probability, as defined in the text

## 14.4    Interaction of an Atom with a Quasi-classical State

The atom, initially in the state $|e\rangle$, is now sent into a cavity where a quasiclassical state $|\alpha\rangle$ of the field has been prepared. At time $t = 0$ the atom enters the cavity and the state of the system is $|e\rangle \otimes |\alpha\rangle$.

**14.4.1**  Calculate the probability $P_f(T, n)$ to find, at time $T$, the atom in the state $|f\rangle$ and the field in the state $|n + 1\rangle$, for $n \geq 0$. What is the probability to find the atom in the state $|f\rangle$ and the field in the state $|0\rangle$?

**14.4.2**  Write the probability $P_f(T)$ to find the atom in the state $|f\rangle$, independently of the state of the field, as an infinite sum of oscillating functions.

**14.4.3**  On Fig. 14.2 are plotted an experimental measurement of $P_f(T)$ and the real part of its Fourier transform $J(\nu)$. The cavity used for this measurement is the same as in Fig. 14.1, but the field has been prepared in a quasi-classical state before the atom is sent in.

(a) Determine the three frequencies $\nu_0$, $\nu_1$, $\nu_2$ which contribute most strongly to $P_f(T)$.
(b) Do the ratios $\nu_1/\nu_0$ and $\nu_2/\nu_0$ have the expected values?
(c) From the values $J(\nu_0)$ and $J(\nu_1)$, determine an approximate value for the mean number of photons $|\alpha|^2$ in the cavity.

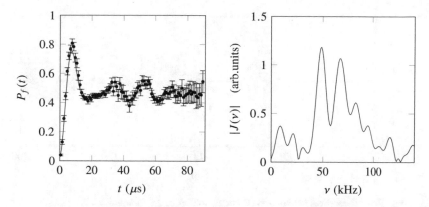

**Fig. 14.2** Left: Probability $P_f(T)$ of measuring the atom in the ground state after the atom has passed through a cavity containing a quasi-classical state of the electromagnetic field. Right: Fourier transform of this probability

## 14.5   Large Numbers of Photons: Damping and Revivals

Consider a quasi-classical state $|\alpha\rangle$ of the field corresponding to a large mean number of photons: $|\alpha|^2 \simeq n_0 \gg 1$, where $n_0$ is an integer. In this case, the probability $\pi(n)$ to find $n$ photons can be cast, in good approximation, in the form:

$$\pi(n) = e^{-|\alpha|^2} \frac{|\alpha|^{2n}}{n!} \simeq \frac{1}{\sqrt{2\pi n_0}} \exp\left(-\frac{(n-n_0)^2}{2n_0}\right). \tag{14.20}$$

This Gaussian limit of the Poisson distribution can be obtained by using the Stirling formula $n! \sim n^n e^{-n} \sqrt{2\pi n}$ and expanding $\ln \pi(n)$ in the vicinity of $n = n_0$.

**14.5.1** Show that this probability takes significant values only if $n$ lies in a neighborhood $\delta n$ of $n_0$. Give the relative value $\delta n / n_0$.

**14.5.2** For such a quasi-classical state, one tries to evaluate the probability $P_f(T)$ of detecting the atom in the state $f$ after its interaction with the field. In order to do so,

- one linearizes the dependence of $\Omega_n$ on $n$ in the vicinity of $n_0$:

$$\Omega_n \simeq \Omega_{n_0} + \Omega_0 \frac{n - n_0}{2\sqrt{n_0 + 1}}, \tag{14.21}$$

- one replaces the discrete summation in $P_f(T)$ by an integral.

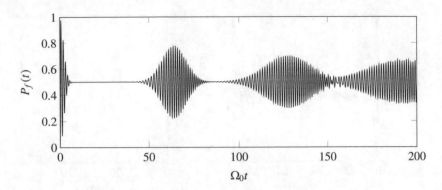

**Fig. 14.3** Exact theoretical calculation of $P_f(T)$ for $\langle n \rangle \simeq 25$ photons

(a) Show that, under these approximations, $P_f(T)$ is an oscillating function of $T$ for short times, but that this oscillation is damped away after a characteristic time $T_D$. Give the value of $T_D$.

We recall that

$$\int_{-\infty}^{\infty} \frac{1}{\sigma\sqrt{2\pi}} e^{-(x-x_0)^2/2\sigma^2} \cos(\alpha x)\, dx = e^{-\alpha^2\sigma^2/2} \cos(\alpha x_0). \qquad (14.22)$$

(b) Does this damping time depend on the mean value of the number of photons $n_0$?

(c) Give a qualitative explanation for this damping.

**14.5.3** If one keeps the expression of $P_f(T)$ as a discrete sum, an exact numerical calculation shows that one expects a revival of the oscillations of $P_f(T)$ for certain times $T_R$ large compared to $T_D$, as shown in Fig. 14.3. This phenomenon is called *quantum revival* and it is currently being studied experimentally.

Keeping the discrete sum but using the approximation (14.21), can you explain the revival qualitatively? How does the time of the first revival depend on $n_0$?

## 14.6   Solutions

### 14.6.1  Quantization of a Mode of the Electromagnetic Field

**14.6.1.1 (a)** The pair of Maxwell equations $\nabla \cdot \boldsymbol{E} = 0$ and $\nabla \cdot \boldsymbol{B} = 0$ are satisfied whatever the values of the functions $e(t)$ and $b(t)$. The equations $\nabla \wedge \boldsymbol{E} = -(\partial \boldsymbol{B}/\partial t)$ and $c^2 \nabla \wedge \boldsymbol{B} = -(\partial \boldsymbol{E}/\partial t)$ require that:

$$\frac{de}{dt} = c^2 k b(t) \qquad \frac{db}{dt} = -k e(t). \qquad (14.23)$$

**(b)** The electromagnetic energy can be written as:

$$U(t) = \int_V \left( \frac{\epsilon_0}{2} e^2(t) \sin^2 kz + \frac{1}{2\mu_0} b^2(t) \cos^2 kz \right) d^3 r$$

$$= \frac{\epsilon_0 V}{4} e^2(t) + \frac{V}{2\mu_0} b^2(t). \tag{14.24}$$

**(c)** Under the change of functions suggested in the text, we obtain:

$$\begin{cases} \dot{\chi} = \omega \Pi \\ \dot{\Pi} = -\omega \chi \end{cases} \qquad U(t) = \frac{\hbar \omega}{2} (\chi^2(t) + \Pi^2(t)). \tag{14.25}$$

These two equations are formally identical to the equations of motion of a particle in a harmonic oscillator potential.

**14.6.1.2** **(a)** From $[\hat{\chi}, \hat{\Pi}] = i$, we deduce that:

$$[\hat{a}, \hat{a}^\dagger] = \frac{1}{2} [\hat{\chi} + i\hat{\Pi}, \hat{\chi} - i\hat{\Pi}] = 1. \tag{14.26}$$

In addition, $\hat{\chi} = (\hat{a} + \hat{a}^\dagger)/\sqrt{2}$ and $\hat{\Pi} = i(\hat{a}^\dagger - \hat{a})/\sqrt{2}$, i.e.:

$$\hat{H}_C = \frac{\hbar \omega}{2} (\hat{a}\hat{a}^\dagger + \hat{a}^\dagger \hat{a}) = \hbar \omega \left( \hat{a}^\dagger \hat{a} + \frac{1}{2} \right), \tag{14.27}$$

or $\hat{H}_C = \hbar \omega \left( \hat{N} + \frac{1}{2} \right)$.

**(b)** For an $n$ photon state, we find $\langle n | \hat{a} | n \rangle = \langle n | \hat{a}^\dagger | n \rangle = 0$, which results in

$$\langle \boldsymbol{E}(\boldsymbol{r}) \rangle = 0 \quad \langle \boldsymbol{B}(\boldsymbol{r}) \rangle = 0. \tag{14.28}$$

The state $|n\rangle$ is an eigenstate of $\hat{H}_C$ with eigenvalue $(n + 1/2)\hbar\omega$, i.e.

$$\langle H_C \rangle = \left( n + \frac{1}{2} \right) \hbar \omega. \tag{14.29}$$

**14.6.1.3** **(a)** The action of $\hat{a}$ on $|\alpha\rangle$ gives

$$\hat{a}|\alpha\rangle = e^{-|\alpha|^2/2} \sum_{n=1}^{\infty} \frac{\alpha^n}{\sqrt{n!}} \sqrt{n} |n-1\rangle$$

$$= \alpha e^{-|\alpha|^2/2} \sum_{n=1}^{\infty} \frac{\alpha^{n-1}}{\sqrt{(n-1)!}} |n-1\rangle = \alpha |\alpha\rangle. \tag{14.30}$$

The vector $|\alpha\rangle$ is normalized:

$$\langle\alpha|\alpha\rangle = e^{-|\alpha|^2} \sum_{n=0}^{\infty} \frac{(\alpha^*)^n \alpha^n}{n!} = 1. \tag{14.31}$$

The expectation value of the number of photons in that state is:

$$\langle n\rangle = \langle\alpha|\hat{N}|\alpha\rangle = \langle\alpha|\hat{a}^{\dagger}\hat{a}|\alpha\rangle = ||\hat{a}|\alpha\rangle||^2 = |\alpha|^2. \tag{14.32}$$

**(b)** The time evolution of $|\psi(t)\rangle$ is given by

$$|\psi(t)\rangle = e^{-|\alpha|^2/2} \sum_{n=0}^{\infty} \frac{\alpha^n}{\sqrt{n!}} e^{-i\omega(n+1/2)t}|n\rangle$$

$$= e^{-i\omega t/2} e^{-|\alpha|^2/2} \sum_{n=0}^{\infty} \frac{(\alpha e^{-i\omega t})^n}{\sqrt{n!}}|n\rangle = e^{-i\omega t/2}|(\alpha e^{-i\omega t})\rangle. \tag{14.33}$$

**(c)** The expectation values of the electric and magnetic fields are

$$\langle E(r)\rangle_t = 2\alpha \cos\omega t \sin kz \sqrt{\frac{\hbar\omega}{\epsilon_0 V}} u_x \tag{14.34}$$

$$\langle B(r)\rangle_t = -2\alpha \sin\omega t \cos kz \sqrt{\frac{\hbar\omega\mu_0}{V}} u_y. \tag{14.35}$$

**(d)** These fields are of the same type as the classical fields considered at the beginning of the problem, with

$$e(t) = 2\alpha \sqrt{\frac{\hbar\omega}{\epsilon_0 V}} \cos\omega t \qquad b(t) = -2\alpha \sqrt{\frac{\hbar\omega\mu_0}{V}} \sin\omega t. \tag{14.36}$$

Given the relation $\epsilon_0\mu_0 c^2 = 1$, we verify that $\dot{e}(t) = c^2 k b(t)$ and $\dot{b} = -ke(t)$. Therefore the expectation values of the field operators satisfy Maxwell's equations.

**(e)** The energy of the classical field can be calculated using the previous results. Since $\cos^2\omega t + \sin^2\omega t = 1$, we find $U(t) = \hbar\omega\alpha^2$. This "classical" energy is therefore time-independent. The expectation value of $\hat{H}_C$ is:

$$\langle H_C\rangle = \langle\hbar\omega(N + 1/2)\rangle = \hbar\omega(\alpha^2 + 1/2). \tag{14.37}$$

It is also time independent (Ehrenfest's theorem).

**(f)** For $|\alpha|$ much larger than 1, the ratio $U(t)/\langle H_C\rangle$ is close to 1. More generally, the expectation value of a physical quantity as calculated for a *quantum* field in the state $|\alpha\rangle$, will be close to the value calculated for a *classical* field such that $E_{cl}(r, t) = \langle E(r)\rangle_t$ and $B_{cl}(r, t) = \langle B(r)\rangle_t$.

## 14.6.2 The Coupling of the Field with an Atom

**14.6.2.1** One checks that

$$\hat{H}_0|f, n\rangle = \left(-\frac{\hbar\omega_A}{2} + \left(n + \frac{1}{2}\right)\hbar\omega\right)|f, n\rangle,$$

$$\hat{H}_0|e, n\rangle = \left(\frac{\hbar\omega_A}{2} + \left(n + \frac{1}{2}\right)\hbar\omega\right)|e, n\rangle. \tag{14.38}$$

**14.6.2.2** For a cavity which resonates at the atom's frequency, i.e. if $\omega = \omega_A$, the couple of states $|f, n + 1\rangle$, $|e, n\rangle$ are degenerate. The first five levels of $H_0$ are shown in Fig. 14.4a. Only the ground state $|f, 0\rangle$ of the atom+field system is non-degenerate.

**14.6.2.3** (a) The action of $\hat{W}$ on the basis vectors of $H_0$ is given by:

$$\hat{W}|f, n\rangle = \sqrt{n}\gamma\,|e, n - 1\rangle \quad \text{if} \quad n \geq 1$$

$$= 0 \quad \text{if} \quad n = 0 \tag{14.39}$$

$$\hat{W}|e, n\rangle = \sqrt{n + 1}\,\gamma\,|f, n + 1\rangle. \tag{14.40}$$

The coupling under consideration corresponds to an electric dipole interaction of the form $-\hat{D} \cdot \hat{E}(r)$, where $\hat{D}$ is the observable electric dipole moment of the atom.

**(b)** $\hat{W}$ couples the two states of each degenerate pair. The term $\hat{a}\hat{\sigma}_+$ corresponds to the absorption of a photon by the atom, which undergoes a transition from the ground state to the excited state. The term $\hat{a}^\dagger\hat{\sigma}_-$ corresponds to the emission of a photon by the atom, which undergoes a transition from the excited state to the ground state.

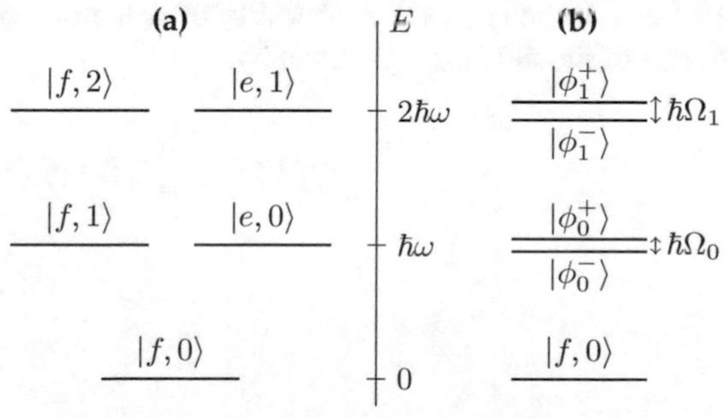

**Fig. 14.4** (a) Positions of the five first energy levels of $H_0$. (b) Positions of the five first energy levels of $\hat{H} = \hat{H}_0 + \hat{W}$

**14.6.2.4** The operator $\hat{W}$ is block-diagonal in the eigenbasis of $\hat{H}_0\{|f,n\rangle, |e,n\rangle\}$. Therefore:

- The state $|f,0\rangle$ is an eigenstate of $\hat{H}_0 + \hat{W}$ with the eigenvalue 0.
- In each eigen-subspace of $\hat{H}_0$ generated by $\{|f,n+1\rangle, |e,n\rangle\}$ with $n \geq 0$, one must diagonalize the $2 \times 2$ matrix:

$$\begin{pmatrix} (n+1)\hbar\omega & \hbar\Omega_n/2 \\ \hbar\Omega_n/2 & (n+1)\hbar\omega \end{pmatrix} \tag{14.41}$$

whose eigenvectors and corresponding eigenvalues are ($n \geq 0$):

$$|\phi_n^+\rangle \quad \text{corresponding to} \quad E_n^+ = (n+1)\hbar\omega + \frac{\hbar\Omega_n}{2} \tag{14.42}$$

$$|\phi_n^-\rangle \quad \text{corresponding to} \quad E_n^- = (n+1)\hbar\omega - \frac{\hbar\Omega_n}{2}. \tag{14.43}$$

The first five energy levels of $\hat{H}_0 + \hat{W}$ are shown in Fig. 14.4b.

### 14.6.3  Interaction of the Atom and an "Empty" Cavity

**14.6.3.1** We expand the initial state on the eigenbasis of $\hat{H}$:

$$|\psi(0)\rangle = |e,0\rangle = \frac{1}{\sqrt{2}}(|\phi_0^+\rangle - |\phi_0^-\rangle). \tag{14.44}$$

The time evolution of the state vector is therefore given by:

$$|\psi(t)\rangle = \frac{1}{\sqrt{2}}\left(e^{-iE_0^+ t/\hbar}|\phi_0^+\rangle - e^{-iE_0^- t/\hbar}|\phi_0^-\rangle\right)$$

$$= \frac{e^{-i\omega t}}{\sqrt{2}}\left(e^{-i\Omega_0 t/2}|\phi_0^+\rangle - e^{i\Omega_0 t/2}|\phi_0^-\rangle\right). \tag{14.45}$$

**14.6.3.2** In general, the probability of detecting the atom in the state $f$, independently of the field state, is given by:

$$P_f(T) = \sum_{n=0}^{\infty} |\langle f,n|\psi(T)\rangle|^2. \tag{14.46}$$

In the particular case of an initially empty cavity, only the term $n = 1$ contributes to the sum. Using $|f, 1\rangle = (|\phi_0^+\rangle + |\phi_0^-\rangle)/\sqrt{2}$, we find

$$P_f(T) = \sin^2 \frac{\Omega_0 T}{2} = \frac{1}{2}(1 - \cos \Omega_0 T). \tag{14.47}$$

It is indeed a periodic function of $T$, with angular frequency $\Omega_0$.

**14.6.3.3** Experimentally, one measures an oscillation of frequency $\nu_0 = 47\,\text{kHz}$. This result corresponds to the expected value:

$$\nu_0 = \frac{1}{2\pi} \frac{2d}{\hbar} \sqrt{\frac{\hbar\omega}{\epsilon_0 V}}. \tag{14.48}$$

## 14.6.4 Interaction of an Atom with a Quasi-classical State

**14.6.4.1** Again, we expand the initial state on the eigenbasis of $\hat{H}_0 + \hat{W}$:

$$|\psi(0)\rangle = |e\rangle \otimes |\alpha\rangle = e^{-|\alpha|^2/2} \sum_{n=0}^{\infty} \frac{\alpha^n}{\sqrt{n!}} |e, n\rangle$$

$$= e^{-|\alpha|^2/2} \sum_{n=0}^{\infty} \frac{\alpha^n}{\sqrt{n!}} \frac{1}{\sqrt{2}} (|\phi_n^+\rangle - |\phi_n^-\rangle). \tag{14.49}$$

At time $t$ the state vector is

$$|\psi(t)\rangle = e^{-|\alpha|^2/2} \sum_{n=0}^{\infty} \frac{\alpha^n}{\sqrt{n!}} \frac{1}{\sqrt{2}} \left( e^{-iE_n^+ t/\hbar} |\phi_n^+\rangle - e^{-iE_n^- t/\hbar} |\phi_n^-\rangle \right). \tag{14.50}$$

We therefore observe that:

- the probability to find the atom in the state $|f\rangle$ and the field in the state $|0\rangle$ vanishes for all values of $T$,
- the probability $P_f(T, n)$ can be obtained from the scalar product of $|\psi(t)\rangle$ and $|f, n+1\rangle = (|\phi_n^+\rangle + |\phi_n^-\rangle)/\sqrt{2}$:

$$P_f(T, n) = \frac{1}{4} e^{-|\alpha|^2} \frac{|\alpha|^{2n}}{n!} \left| e^{-iE_n^+ t/\hbar} - e^{-iE_n^- t/\hbar} \right|^2$$

$$= e^{-|\alpha|^2} \frac{|\alpha|^{2n}}{n!} \sin^2 \frac{\Omega_n T}{2} = \frac{1}{2} e^{-|\alpha|^2} \frac{|\alpha|^{2n}}{n!} (1 - \cos \Omega_n T). \tag{14.51}$$

**14.6.4.2** The probability $P_f(T)$ is simply the sum of all probabilities $P_f(T, n)$:

$$P_f(T) = \sum_{n=0}^{\infty} P_f(T, n) = \frac{1}{2} - \frac{e^{-|\alpha|^2}}{2} \sum_{n=0}^{\infty} \frac{|\alpha|^{2n}}{n!} \cos \Omega_n T. \qquad (14.52)$$

**14.6.4.3** (a)  The three most prominent peaks of $J(\nu)$ occur at the frequencies $\nu_0 = 47\,\text{kHz}$ (already found for an empty cavity), $\nu_1 = 65\,\text{kHz}$ and $\nu_2 = 81\,\text{kHz}$.

(b)  The ratios of the measured frequencies are very close to the theoretical predictions: $\nu_1/\nu_0 = \sqrt{2}$ and $\nu_2/\nu_0 = \sqrt{3}$.

(c)  The ratio $J(\nu_1)/J(\nu_0)$ is of the order of 0.9. Assuming the peaks have the same widths, and that these widths are small compared to the splitting $\nu_1 - \nu_0$, this ratio corresponds to the average number of photons $|\alpha|^2$ in the cavity.

Actually, the peaks overlap, which makes this determination somewhat inaccurate. If one performs a more sophisticated analysis, taking into account the widths of the peaks, one obtains $|\alpha|^2 = 0.85 \pm 0.04$ (see the reference at end of this chapter).

Comment: One can also determine $|\alpha|^2$ from the ratio $J(\nu_2)/J(\nu_1)$ which should be equal to $|\alpha|^2/2$. However, the inaccuracy due to the overlap of the peaks is greater than for $J(\nu_1)/J(\nu_0)$, owing to the smallness of $J(\nu_2)$.

## 14.6.5  Large Numbers of Photons: Damping and Revivals

**14.6.5.1**  The probability $\pi(n)$ takes significant values only if $(n - n_0)^2/(2n_0)$ is not much larger than 1, i.e. for integer values of $n$ in a neighborhood of $n_0$ of relative extension of the order of $1/\sqrt{n_0}$. For $n_0 \gg 1$, the distribution $\pi(n)$ is therefore peaked around $n_0$.

**14.6.5.2** (a)  Consider the result of question 14.6.4.2, where we replace $\Omega_n$ by its approximation (14.21):

$$P_f(T) = \frac{1}{2} - \frac{1}{2} \sum_{n=0}^{\infty} \pi(n) \cos\left[\left(\Omega_{n_0} + \Omega_0 \frac{n - n_0}{2\sqrt{n_0 + 1}}\right) T\right] \qquad (14.53)$$

We now replace the discrete sum by an integral:

$$P_f(T) = \frac{1}{2} - \frac{1}{2} \int_{-\infty}^{\infty} \frac{e^{-u^2/(2n_0)}}{\sqrt{2\pi n_0}} \cos\left[\left(\Omega_{n_0} + \Omega_0 \frac{u}{2\sqrt{n_0 + 1}}\right) T\right] du. \qquad (14.54)$$

We have extended the lower integration bound from $-n_0$ down to $-\infty$, using the fact that the width of the gaussian is $\sqrt{n_0} \ll n_0$. We now develop the expression to be integrated upon:

$$\cos\left[\left(\Omega_{n_0} + \Omega_0 \frac{u}{2\sqrt{n_0+1}}\right)T\right] = \cos(\Omega_{n_0}T)\cos\left(\frac{\Omega_0 uT}{2\sqrt{n_0+1}}\right)$$
$$- \sin(\Omega_{n_0}T)\sin\left(\frac{\Omega_0 uT}{2\sqrt{n_0+1}}\right). \quad (14.55)$$

The sine term does not contribute to the integral (odd function) and we find:

$$P_f(T) = \frac{1}{2} - \frac{1}{2}\cos(\Omega_{n_0}T)\exp\left(-\frac{\Omega_0^2 T^2 n_0}{8(n_0+1)}\right). \quad (14.56)$$

For $n_0 \gg 1$, the argument of the exponential simplifies, and we obtain:

$$P_f(T) = \frac{1}{2} - \frac{1}{2}\cos(\Omega_{n_0}T)\exp\left(-\frac{T^2}{T_D^2}\right) \quad (14.57)$$

with $T_D = 2\sqrt{2}/\Omega_0$.

**(b)** In this approximation, the oscillations are damped out in a time $T_D$ which is independent of the number of photons $n_0$. For a given atomic transition (for fixed values of $d$ and $\omega$), this time $T_D$ increases like the square root of the volume of the cavity. In the limit of an infinite cavity, i.e. an atom in empty space, this damping time becomes infinite: we recover the usual Rabi oscillation. For a cavity of finite size, the number of visible oscillations of $P_f(T)$ is roughly $v_{n_0}T_D \sim \sqrt{n_0}$.

**(c)** The function $P_f(T)$ is made up of a large number of oscillating functions with similar frequencies. Initially, these different functions are in phase, and their sum $P_f(T)$ exhibits marked oscillations. After a time $T_D$, the various oscillations are no longer in phase with one another and the resulting oscillation of $P_f(T)$ is damped. One can find the damping time by simply estimating the time for which the two frequencies at half width on either side of the maximum of $\pi(n)$ are out of phase by $\pi$:

$$\Omega_{n_0+\sqrt{n_0}}T_D \sim \Omega_{n_0-\sqrt{n_0}}T_D + \pi \quad \text{and} \quad \sqrt{n_0 \pm \sqrt{n_0}} \simeq \sqrt{n_0} \pm \frac{1}{2}$$

$$\Rightarrow \quad \Omega_0 T_D \sim \pi.$$

**14.6.5.3** Within the approximation (14.21) suggested in the text, equation (14.53) above corresponds to a periodic evolution of period

$$T_R = \frac{4\pi}{\Omega_0}\sqrt{n_0 + 1}. \tag{14.58}$$

Indeed

$$\left(\Omega_{n_0} + \Omega_0 \frac{n - n_0}{2\sqrt{n_0 + 1}}\right) T_R = 4\pi(n_0 + 1) + 2\pi(n - n_0). \tag{14.59}$$

We therefore expect that all the oscillating functions which contribute to $P_f(T)$ will reset in phase at times $T_R, 2T_R, \ldots$ The time of the first revival, measured in Fig. 14.3, is $\Omega_0 T \simeq 64$, in excellent agreement with this prediction. Notice that $T_R \sim 4\sqrt{n_0}T_D$, which means that the revival time is always large compared to the damping time.

Actually, one can see from the result of Fig. 14.3 that the functions are only partly in phase. This comes from the fact that the numerical calculation has been done with the exact expression of $\Omega_n$. In this case, the difference between two consecutive frequencies $\Omega_{n+1} - \Omega_n$ is not exactly a constant, contrary to what happens in approximation (14.21); the function $P_f(T)$ is not really periodic. After a few revivals, one obtains a complicated behavior of $P_f(T)$, which can be analysed with the techniques developed for the study of chaos.

## 14.6.6 Comments

The damping phenomenon that we observed above is "classical": one would obtain it within a classical description of the interaction of the field and the atom, by considering a field whose intensity is not well defined (this would be the analog of a distribution $\pi(n)$ of the number of photons). On the other hand, the revival comes from the fact that the set of frequencies $\Omega_n$ is discrete. It is a direct consequence of the quantization of the electromagnetic field, in the same way as the occurrence of frequencies $\nu_0\sqrt{2}, \nu_0\sqrt{3}, \ldots$ in the evolution of $P_f(T)$ (Sect. 14.4).

The experiments described in this chapter have been performed in Paris, at the Laboratoire Kastler Brossel. The pair of levels $(f, e)$ correspond to very excited levels of rubidium, which explains the large value of the electric dipole moment $d$. The field is confined in a superconducting niobium cavity (quality factor of $\sim 10^8$), cooled down to 0.8 K in order to avoid perturbations from the thermal black body radiation (M. Brune, F. Schmidt-Kaler, A. Maali, J. Dreyer, E. Hagley, J.-M. Raimond, and S. Haroche, Phys. Rev. Lett. **76**, 1800 (1996)). A detailed analysis of experiments performed with atoms and photons interacting in electromagnetic cavities can be found in *Exploring the Quantum*, by S. Haroche and J.-M. Raimond (Oxford University Press, 2006).

Cavity quantum electrodynamics provides a paradigmatic example of the possibility to manipulate and measure individual particles (matter or light), while preserving their quantum coherence. The 2012 Nobel prize in Physics was awarded to S. Haroche and D. Wineland for the development of such techniques, which are now at the basis of a vast field of research labeled as *quantum technologies*. This field encompasses in particular the possibility to use individual quantum states of atoms, ions or light fields for communication, computation and for simulating other quantum systems; for a general account of this emerging field, see e.g. A. Acin *et al.*, New Jour. Physics **20**, 080201 (2018).

# Schrödinger's Cat

**15**

The superposition principle states that if $|\phi_a\rangle$ and $|\phi_b\rangle$ are two possible states of a quantum system, the quantum superposition $(|\phi_a\rangle + |\phi_b\rangle)/\sqrt{2}$ is also an allowed state for this system. This principle is essential in explaining interference phenomena. However, when it is applied to "large" objects, it leads to paradoxical situations where a system can be in a superposition of states which is classically self-contradictory (antinomic).

The most famous example is the Schrödinger "cat paradox" where the cat is in a superposition of the "dead" and "alive" states. The purpose of this chapter is to show that such superpositions of macroscopic states are not detectable in practice. They are extremely fragile, and a very weak coupling to the environment suffices to destroy the coherent superposition between the two states $|\phi_a\rangle$ and $|\phi_b\rangle$.

## 15.1 The Quasi-classical States of a Harmonic Oscillator

In this chapter, we shall consider high energy excitations of a one-dimensional harmonic oscillator, of mass $m$ and frequency $\omega$. The Hamiltonian is written

$$\hat{H} = \frac{\hat{p}^2}{2m} + \frac{1}{2}m\omega^2\hat{x}^2. \tag{15.1}$$

We denote the eigenstates of $\hat{H}$ by $\{|n\rangle\}$. The energy of the state $|n\rangle$ is $E_n = (n + 1/2)\hbar\omega$.

**15.1.1** Preliminary questions. We introduce the operators $\hat{X} = \hat{x}\sqrt{m\omega/\hbar}$, $\hat{P} = \hat{p}/\sqrt{m\hbar\omega}$ and the annihilation and creation operators

$$\hat{a} = \frac{1}{\sqrt{2}}\left(\hat{X} + i\hat{P}\right) \quad \hat{a}^\dagger = \frac{1}{\sqrt{2}}\left(\hat{X} - i\hat{P}\right) \quad \hat{N} = \hat{a}^\dagger\hat{a}. \tag{15.2}$$

© Springer Nature Switzerland AG 2019
J.-L. Basdevant, J. Dalibard, *The Quantum Mechanics Solver*,
https://doi.org/10.1007/978-3-030-13724-3_15

We recall the commutators: $[\hat{X}, \hat{P}] = i, [\hat{a}, \hat{a}^\dagger] = 1$, and the relations: $\hat{H} = \hbar\omega(\hat{N} + 1/2)$ and $\hat{N}|n\rangle = n|n\rangle$.

(a) Check that if one works with functions of the dimensionless variables $X$ and $P$, one has

$$\hat{P} = -i\frac{\partial}{\partial X} \qquad \hat{X} = i\frac{\partial}{\partial P}. \tag{15.3}$$

(b) Evaluate the commutator $[\hat{N}, \hat{a}]$, and prove that

$$\hat{a}|n\rangle = \sqrt{n}|n - 1\rangle \tag{15.4}$$

up to a phase factor which we set equal to 1 in what follows.

(c) Using (15.4) for $n = 0$ and expressing $\hat{a}$ in terms of $\hat{X}$ and $\hat{P}$, calculate the wave function of the ground state $\psi_0(X)$ and its Fourier transform $\varphi_0(P)$. It is not necessary to normalize the result.

**15.1.2  The Quasi-classical States.** The eigenstates of the operator $\hat{a}$ are called *quasi-classical* states, for reasons which we now examine.

Consider an arbitrary complex number $\alpha$. Show that the following state

$$|\alpha\rangle = e^{-|\alpha|^2/2} \sum_n \frac{\alpha^n}{\sqrt{n!}}|n\rangle \tag{15.5}$$

is a normalized eigenstate of $\hat{a}$ with eigenvalue $\alpha : \hat{a}|\alpha\rangle = \alpha|\alpha\rangle$.

**15.1.3** Calculate the expectation value of the energy in a quasi-classical state $|\alpha\rangle$. Calculate also the expectation values $\langle x \rangle$ and $\langle p \rangle$ and the root mean square deviations $\Delta x$ and $\Delta p$ for this state. Show that one has $\Delta x \, \Delta p = \hbar/2$.

**15.1.4** Following a similar procedure as in question 15.1.1 (c) above, determine the wave function $\psi_\alpha(X)$ of the quasi-classical state $|\alpha\rangle$, and its Fourier transform $\varphi_\alpha(P)$. Again, it is not necessary to normalize the result.

**15.1.5** Suppose that at time $t = 0$, the oscillator is in a quasi-classical state $|\alpha_0\rangle$ with $\alpha_0 = \rho e^{i\phi}$ where $\rho$ is a real positive number.

(a) Show that at any later time $t$ the oscillator is also in a quasi-classical state which can be written as $e^{-i\omega t/2}|\alpha(t)\rangle$. Determine the value of $\alpha(t)$ in terms of $\rho, \phi, \omega$ and $t$.

(b) Evaluate $\langle x \rangle_t$ and $\langle p \rangle_t$. Assuming that $|\alpha| \gg 1$, justify briefly why these states are called "quasi-classical".

**15.1.6 Numerical Example.** Consider a simple pendulum of length 1 meter and of mass 1 gram. Assume the state of this pendulum can be described by a quasi-classical state. At time $t = 0$ the pendulum is at $\langle x_0 \rangle = 1$ micrometer from its classical equilibrium position, with zero mean velocity.

(a) What is the corresponding value of $\alpha(0)$?
(b) What is the relative uncertainty on its position $\Delta x / x_0$?
(c) What is the value of $\alpha(t)$ after 1/4 period of oscillation?

## 15.2   Construction of a Schrödinger-Cat State

During the time interval $[0, T]$, one adds to the harmonic potential, the coupling

$$\hat{W} = \hbar g (\hat{a}^\dagger \hat{a})^2. \tag{15.6}$$

We assume that $g$ is much larger than $\omega$ and that $\omega T \ll 1$. Hence, we can make the approximation that, during the interval $[0, T]$, the Hamiltonian of the system is simply $\hat{W}$. At time $t = 0$, the system is in a quasi-classical state $|\psi(0)\rangle = |\alpha\rangle$.

**15.2.1** Show that the states $|n\rangle$ are eigenstates of $\hat{W}$, and write the expansion of the state $|\psi(T)\rangle$ at time $T$ on the basis $\{|n\rangle\}$.

**15.2.2** How does $|\psi(T)\rangle$ simplify in the particular cases $T = 2\pi/g$ and $T = \pi/g$?

**15.2.3** One now chooses $T = \pi/2g$. Show that this gives

$$|\psi(T)\rangle = \frac{1}{\sqrt{2}} \left( e^{-i\pi/4} |\alpha\rangle + e^{i\pi/4} |-\alpha\rangle \right). \tag{15.7}$$

**15.2.4** Suppose $\alpha$ is pure imaginary: $\alpha = i\rho$.

(a) Discuss qualitatively the physical properties of the state (15.7).
(b) Consider a value of $|\alpha|$ of the same order of magnitude as in 1.6. In what sense can this state be considered a concrete example of the "Schrödinger cat" type of state mentioned in the introduction?

## 15.3   Quantum Superposition Versus Statistical Mixture

We now study the properties of the state (15.7) in a "macroscopic" situation $|\alpha| \gg 1$. We choose $\alpha$ pure imaginary, $\alpha = i\rho$, and we set $p_0 = \rho\sqrt{2m\hbar\omega}$.

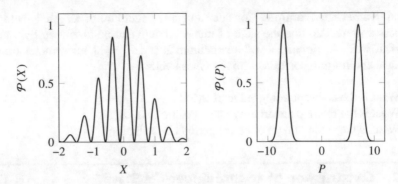

**Fig. 15.1** Probability distributions for the position (left) and for the momentum (right) of a system in the state (15.7) for $\alpha = 5i$. The quantities $X$ and $P$ are the dimensionless variables introduced in the first part of the problem. The vertical scale is arbitrary

**15.3.1** Consider a quantum system in the state (15.7). Write the (non-normalized) probability distributions for the position and for the momentum of the system. These probability distributions are represented in Fig. 15.1 for $\alpha = 5i$. Interpret these distributions physically.

**15.3.2** A physicist (Alice) prepares $N$ independent systems all in the state (15.7) and measures the momentum of each of these systems. The measuring apparatus has a resolution $\delta p$ such that:

$$\sqrt{m\hbar\omega} \ll \delta p \ll p_0. \tag{15.8}$$

For $N \gg 1$, draw qualitatively the histogram of the results of the $N$ measurements.

**15.3.3** The state (15.7) represents the quantum superposition of two states which are macroscopically different, and therefore leads to the paradoxical situations mentioned in the introduction. Another physicist (Bob) claims that the measurements done by Alice have not been performed on $N$ quantum systems in the state (15.7), but that Alice is actually dealing with a non-paradoxical statistical mixture", that is to say that half of the $N$ systems are in the state $|\alpha\rangle$ and the other half in the state $|-\alpha\rangle$. Assuming this is true, does one obtain the same probability distribution as for the previous question for the $N$ *momentum* measurements?

**15.3.4** In order to settle the matter, Alice now measures the position of each of $N$ independent systems, all prepared in the state (15.7). Draw the shape of the resulting distribution of events, assuming that the resolution $\delta x$ of the measuring apparatus is such that:

$$\delta x \ll \frac{1}{|\alpha|}\sqrt{\frac{\hbar}{m\omega}}. \tag{15.9}$$

**15.3.5** Can Bob obtain the same result concerning the $N$ position measurements assuming he is dealing with a statistical mixture?

**15.3.6** Considering the numerical value obtained in the case of a simple pendulum in question 15.1.6, evaluate the resolution $\delta x$ which is necessary in order to tell the difference between a set of $N$ systems in the quantum superposition (15.7), and a statistical mixture consisting in $N/2$ pendulums in the state $|\alpha\rangle$ and $N/2$ pendulums in the state $|-\alpha\rangle$.

## 15.4 The Fragility of a Quantum Superposition

In a realistic physical situation, one must take into account the coupling of the oscillator with its environment, in order to estimate how long one can discriminate between the quantum superposition (15.7) (that is to say the Schrödinger cat" which is "alive *and* dead") and a simple statistical mixture (i.e. a set of cats (systems), half of which are alive, the other half being dead; each cat being *either* alive *or* dead.)

If the oscillator is initially in the quasi-classical state $|\alpha_0\rangle$ and if the environment is in a state $|\chi_e(0)\rangle$, the wave function of the total system is the product of the individual wave functions, and the state vector of the total system can be written as the (tensor) product of the state vectors of the two subsystems:

$$|\Phi(0)\rangle = |\alpha_0\rangle|\chi_e(0)\rangle. \tag{15.10}$$

The coupling is responsible for the damping of the oscillator's amplitude. At a later time $t$, the state vector of the total system becomes:

$$|\Phi(t)\rangle = |\alpha_1\rangle|\chi_e(t)\rangle \tag{15.11}$$

with $\alpha_1 = \alpha(t)e^{-\gamma t}$; the number $\alpha(t)$ corresponds to the quasi-classical state one would find in the absence of damping (question 15.1.4(a)) and $\gamma$ is a real positive number.

**15.4.1** Using the result 1.3, give the expectation value of the energy of the oscillator at time $t$, and the energy acquired by the environment when $2\gamma t \ll 1$.

**15.4.2** For initial states of the "Schrödinger cat" type for the oscillator, the state vector of the total system is, at $t = 0$,

$$|\Phi(0)\rangle = \frac{1}{\sqrt{2}} \left( e^{-i\pi/4}|\alpha_0\rangle + e^{i\pi/4}|-\alpha_0\rangle \right) |\chi_e(0)\rangle \tag{15.12}$$

and, at a later time $t$,

$$|\Phi(t)\rangle = \frac{1}{\sqrt{2}} \left( e^{-i\pi/4}|\alpha_1\rangle|\chi_e^{(+)}(t)\rangle + e^{i\pi/4}|-\alpha_1\rangle|\chi_e^{(-)}(t)\rangle \right) \qquad (15.13)$$

still with $\alpha_1 = \alpha(t)e^{-\gamma t}$. We choose $t$ such that $\alpha_1$ is pure imaginary, with $|\alpha_1| \gg 1$. $|\chi_e^{(+)}(t)\rangle$ and $|\chi_e^{(-)}(t)\rangle$ are two normalized states of the environment that are a priori different (but not orthogonal).

The probability distribution of the oscillator's position, measured independently of the state of the environment, is then

$$\mathcal{P}(x) = \frac{1}{2}[|\psi_{\alpha_1}(x)|^2 + |\psi_{-\alpha_1}(x)|^2 + 2\mathcal{R}e(i\psi_{\alpha_1}^*(x)\psi_{-\alpha_1}(x)\langle\chi_e^{(+)}(t)|\chi_e^{(-)}(t)\rangle)].$$
$$(15.14)$$

Setting $\eta = \langle\chi_e^{(+)}(t)|\chi_e^{(-)}(t)\rangle$ with $0 \le \eta \le 1$ ($\eta$ is supposed to be real) and using the results of Sect.15.3, describe without any calculation, the result of:

(a) $N$ independent position measurements,
(b) $N$ independent momentum measurements.

Which condition on $\eta$ allows one to distinguish between a quantum superposition and a statistical mixture?

**15.4.3** In a very simple model, the environment is represented by a second oscillator, of same mass and frequency as the first one. We assume that this second oscillator is initially in its ground state $|\chi_e(0)\rangle = |0\rangle$. If the coupling between the two oscillators is quadratic, we will take for granted that

- the states $|\chi_e^{(\pm)}(t)\rangle$ are quasi-classical states: $\left|\chi_e^{(\pm)}(t)\rangle = \right| \pm \beta\rangle$,
- and that, for short times ($\gamma t \ll 1$) : $|\beta|^2 = 2\gamma t|\alpha_0|^2$.

  (a) From the expansion (15.5), show that $\eta = \langle\beta|-\beta\rangle = \exp(-2|\beta|^2)$.
  (b) Using the expression found in question 15.5.4.1 for the energy of the first oscillator, determine the typical energy transfer between the two oscillators, above which the difference between a quantum superposition and a statistical mixture becomes unobservable.

**15.4.4** Consider again the simple pendulum described above. Assume the damping time is one year (a pendulum in vacuum with reduced friction). Using the result of the previous question, evaluate the time during which a "Schrödinger cat" state can be observed. Comment and conclude.

## 15.5  Solutions

### 15.5.1 The Quasi-classical States of a Harmonic Oscillator

**15.5.1.1 (a)** A simple change of variables gives

$$\hat{P} = \frac{\hat{p}}{\sqrt{m\hbar\omega}} = \frac{1}{\sqrt{m\hbar\omega}} \frac{\hbar}{i} \frac{\partial}{\partial x} = -i\sqrt{\frac{\hbar}{m\omega}} \frac{\partial}{\partial x} = -i\frac{\partial}{\partial X} \qquad (15.15)$$

$$\hat{X} = \sqrt{\frac{m\omega}{\hbar}} \hat{x} = \sqrt{\frac{m\omega}{\hbar}} i\hbar \frac{\partial}{\partial p} = i\sqrt{m\hbar\omega} \frac{\partial}{\partial p} = i\frac{\partial}{\partial P} \qquad (15.16)$$

**(b)** We have the usual relations $[\hat{N}, \hat{a}] = [\hat{a}^\dagger\hat{a}, \hat{a}] = [\hat{a}^\dagger, \hat{a}]\hat{a} = -\hat{a}$. Consequently:

$$[\hat{N}, \hat{a}]|n\rangle = -\hat{a}|n\rangle \quad \Rightarrow \quad \hat{N}\hat{a}|n\rangle = (n-1)\hat{a}|n\rangle, \qquad (15.17)$$

and $\hat{a}|n\rangle$ is an eigenvector of $\hat{N}$ corresponding to the eigenvalue $n-1$. We know from the theory of the one-dimensional harmonic oscillator that the energy levels are not degenerated. Therefore we find that $\hat{a}|n\rangle = \mu|n-1\rangle$, where the coefficient $\mu$ is determined by calculating the norm of $\hat{a}|n\rangle$:

$$\|\hat{a}|n\rangle\|^2 = \langle n|\hat{a}^\dagger\hat{a}|n\rangle = n \Rightarrow \mu = \sqrt{n} \qquad (15.18)$$

up to an arbitrary phase.

**(c)** The equation $\hat{a}|0\rangle = 0$ corresponds to $(\hat{X} + i\hat{P})|0\rangle = 0$.
  In real space: $\left(X + \frac{\partial}{\partial X}\right)\psi_0(X) = 0 \Rightarrow \psi_0(X) \propto \exp\left(-X^2/2\right)$.
  In momentum space: $\left(P + \frac{\partial}{\partial P}\right)\varphi_0(P) = 0 \Rightarrow \varphi_0(P) \propto \exp\left(-P^2/2\right)$.

**15.5.1.2** One can check directly the relation $\hat{a}|\alpha\rangle = \alpha|\alpha\rangle$:

$$\hat{a}|\alpha\rangle = e^{-|\alpha|^2/2} \sum_n \frac{\alpha^n}{\sqrt{n!}} \hat{a}|n\rangle = e^{-|\alpha|^2/2} \sum_n \frac{\alpha^n}{\sqrt{n!}} \sqrt{n}|n-1\rangle$$

$$= \alpha e^{-|\alpha|^2/2} \sum_n \frac{\alpha^n}{\sqrt{n!}}|n\rangle = \alpha|\alpha\rangle \qquad (15.19)$$

The calculation of the norm of $|\alpha\rangle$ yields: $\langle\alpha|\alpha\rangle = e^{-|\alpha|^2} \sum_n \frac{|\alpha|^{2n}}{n!} = 1$.

**15.5.1.3** The expectation value of the energy is:

$$\langle E\rangle = \langle\alpha|\hat{H}|\alpha\rangle = \hbar\omega\langle\alpha|\hat{N} + 1/2|\alpha\rangle = \hbar\omega(|\alpha|^2 + 1/2). \qquad (15.20)$$

For $\langle x \rangle$, and $\langle p \rangle$, we use

$$\langle x \rangle = \sqrt{\frac{\hbar}{2m\omega}} \langle \alpha | \hat{a} + \hat{a}^\dagger | \alpha \rangle = \sqrt{\frac{\hbar}{2m\omega}} (\alpha + \alpha^*) \tag{15.21}$$

$$\langle p \rangle = -i\sqrt{\frac{m\hbar\omega}{2}} \langle \alpha | \hat{a} - \hat{a}^\dagger | \alpha \rangle = i\sqrt{\frac{m\omega\hbar}{2}} (\alpha^* - \alpha) \tag{15.22}$$

$$\Delta x^2 = \frac{\hbar}{2m\omega} \langle \alpha | (\hat{a} + \hat{a}^\dagger)^2 | \alpha \rangle - \langle x \rangle^2 = \frac{\hbar}{2m\omega} ((\alpha + \alpha^*)^2 + 1) - \langle x \rangle^2. \tag{15.23}$$

Therefore $\Delta x = \sqrt{\hbar/2m\omega}$, which is independent of $\alpha$.
   Similarly

$$\Delta p^2 = -\frac{m\hbar\omega}{2} \langle \alpha | (\hat{a} - \hat{a}^\dagger)^2 | \alpha \rangle - \langle p \rangle^2 = -\frac{m\hbar\omega}{2} ((\alpha - \alpha^*)^2 - 1) - \langle p \rangle^2 \tag{15.24}$$

Therefore $\Delta p = \sqrt{m\hbar\omega/2}$. The Heisenberg inequality becomes in this case an equality $\Delta x \, \Delta p = \hbar/2$, independently of the value of $\alpha$.

**15.5.1.4**  With the $X$ variable, we have

$$\frac{1}{\sqrt{2}} \left( X + \frac{\partial}{\partial X} \right) \psi_\alpha(X) = \alpha \psi_\alpha(X) \quad \Rightarrow \quad \psi_\alpha(X) = C \exp\left( -\frac{(X - \alpha\sqrt{2})^2}{2} \right) \tag{15.25}$$

Similarly, with the $P$ variable,

$$\frac{i}{\sqrt{2}} \left( P + \frac{\partial}{\partial P} \right) \varphi_\alpha(P) = \alpha \varphi_\alpha(P) \quad \Rightarrow \quad \varphi_\alpha(P) = C' \exp\left( -\frac{(P + i\alpha\sqrt{2})^2}{2} \right). \tag{15.26}$$

**15.5.1.5**  **(a)**  Setting $\alpha(t) = \alpha_0 \, e^{-i\omega t} = \rho \, e^{-i(\omega t - \phi)}$, we find

$$|\psi(0)\rangle = |\alpha_0\rangle \tag{15.27}$$

$$|\psi(t)\rangle = e^{-|\alpha|^2/2} \sum_n \frac{\alpha_0^n}{\sqrt{n!}} e^{-iE_n t/\hbar} |n\rangle = e^{-|\alpha|^2/2} e^{-i\omega t/2} \sum_n \frac{\alpha_0^n}{\sqrt{n!}} e^{-in\omega t} |n\rangle$$

$$= e^{-i\omega t/2} |\alpha(t)\rangle. \tag{15.28}$$

**(b)**  Setting $x_0 = \rho\sqrt{2\hbar/(m\omega)}$ and $p_0 = \rho\sqrt{2m\hbar\omega}$ we find

$$\langle x \rangle_t = x_0 \cos(\omega t - \phi), \qquad \langle p \rangle_t = -p_0 \sin(\omega t - \phi). \tag{15.29}$$

These are the equations of motions of a classical oscillator. Using the result of question 15.5.1.3, we obtain

$$\frac{\Delta x}{x_0} = \frac{1}{2\rho} \ll 1, \quad \frac{\Delta p}{p_0} = \frac{1}{2\rho} \ll 1. \tag{15.30}$$

The relative uncertainties on the position and on the momentum of the oscillator are quite accurately defined at any time. Hence the name "quasi-classical state".

**15.5.1.6 (a)** The appropriate choice is $\langle x \rangle_0 = x_0$ and $\langle p \rangle_0 = 0$, i.e. $\phi = 0$

$$\omega = 2\pi \nu = \sqrt{\frac{g}{\ell}} = 3.13 \text{ s}^{-1} \quad \Rightarrow \quad \alpha(0) = 3.9 \; 10^9 \tag{15.31}$$

**(b)** $\Delta x / x_0 = 1/(2\alpha(0)) = 1.3 \; 10^{-10}$.

**(c)** After 1/4 period, $e^{i\omega t} = e^{i\pi/2} = i \Rightarrow \alpha(T/4) = -i \; 3.9 \; 10^9$

## 15.5.2 Construction of a Schrödinger-Cat State

**15.5.2.1** The eigenvectors of $\hat{W}$ are simply the previous $|n\rangle$, therefore:

$$\hat{W} |n\rangle = \hbar g \, n^2 |n\rangle \tag{15.32}$$

and

$$|\psi(0)\rangle = |\alpha\rangle \quad \Rightarrow \quad |\psi(T)\rangle = e^{-|\alpha|^2/2} \sum_n \frac{\alpha^n}{\sqrt{n!}} e^{-ign^2 T} |n\rangle. \tag{15.33}$$

**15.5.2.2** If $T = 2\pi/g$, then $e^{-ign^2 T} = e^{-2i\pi n^2} = 1$ and

$$|\psi(T)\rangle = |\alpha\rangle. \tag{15.34}$$

If $T = \pi/g$, then $e^{-ign^2 T} = e^{-i\pi n^2} = 1$ if $n$ is even, $-1$ if $n$ is odd, therefore

$$e^{-ign^2 T} = (-1)^n \quad \Rightarrow \quad |\psi(T)\rangle = |-\alpha\rangle \tag{15.35}$$

**15.5.2.3** If $T = \pi/2g$, then $e^{-ign^2 T} = e^{-i\frac{\pi}{2}n^2} = 1$ for $n$ even, and $e^{-ign^2 T} = -i$ if $n$ is odd.

We can rewrite this relation as

$$e^{-ign^2 T} = \frac{1}{2}(1 - i + (1 + i)(-1)^n) = \frac{1}{\sqrt{2}}(e^{-i\frac{\pi}{4}} + e^{i\frac{\pi}{4}}(-1)^n) \tag{15.36}$$

or, equivalently,

$$|\psi(T)\rangle = \frac{1}{\sqrt{2}}(e^{-i\pi/4}|\alpha\rangle + e^{i\pi/4}|-\alpha\rangle). \tag{15.37}$$

**15.5.2.4 (a)** For $\alpha = i\rho$, in the state $|\alpha\rangle$, the oscillator has a zero mean position and a positive velocity. In the state $|-\alpha\rangle$, the oscillator also has a zero mean position, but a negative velocity. The state 15.3 is a quantum superposition of these two situations.

**(b)** If $|\alpha| \gg 1$, the states $|\alpha\rangle$ and $|-\alpha\rangle$ are macroscopically different (antinomic). The state 15.3 is a quantum superposition of such states. It therefore constitutes a (peaceful) version of Schrödinger's cat, where we represent "dead" or "alive" cats by simple vectors of Hilbert space.

### 15.5.3 Quantum Superposition Versus Statistical Mixture

**15.5.3.1** The probability distributions of the position and of the momentum are

$$\mathcal{P}(X) \propto |e^{-i\pi/4}\psi_\alpha(X) + e^{i\pi/4}\psi_{-\alpha}(X)|^2$$

$$\propto \left| e^{-i\pi/4} \exp\left(-\frac{1}{2}(X - i\rho\sqrt{2})^2\right) + e^{i\pi/4}\exp\left(-\frac{1}{2}(X + i\rho\sqrt{2})^2\right) \right|^2$$

$$\propto e^{-X^2} \cos^2\left(X\rho\sqrt{2} - \frac{\pi}{4}\right) \tag{15.38}$$

$$\mathcal{P}(P) \propto |e^{-i\pi/4}\varphi_\alpha(P) + e^{i\pi/4}\varphi_{-\alpha}(P)|^2$$

$$\simeq \exp(-(P - \rho\sqrt{2})^2) + \exp(-(P + \rho\sqrt{2})^2). \tag{15.39}$$

In the latter equation, we have used the fact that, for $\rho \gg 1$, the two Gaussians centered at $\rho\sqrt{2}$ and $-\rho\sqrt{2}$ have a negligible overlap.

**15.5.3.2** Alice will find two peaks, each of which contains roughly half of the events, centered respectively at $p_0$ and $-p_0$.

**15.5.3.3** The statistical mixture of Bob leads to the same momentum distribution as that measured by Alice: the $N/2$ oscillators in the state $|\alpha\rangle$ all lead to a mean momentum $+p_0$, and the $N/2$ oscillators in the state $|-\alpha\rangle$ to $-p_0$. Up to this point, there is therefore no difference, and no paradoxical behavior related to the quantum superposition (15.7).

**15.5.3.4** In the $X$ variable, the resolution of the detector satisfies

$$\delta X \ll \frac{1}{|\alpha|} = \frac{1}{\rho} \tag{15.40}$$

Alice therefore has a sufficient resolution to observe the oscillations of the function $\cos^2(X\rho\sqrt{2} - \pi/4)$ in the distribution $\mathcal{P}(X)$. The shape of the distribution will therefore reproduce the probability law of $X$ drawn on Fig. 15.1, i.e. a modulation of period $(\hbar\pi^2/(2m\alpha^2\omega))^{1/2}$, with a Gaussian envelope.

**15.5.3.5** If Bob performs a position measurement on the $N/2$ systems in the state $|\alpha\rangle$, he will find a Gaussian distribution corresponding to the probability law $\mathcal{P}(X) \propto |\psi_\alpha(X)|^2 \propto \exp(-X)^2$. He will find the same distribution for the $N/2$ systems in the state $|-\alpha\rangle$. The sum of his results will be a Gaussian distribution, which is quite different from the result expected by Alice. The position measurement should, in principle, allow one to discriminate between the quantum superposition and the statistical mixture.

**15.5.3.6** The necessary resolution is $\delta x \ll \frac{1}{|\alpha|}\sqrt{\frac{\hbar}{m\omega}} \sim 5\,10^{-26}$ m. Unfortunately, it is impossible to attain such a resolution in practice.

## 15.5.4 The Fragility of a Quantum Superposition

**15.5.4.1** We have $E(t) = \hbar\omega(|\alpha_0|^2 e^{-2\gamma t} + 1/2)$: this energy decreases with time. After a time much longer than $\gamma^{-1}$, the oscillator is in its ground state. This dissipation model corresponds to a zero temperature environment. The mean energy acquired by the environment $E(0) - E(t)$ is, for $2\gamma t \ll 1$, $\Delta E(t) \simeq 2\hbar\omega|\alpha_0|^2\gamma t$.

**15.5.4.2** **(a)** The probability distribution of the position keeps its Gaussian envelope, but the contrast of the oscillations is reduced by a factor $\eta$.

**(b)** The probability distribution of the momentum is given by

$$\mathcal{P}(p) = \frac{1}{2}(|\varphi_{\alpha_1}(p)|^2 + |\varphi_{-\alpha_1}(p)|^2 + 2\eta\,\mathcal{R}e(i\varphi^*_{-\alpha_1}(p)\varphi_{\alpha_1}(p))) \tag{15.41}$$

Since the overlap of the two Gaussians $\varphi_{\alpha_1}(p)$ and $\varphi_{-\alpha_1}(p)$ is negligible for $|\alpha_1| \gg 1$, the crossed term, which is proportional to $\eta$ does not contribute significantly. One recovers two peaks centered at $\pm|\alpha_1|\sqrt{2m\hbar\omega}$.

The difference between a quantum superposition and a statistical mixture can be made by position measurements. The quantum superposition leads to a modulation of spatial period $(\hbar\pi^2/(2m\alpha^2\omega))^{1/2}$ with a Gaussian envelope, whereas only the Gaussian is observed on a statistical mixture. In order to see this modulation, it must not be too small, say

$$\eta \geq 1/10. \tag{15.42}$$

**15.5.4.3 (a)** A simple calculation gives

$$\langle \beta | - \beta \rangle = e^{-|\beta|^2} \sum_n \frac{\beta^{*n}(-\beta)^n}{n!} = e^{-|\beta|^2} e^{-|\beta|^2} = e^{-2|\beta|^2} \tag{15.43}$$

**(b)** From the previous considerations, we must have $e^{-2|\beta|^2} \geq 1/10$, i.e. $|\beta| \leq 1$.

For times shorter than $\gamma^{-1}$, the energy of the first oscillator is

$$E(t) = E(0) - 2\gamma t |\alpha_0|^2 \hbar\omega. \tag{15.44}$$

The energy of the second oscillator is

$$E'(t) = \hbar\omega(|\beta(t)|^2 + 1/2) = \hbar\omega/2 + 2\gamma t |\alpha_0|^2 \hbar\omega. \tag{15.45}$$

The total energy is conserved; the energy transferred during time $t$ is $\Delta E(t) = 2\gamma t |\alpha_0|^2 \hbar\omega = \hbar\omega|\beta|^2$. In order to distinguish between a quantum superposition and a statistical mixture, we must have $\Delta E \leq \hbar\omega$. In other words, if a single energy quantum $\hbar\omega$ is transferred, it becomes problematic to tell the difference.

**15.5.4.4** With $1/2\gamma = 1$ year $= 3 \times 10^7$ seconds, the time it takes to reach $|\beta| = 1$ is $(2\gamma|\alpha_0|^2)^{-1} \simeq 2 \times 10^{-12}$ seconds!

### 15.5.5 Comments

Even for a system well protected from the environment, as the pendulum considered above, the quantum superpositions of macroscopic states are unobservable. After a very short time, all measurements that one can make on a system initially prepared in such a state coincide with those made on a statistical mixture. It is therefore not possible, at present, to observe the effects related to the paradoxical character of a macroscopic quantum superposition.

However, it is quite possible to observe "mesoscopic" kittens, for systems which have a limited number of degrees of freedom and are well isolated. The idea developed here, which is oriented towards quantum optics, was proposed by Bernard Yurke and David Stoler, Phys. Rev. Lett. **57**, p. 13 (1986). One of the earliest conclusive results were obtained at the Ecole Normale Supérieure in Paris, on microwave photons (50 GHz) stored in a superconducting cavity (M. Brune, E. Hagley, J. Dreyer, X. Maitre, A. Maali, C. Wunderlich, J.-M. Raimond, and S. Haroche, Phys. Rev. Lett. **77**, 4887 (1996)). In this case, the field stored in the cavity is a quasi-perfect harmonic oscillator. The preparation of the kitten with 5 to 10 photons, i.e. $|\alpha|^2 = 5$ or $10$ (Sect. 15.2), is then accomplished by sending atoms through the cavity. Dissipation (Sect. 15.4) corresponds to the very weak residual absorption by the walls of the superconducting cavity (for more details, see e.g. *Exploring the Quantum*, by S. Haroche and J.-M. Raimond (Oxford University Press, 2006)).

During the last decade, Schrödinger cat states have been implemented on several different platforms, such as trapped ions, superconducting quantum interference devices (SQUIDs), optical photons and circuit Quantum Electrodynamics devices. A survey of these implementations, discussing both the foundational questions that can be addressed and the practical applications of such states within the framework of Quantum Technologies, can be found in the review article of F. Fröwis, P. Sekatski, W. Dür, N. Gisin, and N. Sangouard, *Macroscopic quantum states: Measures, fragility, and implementations*, Rev. Mod. Phys. **90**, 025004 (2018).

# Quantum Cryptography

**16**

Cryptography consists in sending an encrypted message to a correspondent and in minimizing the risk for this message to be intercepted and deciphered by an unwanted outsider. The present chapter shows how quantum mechanics can provide a procedure to achieve this goal. We assume here that Alice (A) wants to send Bob (B) some information which may be coded in the binary system, for instance

$$+ + - - - + + - \cdots \tag{16.1}$$

We denote the number of bits of this message by $n$. Alice wants to send this message to Bob only if she has made sure that no "spy" is listening to the communication.

## 16.1 Preliminaries

Consider a spin 1/2 particle. The spin operator is $\hat{S} = (\hbar/2)\hat{\sigma}$ where the set $\hat{\sigma}_i$, $i = x, y, z$ are the Pauli matrices. We write $|\sigma_z = +1\rangle$ and $|\sigma_z = -1\rangle$ for the eigenstates of $\hat{S}_z$ with respective eigenvalues $+\hbar/2$ and $-\hbar/2$.

Consider a particle in the state $|\sigma_z = +1\rangle$. One measures the component of the spin along an axis $u$ in the $(x, z)$ plane, defined by the unit vector

$$e_u = \cos\theta\, e_z + \sin\theta\, e_x, \tag{16.2}$$

where $e_z$ and $e_x$ are the unit vectors along the $z$ and $x$ axes respectively. We recall that the corresponding operator is

$$\hat{S} \cdot e_u = \frac{\hbar}{2}(\cos\theta\, \hat{\sigma}_z + \sin\theta\, \hat{\sigma}_x). \tag{16.3}$$

**16.1.1** Show that the possible results of the measurement are $+\hbar/2$ and $-\hbar/2$.

© Springer Nature Switzerland AG 2019
J.-L. Basdevant, J. Dalibard, *The Quantum Mechanics Solver*,
https://doi.org/10.1007/978-3-030-13724-3_16

**16.1.2** Show that the eigenstates of the observable (16.3) are (up to a multiplicative constant):

$$|\sigma_u = +1\rangle = \quad \cos\phi\, |\sigma_z = +1\rangle + \sin\phi\, |\sigma_z = -1\rangle \qquad (16.4)$$

$$|\sigma_u = -1\rangle = -\sin\phi\, |\sigma_z = +1\rangle + \cos\phi\, |\sigma_z = -1\rangle \qquad (16.5)$$

and express $\phi$ in terms of $\theta$. Write the probabilities $p_u^{\pm}$ of finding $+\hbar/2$ and $-\hbar/2$ when measuring the projection of the spin along the $u$ axis.

**16.1.3** What are the spin states after measurements that give the results $+\hbar/2$ and $-\hbar/2$ along $u$?

**16.1.4** Immediately after such a measurement, one measures the $z$ component of the spin.

(a) What are the possible results and what are the probabilities of finding these results in terms of the results found previously along the $u$ axis (observable (16.3)).

(b) Show that the probability to recover the same value $S_z = +\hbar/2$ as in the initial state $|\sigma_z = +1\rangle$ is

$$P_{++}(\theta) = (1 + \cos^2\theta)/2. \qquad (16.6)$$

(c) Assuming now that the initial state is $|\sigma_z = -1\rangle$, what is, for the same sequence of measurements, the probability $P_{--}(\theta)$ to recover $S_z = -\hbar/2$ in the last measurement?

## 16.2   Correlated Pairs of Spins

A source produces a pair $(a, b)$ of spin-1/2 particles (Fig. 16.1), prepared in the state $|\psi\rangle = \phi(r_a, r_b)|\Sigma\rangle$ where the spin state of the two particles is

$$|\Sigma\rangle = \frac{1}{\sqrt{2}}(|\sigma_z^a = +1\rangle \otimes |\sigma_z^b = +1\rangle + |\sigma_z^a = -1\rangle \otimes |\sigma_z^b = -1\rangle). \qquad (16.7)$$

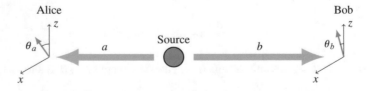

**Fig. 16.1** A source emits a pair $(a, b)$ of spin-1/2 particles. Alice measures the component of the spin of $a$ along a direction $\theta_a$ and Bob measures the component of the spin of $b$ along a direction $\theta_b$

In other words, the spin variables are decoupled from the space variables $(\boldsymbol{r}_a, \boldsymbol{r}_b)$. In (16.7), $|\sigma_u^a = \pm\rangle$ (specifically $u = z$) are the eigenstates of the $u$ component of the spin of particle $a$, and similarly for $b$.

**16.2.1** Show that this state can also be written as:

$$|\Sigma\rangle = \frac{1}{\sqrt{2}}(|\sigma_x^a = +1\rangle \otimes |\sigma_x^b = +1\rangle + |\sigma_x^a = -1\rangle \otimes |\sigma_x^b = -1\rangle). \qquad (16.8)$$

**16.2.2** The pair of particles $(a, b)$ is prepared in the spin state (16.7) and (16.8). As the two particles move away from each another, this spin state remains unchanged (unless a measurement is made).

(a) Alice *first* measures the spin component of $a$ along an axis $u_a$ of angle $\theta_a$. What are the possible results and the corresponding probabilities in the two cases $\theta_a = 0$, i.e. the $z$ axis, and $\theta_a = \pi/2$, i.e. the $x$ axis?
(b) Show that, after Alice's measurement, the spin state of the two particles depends as follows on the measurement and its result

| Axis | Result | State |
|------|--------|-------|
| $z$ | $+\hbar/2$ | $|\sigma_z^a = +1\rangle \otimes |\sigma_z^b = -1\rangle$ |
| $z$ | $+\hbar/2$ | $|\sigma_x^a = +1\rangle \otimes |\sigma_x^b = +1\rangle$ |
| $x$ | $-\hbar/2$ | $|\sigma_x^a = -1\rangle \otimes |\sigma_x^b = -1\rangle$ |

From then on, why can one ignore particle $a$ as far as spin measurements on $b$ are concerned?
(We recall that if $|\psi\rangle = |u\rangle \otimes |v\rangle$ is a factorized state and $\hat{C} = \hat{A} \otimes \hat{B}$, where $\hat{A}$ and $\hat{B}$ act respectively on the spaces of $|u\rangle$ and $|v\rangle$, then $\langle\psi|\hat{C}|\psi\rangle = \langle u|\hat{A}|u\rangle\langle v|\hat{B}|v\rangle$.)

**16.2.3** *After* Alice's measurement, Bob measures the spin of particle $b$ along an axis $u_b$ of angle $\theta_b$.

Give the possible results of Bob's measurement and their probabilities in terms of Alice's results in the four following configurations:

(a) $\theta_a = 0, \theta_b = 0$;
(b) $\theta_a = 0, \theta_b = \pi/2$;
(c) $\theta_a = \pi/2, \theta_b = 0$;
(d) $\theta_a = \pi/2, \theta_b = \pi/2$.

In which cases do the measurements on $a$ and $b$ give with certainty the same result?

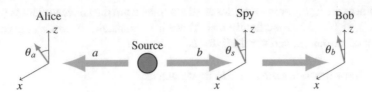

**Fig. 16.2**   A spy, sitting between the source and Bob, measures the component of the $b$ spin along an axis $\theta_s$

**16.2.4** Consider the situation $\theta_a = 0$. Suppose that a "spy" sitting between the source and Bob measures the spin of particle $b$ along an axis $u_s$ of angle $\theta_s$ as sketched in Fig. 16.2.

(a) What are, in terms of $\theta_s$ and of Alice's findings, the results of the spy's measurements and their probabilities?
(b) After the spy's measurement, Bob measures the spin of $b$ along the axis defined by $\theta_b = 0$. What does Bob find, and with what probabilities, in terms of the spy's results?
(c) What is the probability $P(\theta_s)$ that Alice and Bob find the same results after the spy's measurement?
(d) What is the expectation value of $P(\theta_s)$ if the spy chooses $\theta_s$ at random in the interval $[0, 2\pi]$ with uniform probability?
   What is this expectation value if the spy chooses only the two values $\theta_s = 0$ and $\theta_s = \pi/2$ each with the same probability $p = 1/2$?

## 16.3   The Quantum Cryptography Procedure

In order to transmit confidential information, Alice and Bob use the procedure outlined in Fig. 16.3. Comment on this procedure, and answer the following questions.

**16.3.1** How can Alice be sure that a spy is present?

**16.3.2** What is the probability that an operating spy will escape being detected? Calculate this probability for $FN = 200$.

**16.3.3** Does the spy become more "invisible" if he knows the system of axes $(x, z)$ chosen by Alice and Bob to perform their measurements?

**16.3.4** Comment on the two "experiments" whose results are given in Tables 16.1 and 16.2. Show that a spy has certainly listened to communication 2. What is the probability that a spy listened to communication 1, but remained undetected?

1. Alice and Bob decide along which axes $x$ and $z$ they will make their measurements.

2. Alice, who controls the source $S$, prepares an ordered sequence of $N \gg n$ pairs of spins in the state (16.7) ($n$ is the number of bits of the message). She sends the $b$ spins to Bob and keeps the $a$ spins.

3. For each spin that they collect, Alice and Bob measure either the $x$ or the $z$ component. Each of them chooses the $x$ or $z$ direction at random with probability $p = 1/2$. There is no correlation, for a given pair of spins $(a, b)$, between the axis chosen by Alice and the one chosen by Bob. They both register all their results.

4. Bob selects a subset $FN$ of his measurements. He communicates openly to Alice (by radio, www, etc.) the axis and the result of the measurement for *each* event of this subset. In practice $F \sim 0.5$.

5. Alice compares, for this subset $FN$, her axes and her results with those just communicated by Bob. By doing so, she can tell whether or not a spy is present. If a spy is spotted, the procedure stops and a "physical" search for the spy must be undertaken. Otherwise:

6. Alice makes a public announcement that she is convinced not to have been spied upon, and Bob, still openly, communicates his *axes* of measurements for the remaining spins. However, he does not communicate the corresponding results.

7. . . .

**Fig. 16.3** The procedure for quantum cryptography

**Table 16.1** Experiment 1, performed with 12 pairs of spins. Top: set of axes and results obtained by Alice. Bottom: choices of axes and results publicly communicated by Bob

| A | Spin # | 1 | 2 | 3 | 4 | 5 | 6 | 7 | 8 | 9 | 10 | 11 | 12 |
|---|--------|---|---|---|---|---|---|---|---|---|----|----|----|
| A | Axis | $x$ | $x$ | $z$ | $x$ | $z$ | $z$ | $x$ | $z$ | $z$ | $z$ | $x$ | $x$ |
| A | Result | + | − | + | + | − | − | + | + | + | − | + | − |
| B | Spin # | 1 | 2 | 3 | 4 | 5 | 6 | 7 | 8 | 9 | 10 | 11 | 12 |
| B | Axis | $x$ | | $x$ | $z$ | | | $x$ | | | $x$ | $x$ | |
| B | Result | + | | − | − | | | + | | | + | + | |

**Table 16.2** Experiment 2, performed with 12 pairs of spins. Top: set of axes and results obtained by Alice. Bottom: choices of axes and results publicly communicated by Bob

| A | Spin # | 1 | 2 | 3 | 4 | 5 | 6 | 7 | 8 | 9 | 10 | 11 | 12 |
|---|--------|---|---|---|---|---|---|---|---|---|----|----|----|
| A | Axis | $x$ | $z$ | $z$ | $z$ | $x$ | $x$ | $z$ | $x$ | $x$ | $z$ | $x$ | $z$ |
| A | Result | + | + | − | + | + | − | + | + | − | − | + | + |
| B | Spin # | 1 | 2 | 3 | 4 | 5 | 6 | 7 | 8 | 9 | 10 | 11 | 12 |
| B | Axis | | $x$ | | | $x$ | | | $x$ | $z$ | | $z$ | $z$ |
| B | Result | | + | | | + | | | − | + | | + | − |

**Table 16.3** Choice of axes publicly communicated by Bob in the framework of experiment 1, after Alice has said she is convinced that she is not being spied upon

| Spin # | 2 | 5 | 6 | 8 | 9 | 12 |
|--------|---|---|---|---|---|----|
| Axis   | $x$ | $x$ | $x$ | $z$ | $x$ | $x$ |

**16.3.5** Complete the missing item (number 7 in the above procedure), and indicate how Alice can send her message (16.1) to Bob without using any other spin pairs than the $N$ pairs which Bob and her have already analyzed. Using Table 16.3, tell how, in experiment 1, Alice can send to Bob the message $(+, -)$.

---

## 16.4  Solutions

### 16.4.1 Preliminaries

**16.4.1.1** The spin observable along the $u$ axis is

$$\hat{S} \cdot \hat{e}_u = \frac{\hbar}{2} \begin{pmatrix} \cos\theta & \sin\theta \\ \sin\theta & -\cos\theta \end{pmatrix}. \tag{16.9}$$

The possible results of the measurement are the eigenvalues of $\hat{S} \cdot \hat{e}_u$, i.e. $\pm\hbar/2$.

**16.4.1.2** The corresponding eigenvectors are

$$|\sigma_u = +1\rangle = \cos(\theta/2)|\sigma_z = +1\rangle + \sin(\theta/2)|\sigma_z = -1\rangle \tag{16.10}$$

$$|\sigma_u = -1\rangle = -\sin(\theta/2)|\sigma_z = +1\rangle + \cos(\theta/2)|\sigma_z = -1\rangle, \tag{16.11}$$

therefore $\phi = \theta/2$. The probabilities follow directly:

$$p_u^\pm = |\langle \sigma_u = \pm 1 | \sigma_z = +1\rangle|^2, \quad p_u^+ = \cos^2(\theta/2), \quad p_u^- = \sin^2(\theta/2). \tag{16.12}$$

**16.4.1.3** The state after a measurement with the result $+\hbar/2$ (or $-\hbar/2$) is $|\sigma_u = +1\rangle$ (or $|\sigma_u = -1\rangle$).

**16.4.1.4  (a)** If the measurement along $u$ has given $+\hbar/2$, then the probabilities for the second measurement are:

$$p_z^+(\pm\hbar/2) = |\langle \sigma_z = \pm 1 | \sigma_u = +1\rangle|^2 \tag{16.13}$$

with

$$p_z^+(+\hbar/2) = \cos^2(\theta/2), \quad p_z^+(-\hbar/2) = \sin^2(\theta/2). \tag{16.14}$$

If the measurement along $u$ has given $-\hbar/2$, then

$$p_z^-(-\hbar/2) = \cos^2(\theta/2) \, , \; p_z^-(+\hbar/2) = \sin^2(\theta/2). \tag{16.15}$$

**(b)** One recovers $S_z = +\hbar/2$ with probabilities:

(i) $p_u^+ \cdot p_z^+(+\hbar/2) = \cos^4(\theta/2)$ if the measurement along $u$ has given $+\hbar/2$,
(ii) $p_u^- \cdot p_z^-(+\hbar/2) = \sin^4(\theta/2)$ if the measurement along $u$ has given $-\hbar/2$.

Altogether, one has

$$P_{++} = \cos^4\frac{\theta}{2} + \sin^4\frac{\theta}{2} = \frac{1}{2}(1 + \cos^2\theta). \tag{16.16}$$

**(c)** The intermediate results are reversed, but the final probability is the same

$$P_{--} = \frac{1}{2}(1 + \cos^2\theta). \tag{16.17}$$

### 16.4.2 Correlated Pairs of Spins

**16.4.2.1** The $z$ and $x$ eigenstates are related by $|\sigma_x = \pm 1\rangle = (|\sigma_z = +1\rangle \pm |\sigma_z = -1\rangle)/\sqrt{2}$.

If we make the substitution in expression (16.7), we obtain

$$\frac{1}{2\sqrt{2}} \left( (|\sigma_z^a = +1\rangle + |\sigma_z^a = -1\rangle) \otimes (|\sigma_z^b = +1\rangle + |\sigma_z^b = -1\rangle) \right. \tag{16.18}$$

$$\left. + (|\sigma_z^a = +1\rangle - |\sigma_z^a = -1\rangle) \otimes (|\sigma_z^b = +1\rangle - |\sigma_z^b = -1\rangle) \right), \tag{16.19}$$

where the crossed terms disappear. More generally, the state under consideration is actually invariant under rotations around the $y$ axis. In an actual experiment, it would be simpler to work with the singlet state

$$|0, 0\rangle = \frac{1}{\sqrt{2}}(|\sigma_z^a = +1\rangle \otimes |\sigma_z^b = -1\rangle - |\sigma_z^a = -1\rangle \otimes |\sigma_z^b = +1\rangle)/\sqrt{2}, \tag{16.20}$$

where Alice and Bob would simply find results of opposite signs by measuring along the same axis.

**16.4.2.2 (a)** Alice finds $\pm\hbar/2$ with $p = 1/2$ in each case. This result is obtained by noticing that the projector on the eigenstate $|\sigma_z^a = +1\rangle$ is $\hat{P}_+^a = |\sigma_z^a = +1\rangle\langle\sigma_z^a = +1| \otimes \hat{I}^b$ and that $p(+\hbar/2) = \langle\Sigma|\hat{P}_+^a|\Sigma\rangle = 1/2$, (and similarly for $p(-\hbar/2)$).

**(b)** This array of results is a consequence of the reduction of the wave packet. If Alice measures along the $z$ axis, we use (16.7); the normalized projections on the eigenstates of $S_z^a$ are $|\sigma_z^a = +1\rangle \otimes |\sigma_z^b = +1\rangle$ (Alice's result: $+\hbar/2$) and $|\sigma_z^a = -1\rangle \otimes |\sigma_z^b = -1\rangle$ (Alice's result: $-\hbar/2$). A similar formula holds for a measurement along the $x$ axis, because of the invariance property, and its consequence, (16.8).

Any measurement on $b$ (a probability, an expectation value) will imply expectation values of operators of the type $\hat{I}^a \otimes \hat{B}^b$ where $\hat{B}^b$ is a projector or a spin operator. Since the states under consideration are factorized, the corresponding expressions for spin measurements on $b$ will be of the type

$$(\langle \sigma_z^a = +1| \otimes \langle \sigma_z^b = +1|) \hat{I}^a \otimes \hat{B}^b (|\sigma_z^a = +1\rangle \otimes |\sigma_z^b = +1\rangle). \tag{16.21}$$

This reduces to

$$\langle \sigma_z^a = +1|\sigma_z^a = +1\rangle . \langle \sigma_z^b = +1|\hat{B}^b|\sigma_z^b = +1\rangle = \langle \sigma_z^b = +1|\hat{B}^b|\sigma_z^b = +1\rangle \tag{16.22}$$

where the spin state of $a$ is irrelevant.

**16.4.2.3** For the first and second configurations, we can summarize the results as follows:

| $\theta_a$ | $\theta_b$ | Alice | Bob | Probability |
|---|---|---|---|---|
| 0 | 0 | $+\hbar/2$ | $+\hbar/2$ | $p = 1$ |
| 0 | 0 | $-\hbar/2$ | $-\hbar/2$ | $p = 1$ |
| 0 | $\pi/2$ | $+\hbar/2$ | $\pm\hbar/2$ | $p\pm = 1/2$ |
| 0 | $\pi/2$ | $-\hbar/2$ | $\pm\hbar/2$ | $p\pm = 1/2$ |

The results $\theta_a = \pi/2, \theta_b = 0$ are identical to those of $\theta_a = 0, \theta_b = \pi/2$; similarly, the case $\theta_a = \pi/2, \theta_b = \pi/2$ is identical to $\theta_a = 0$, $\theta_b = 0$ (one actually recovers the same result for any $\theta_a = \theta_b$).

In the two cases (a) and (d), where $\theta_a = \theta_b$, i.e. when they measure along the same axis, Alice and Bob are sure to find the same result.

**16.4.2.4** **(a)** Concerning the findings of Alice and of the spy, we have:

| Alice | Spy | Probability |
|---|---|---|
| $+\hbar/2$ | $+\hbar/2$ | $\cos^2(\theta_s/2)$ |
| $+\hbar/2$ | $-\hbar/2$ | $\sin^2(\theta_s/2)$ |
| $-\hbar/2$ | $+\hbar/2$ | $\sin^2(\theta_s/2)$ |
| $-\hbar/2$ | $-\hbar/2$ | $\cos^2(\theta_s/2)$ |

**(b)** Concerning the findings of Bob and of the spy:

| Spy | Bob | Probability |
|---|---|---|
| $+\hbar/2$ | $+\hbar/2$ | $\cos^2(\theta_s/2)$ |
| $+\hbar/2$ | $-\hbar/2$ | $\sin^2(\theta_s/2)$ |
| $-\hbar/2$ | $+\hbar/2$ | $\sin^2(\theta_s/2)$ |
| $-\hbar/2$ | $-\hbar/2$ | $\cos^2(\theta_s/2)$ |

**(c)** The probability that Alice and Bob find the same result has actually been calculated in questions 1.4 (b) and 1.4 (c), we simply have

$$P(\theta_S) = \frac{1}{2}(1 + \cos^2\theta_s). \tag{16.23}$$

**(d)** Amazingly enough, the two expectation values are the same. On one hand, one has $\int_0^{2\pi} P(\theta_S)d\theta_s/(2\pi) = 3/4$. On the other, since $P(0) = 1$ and $P(\pi/2) = 1/2$, on the average $\overline{p} = 3/4$ if the values $\theta_s = 0$ and $\theta_s = \pi/2$ are chosen with equal probabilities.

### 16.4.3 The Quantum Cryptography Procedure

**16.4.3.1** Necessarily, if $\theta_a = \theta_b$, the results of Alice and Bob must be the same. If a *single* measurement done along the *same* axis $\theta_a = \theta_b$ gives *different* results for Alice and Bob, a spy is certainly operating (at least in an ideal experiment). If $\theta_a \neq \theta_b$, on the average half of the results are the same, half have opposite signs.

**16.4.3.2** The only chance for the spy to remain invisible is that Alice and Bob always find the same results when they choose the same axis. For each pair of spins, there is a probability 1/2 that they choose the same axis, and there is in this case a probability 1/4 that they do not find the same result if a spy is operating (question 6.4(d)). Therefore, for each pair of spins, there is a probability 1/8 that the spy is detected, and a probability 7/8 that the spy remains invisible.

This may seem a quite inefficient detection method. However, for a large number of events, the probability $(7/8)^{FN}$ that the spy remains undetected is very small. For $FN = 200$ one has $(7/8)^{200} \approx 2.5 \times 10^{-12}$.

**16.4.3.3** Quite surprisingly, as mentioned above, the spy does not gain anything in finding out which $x$ and $z$ axes Alice and Bob have agreed on in step 1 of the procedure.

**16.4.3.4** Experiment number 2. Measurements 8 and 12, where the axes are the same, give opposite results: rush upon the spy!

In experiment number 1, however, measurements 1, 7 and 11 along the $x$ axis do give the same results and are consistent with the assumption that there is no spy around. However, the number $N = 3$ is quite small in the present case. If a spy is operating, the probability that he remains undetected is $\approx 40\%$.

**16.4.3.5** Among the $(1 - F)N$ remaining measurements, Alice selects a sequence of events where the axes are the same and which reproduces her message. She communicates openly to Bob the labels of these events, and Bob can (at last!) read the message on his own set of data.

In the present case, Alice tells Bob to look at the results #8 and #12, where Bob can read $(+, -)$.

### 16.4.4 References

This procedure is presently being developed in several industrial research laboratories. In practice, one uses photon pairs with correlated polarizations rather than spin 1/2 material particles. See, for instance C. H. Bennett, G. Brassard, and A. Ekert, *Quantum Cryptography*, Scientific American **267**, 26 (October 1992); D. J. Bernstein, J. Buchmann, and E. Dahmen (Eds.) (2009), *Post-quantum cryptography*, Springer Science & Business Media; C. H. Bennett and G. Brassard, *Quantum cryptography: Public key distribution and coin tossing*, Theoretical computer science **560**, 7 (2014).

# Ideal Quantum Measurement

# 17

In 1940, John von Neumann proposed a definition for an optimal, or "ideal" measurement of a quantum physical quantity. In this chapter, we study a practical example of such a procedure. Our ambition is to measure the excitation number of a harmonic oscillator $S$ by coupling it to another oscillator $\mathcal{D}$ whose phase is measured.

We recall that, for $k$ integer:

$$\sum_{n=0}^{N} e^{\frac{2i\pi kn}{N+1}} = N+1 \quad \text{for} \quad k = p(N+1), \ p \text{ integer}; \quad = 0 \quad \text{otherwise.} \quad (17.1)$$

## 17.1 A Von Neumann Detector

We want to measure a physical quantity $A$ on a quantum system $S$. We use a detector $\mathcal{D}$ devised for such a measurement. There are two stages in the measurement process. First, we let $S$ and $\mathcal{D}$ interact. Then, after $S$ and $\mathcal{D}$ get separated and do not interact anymore, we read a result on the detector $\mathcal{D}$. We assume that $\mathcal{D}$ possesses an orthonormal set of states $\{|D_i\rangle\}$ with $\langle D_i|D_j\rangle = \delta_{i,j}$. These states correspond for instance to the set of values which can be read on a digital display.

Let $|\psi\rangle$ be the state of the system $S$ under consideration, and $|D\rangle$ the state of the detector $\mathcal{D}$. Before the measurement, the state of the global system $S \mid \mathcal{D}$ is

$$|\Psi_i\rangle = |\psi\rangle \otimes |D\rangle. \quad (17.2)$$

© Springer Nature Switzerland AG 2019
J.-L. Basdevant, J. Dalibard, *The Quantum Mechanics Solver*,
https://doi.org/10.1007/978-3-030-13724-3_17

Let $a_i$ and $|\phi_i\rangle$ be the eigenvalues and corresponding eigenstates of the observable $\hat{A}$. The state $|\psi\rangle$ of the system $\mathcal{S}$ can be expanded as

$$|\psi\rangle = \sum_i \alpha_i |\phi_i\rangle. \tag{17.3}$$

**17.1.1** Using the axioms of quantum mechanics, what are the probabilities $p(a_i)$ to find the values $a_i$ in a measurement of the quantity $A$ on this state?

**17.1.2** After the interaction of $\mathcal{S}$ and $\mathcal{D}$, the state of the global system is in general of the form

$$|\Psi_f\rangle = \sum_{i,j} \gamma_{ij} |\phi_i\rangle \otimes |D_j\rangle \tag{17.4}$$

We now observe the state of the detector. What is the probability to find the detector in the state $|D_j\rangle$?

**17.1.3** After this measurement, what is the state of the global system $\mathcal{S} + \mathcal{D}$?

**17.1.4** A detector is called *ideal* if the choice of $|D_0\rangle$ and of the $\mathcal{S} - \mathcal{D}$ coupling leads to coefficients $\gamma_{ij}$ which, for any state $|\psi\rangle$ of $\mathcal{S}$, verify: $|\gamma_{ij}| = \delta_{i,j}|\alpha_j|$. Justify this designation.

## 17.2   Phase States of the Harmonic Oscillator

We consider a harmonic oscillator of angular frequency $\omega$. We note $\hat{N}$ the "number" operator, i.e. the Hamiltonian is $\hat{H} = (\hat{N} + 1/2))\hbar\omega$ with eigenstates $|N\rangle$ and eigenvalues $E_N = (N + \frac{1}{2})\hbar\omega$, $N$ integer $\geq 0$.

Let $s$ be a positive integer. The so-called "phase states" are the family of states defined at each time $t$ by:

$$|\theta_m\rangle = \frac{1}{\sqrt{s+1}} \sum_{N=0}^{N=s} e^{-iN(\omega t + \theta_m)}|N\rangle \tag{17.5}$$

where $\theta_m$ can take any of the $2s + 1$ values

$$\theta_m = \frac{2\pi m}{s+1} \quad (m = 0, 1, \dots, s). \tag{17.6}$$

**17.2.1** Show that the states $|\theta_m\rangle$ are orthonormal.

**17.2.2** We consider the subspace of states of the harmonic oscillator such that the number of quanta $N$ is bounded from above by some value $s$. The sets $\{|N\rangle, N = 0, 1, \ldots, s\}$ and $\{|\theta_m\rangle, m = 0, 1, \ldots, s\}$ are two bases in this subspace. Express the vectors $|N\rangle$ in the basis of the phase states.

**17.2.3** What is the probability to find $N$ quanta in a phase state $|\theta_m\rangle$?

**17.2.4** Calculate the expectation value of the position $\hat{x}$ in a phase state, and find a justification for the name "phase state". We recall the relation $\hat{x}|N\rangle = x_0(\sqrt{N+1}|N+1\rangle + \sqrt{N}|N-1\rangle)$, where $x_0$ is the characteristic length of the problem. We set $C_s = \sum_{N=0}^{s} \sqrt{N}$.

---

## 17.3   The Interaction Between the System and the Detector

We want to perform an "ideal" measurement of the number of excitation quanta of a harmonic oscillator. In order to do so, we couple this oscillator $S$ with another oscillator $\mathcal{D}$, which is our detector. Both oscillators have the same angular frequency $\omega$. The eigenstates of $\hat{H}_S = (\hat{n} + \frac{1}{2})\hbar\omega$ are noted $|n\rangle, n = 0, 1, \ldots, s$, those of $\hat{H}_D = (\hat{N} + \frac{1}{2})\hbar\omega$ are noted $|N\rangle, N = 0, 1 \ldots s$ where $\hat{n}$ and $\hat{N}$ are the number operators of $S$ and $\mathcal{D}$.

We assume that both numbers of quanta $n$ and $N$ are bounded from above by $s$. The coupling between $S$ and $\mathcal{D}$ has the form:

$$\hat{V} = \hbar g \,\hat{n}\hat{N} \tag{17.7}$$

*This Hamiltonian is realistic. If the two oscillators are two modes of the electromagnetic field, it originates from the crossed Kerr effect.*

**17.3.1** What are the eigenstates and eigenvalues of the total Hamiltonian

$$\hat{H} = \hat{H}_S + \hat{H}_D + \hat{V} \; ? \tag{17.8}$$

**17.3.2** We assume that the initial state of the global system $S + \mathcal{D}$ is *factorized* as:

$$|\Psi(0)\rangle = |\psi_S\rangle \otimes |\psi_D\rangle, \quad \text{with :} \quad |\psi_S\rangle = \sum_n a_n|n\rangle, \quad |\psi_D\rangle = \sum_N b_N|N\rangle \tag{17.9}$$

where we assume that $|\psi_S\rangle$ and $|\psi_D\rangle$ are normalized. We perform a measurement of $\hat{n}$ in the state $|\Psi(0)\rangle$. What results can one find, with what probabilities? Answer the same question for a measurement of $\hat{N}$.

**17.3.3** During the time interval $[0, t]$, we couple the two oscillators. The coupling is switched off at time $t$. What is the state $|\Psi(t)\rangle$ of the system? Is it also *a priori* factorizable?

**17.3.4** Is the probability law for the couple of random variables $\{n, N\}$ affected by the interaction? Why?

## 17.4   An "Ideal" Measurement

Initially, at time $t = 0$, the oscillator $S$ is in a state $|\psi_S\rangle = \sum_{n=0}^{s} a_n |n\rangle$. The oscillator $\mathcal{D}$ is prepared in the state

$$|\psi_D\rangle = \frac{1}{\sqrt{s+1}} \sum_{N=0}^{s} |N\rangle. \tag{17.10}$$

**17.4.1** We switch on the interaction $\hat{V}$ during the time interval $[0, t]$. Express the state $|\Psi(t)\rangle$ in terms of the phase states $\{|\theta_k\rangle\}$ of the oscillator $\mathcal{D}$.

**17.4.2** We assume the interaction time is $t = t_0 \equiv 2\pi/[g(s+1)]$. Write the state $|\Psi(t_0)\rangle$ of the system.

**17.4.3** What is the probability to find the value $\theta_k$ in a measurement of the phase of the "detector" oscillator $\mathcal{D}$?

**17.4.4** After this measurement has been performed, what is the state of the oscillator $S$? Describe qualitatively what will happen if one were to choose an interaction time $t \neq t_0$.

**17.4.5** Comment on the result. In your opinion, why did J. von Neumann consider this as an "ideal" quantum-measurement process?

## 17.5   Solutions

### 17.5.1 A Von Neumann Detector

**17.5.1.1** Since the state of the system is $|\psi\rangle = \sum_i \alpha_i |\phi_i\rangle$, the probability to find the value $a_j$ in a measurement of $A$ is $p(a_j) = |\alpha_j|^2$.

**17.5.1.2** The state of the global system is

$$|\Psi_1\rangle = \sum_{i,j} \gamma_{ij} |\phi_i\rangle \otimes |D_j\rangle. \tag{17.11}$$

The probability $p_j$ to find the detector in the state $|D_j\rangle$ is the sum of the probabilities $|\gamma_{ij}|^2$:

$$p_j = \sum_i |\gamma_{ij}|^2, \tag{17.12}$$

since the states $|\phi_i\rangle$ are orthogonal.

**17.5.1.3** After this measurement, the state of the global system $\mathcal{S} + \mathcal{D}$ is, after the principle of wave packet reduction,

$$|\Psi\rangle = \frac{1}{\sqrt{p_j}} \left[ \sum_i \gamma_{ij} |\phi_i\rangle \right] \otimes |D_j\rangle. \tag{17.13}$$

**17.5.1.4** For an ideal detector, the probability that the detector is in the state $|D_j\rangle$ is $p_j = |\alpha_j|^2 = p(a_j)$ and the state of the set system + detector, once we know the state of the detector, is $|\phi_j\rangle \otimes |D_j\rangle$. This is the expected result, given the wave packet reduction principle.

## 17.5.2 Phase States of the Harmonic Oscillator

**17.5.2.1** Given the definition of the phase states, one has:

$$\langle \theta_m | \theta_n \rangle = \frac{1}{s+1} \sum_{N=0}^{s} \sum_{N'=0}^{s} e^{iN(\omega t + \theta_m)} e^{-iN'(\omega t + \theta_n)} \langle N | N' \rangle = \frac{1}{s+1} \sum_{N=0}^{s} e^{iN(\theta_m - \theta_n)}$$

which simplifies as

$$\langle \theta_m | \theta_n \rangle = \frac{1}{s+1} \sum_{N=0}^{s} e^{2i\pi N(m-n)/(s+1)} = \delta_{m,n}, \tag{17.14}$$

where the last equality stands because $-s \leq m - n \leq s$.

**17.5.2.2** The scalar product of a state $|N\rangle$ with a phase state is

$$\langle \theta_m | N \rangle = (\langle N | \theta_m \rangle)^* = \frac{1}{\sqrt{s+1}} e^{iN(\omega t + \theta_m)}, \tag{17.15}$$

hence the expansion:

$$|N\rangle = \sum_{m=0}^{s} \langle \theta_m | N \rangle |\theta_m\rangle = \frac{1}{\sqrt{s+1}} \sum_{m=0}^{s} e^{iN(\omega t + \theta_m)} |\theta_m\rangle. \tag{17.16}$$

**17.5.2.3** Given the definition of a phase state, the probability to find $N$ quanta in a state $|\theta_m\rangle$ is

$$p(N, \theta_m) = |\langle N | \theta_m \rangle|^2 = \frac{1}{s+1}. \tag{17.17}$$

**17.5.2.4** One obtains

$$\langle \theta_m | \hat{x} | \theta_m \rangle = 2x_0 \frac{C_s}{s+1} \cos(\omega t + \theta_m). \tag{17.18}$$

The phases of the expectation values of $x$ in two phase states $|\theta_m\rangle$ and $|\theta_n\rangle$ differ by an integer multiple $2(m - n)\pi/(s + 1)$ of the elementary phase $2\pi/(s + 1)$.

### 17.5.3  The Interaction Between the System and the Detector

**17.5.3.1** The factorized states $|n\rangle \otimes |N\rangle$ are eigenstates of the total Hamiltonian

$$\hat{H} = \hat{H}_S + \hat{H}_D + \hat{V} = (\hat{n} + \hat{N} + 1)\hbar\omega + \hbar g\, \hat{n} \otimes \hat{N}, \tag{17.19}$$

with eigenvalues $E_{n,N} = (n + N + 1)\hbar\omega + \hbar g\, nN$.

**17.5.3.2** The results of measurements and the corresponding probabilities are $n = 0, 1, \ldots, s$, $p(n) = |a_n|^2$ and $N = 0, 1, \ldots, s$, $p(N) = |b_N|^2$.

**17.5.3.3** The state of the system at time $t$ is

$$|\Psi(t)\rangle = \sum_n \sum_N a_n\, b_N\, e^{-i[(n+N+1)\omega + gnN]t} |n\rangle \otimes |N\rangle. \tag{17.20}$$

In general, it is not factorized.

**17.5.3.4** The probability law for the couple of random variables $\{n, N\}$ is still $p(n, N) = |a_n|^2 |b_N|^2$. It is not modified by the interaction since $\hat{V}$ commutes with $\hat{n}$ and $\hat{N}$. The quantities $n$ and $N$ are constants of the motion.

### 17.5.4  An "Ideal" Measurement

**17.5.4.1** One has $b_N = 1/\sqrt{s + 1}$, hence

$$|\Psi(t)\rangle = \frac{1}{\sqrt{s+1}} \sum_n \sum_N a_n e^{-i[(n+N+1)\omega + gnN)]t} |n\rangle \otimes |N\rangle. \tag{17.21}$$

Inserting the expansion of the states $|N\rangle$ in terms of the phase states, one obtains

$$|\Psi(t)\rangle = \sum_n \sum_m \left( \sum_N \frac{e^{i(\theta_m - gnt)N}}{s+1} \right) e^{-i(n+1)\omega t} a_n \, |n\rangle \otimes |\theta_m\rangle. \qquad (17.22)$$

**17.5.4.2**  If the interaction time is $t_0 = 2\pi/[g(s+1)]$, this expression reduces to

$$|\Psi(t_0)\rangle = \sum_{n=0}^{s} e^{-i(n+1)\omega t_0} a_n \, |n\rangle \otimes |\theta_n\rangle. \qquad (17.23)$$

**17.5.4.3**  The probability to find the result $\theta_n$ by measuring the phase of the detecting oscillator $\mathcal{D}$ on this state is $p(\theta_n) = |a_n|^2$.

**17.5.4.4**  After this measurement, the state of the oscillator $\mathcal{S}$ is simply $|n\rangle$ (up to an arbitrary phase factor). In the state (17.23), the two systems are perfectly correlated. To a phase state of $\mathcal{D}$ there corresponds only one state of number of quanta of $\mathcal{S}$. If one were to choose a time interval different from $t_0$, this correlation would not be perfect. After a measurement of the phase of $\mathcal{D}$, the state of $\mathcal{S}$ would be a superposition of states with different numbers of quanta.

**17.5.4.5**  We see that this procedure, which supposes a well defined interaction time interval between the system and the detector, gives the value of the probability $p(n) = |a_n|^2$ that $\mathcal{S}$ is in a state with $n$ quanta. In addition, after one has read the result $\theta_n$ on the detector, one is sure that $\mathcal{S}$ is in the state $|n\rangle$, without having to further interact with it (reduction of the wave packet). In this sense, this procedure does follow exactly the axioms of quantum mechanics on measurement. It is therefore an "ideal" measurement of a quantum physical quantity.

## 17.5.5  Comments

We presented in this chapter a concrete example of a measuring device in Quantum Mechanics, which is often used to discuss the general problem of a quantum measurement. Several aspects of this problem are addressed in the book *Quantum theory and measurement*, J.A. Wheeler and W, H, Zurek (Eds.) (Princeton University Press, 2014).

One can extend formally the result of this chapter to other systems than harmonic oscillators. In practice, the case studied here is a simplification of the concrete case where the oscillators $\mathcal{S}$ and $\mathcal{D}$ are modes of the electromagnetic field. The Hamiltonian which is effectively encountered in a optically non-linear crystal, comes from the phenomenon called the *crossed Kerr effect*. In an interferometer, where $\mathcal{D}$ is a laser beam split in two parts by a semi-transparent mirror, one can let the signal oscillator $\mathcal{S}$ interact with one of the beams. The measurement consists in an interferometric measurement when the two beams of $\mathcal{D}$ recombine. This type of experiment has been carried out intensively in recent years. It is also called a "non-destructive" quantum measurement (or QND measurement).

A description of earlier QND realizations can be found in J.-P. Poizat and P. Grangier, *Experimental realization of a quantum optical tap*, Phys. Rev. Lett. **70**, 271 (1993). Recent discussions and implementations are presented by C. Monroe, *Demolishing Quantum Nondemolition*, Physics Today **64**, 8 (2011), and by A. Clerk, M. Devoret, S. M. Girvin, F. Marquardt, and R. J. Schoelkopf, *Introduction to quantum noise, measurement, and amplification*, Rev. Mod. Phys. **82**, 1155 (2010). On a fundamental level, the problem investigated here can also be viewed as a model for the coupling of a quantum system and its environment, which is responsible for decoherence; for more details, see e.g. *Quantum Decoherence*, Poincaré Seminar, B. Duplantier, J.-M. Raimond and V. Rivasseau (Edts) (Birkhäuser Verlag, Basel, 2007).

# The Quantum Eraser

# 18

This chapter deals with a quantum process where the superposition of two probability amplitudes leads to an interference phenomenon. The two amplitudes can be
associated with two quantum paths, as in a double slit interference experiment. We
shall first show that these interferences disappear if an intermediate measurement
gives information about which path has actually been followed. Next, we shall see
how interferences can actually *reappear* if this information is "erased" by a quantum
device.

We consider a beam of neutrons, which are particles of charge zero and spin
1/2, propagating along the $x$ axis with velocity $v$. In all what follows, the motion
of the neutrons in space is treated classically as a uniform linear motion. Only the
evolution of their spin states is treated quantum mechanically.

## 18.1 Magnetic Resonance

The eigenstates of the $z$ component of the neutron spin are noted $|n : +\rangle$ and $|n : -\rangle$.
A constant uniform magnetic field $\boldsymbol{B}_0 = B_0\boldsymbol{u}_z$ is applied along the $z$ axis ( $\boldsymbol{u}_z$ is
the unit vector along the $z$ axis). The magnetic moment of the neutron is denoted
$\hat{\boldsymbol{\mu}}_n = \gamma_n\hat{\boldsymbol{S}}_n$, where $\gamma_n$ is the gyromagnetic ratio and $\hat{\boldsymbol{S}}_n$ the spin operator of the
neutron.

**18.1.1** What are the magnetic energy levels of a neutron in the presence of the field
$\boldsymbol{B}_0$? Express the result in terms of $\omega_0 = -\gamma_n B_0$.

**18.1.2** The neutrons cross a cavity of length $L$ between times $t_0$ and $t_1 = t_0 + L/v$. Inside this cavity, in addition to the constant field $\boldsymbol{B}_0$, a rotating field $\boldsymbol{B}_1(t)$
is applied. The field $\boldsymbol{B}_1(t)$ lies in the $(x, y)$ plane and it has a constant angular

© Springer Nature Switzerland AG 2019
J.-L. Basdevant, J. Dalibard, *The Quantum Mechanics Solver*,
https://doi.org/10.1007/978-3-030-13724-3_18

frequency $\omega$:

$$B_1(t) = B_1(\cos \omega t \, u_x + \sin \omega t \, u_y). \qquad (18.1)$$

Let $|\psi_n(t)\rangle = \alpha_+(t)|n : +\rangle + \alpha_-(t)|n : -\rangle$ be the neutron spin state at time $t$, and consider a neutron entering the cavity at time $t_0$.

(a) Write the equations of evolution for $\alpha_\pm(t)$ when $t_0 \le t \le t_1$. We set hereafter $\omega_1 = -\gamma_n B_1$.

(b) Setting $\alpha_\pm(t) = \beta_\pm(t) \exp[\mp i\omega(t - t_0)/2]$, show that the problem reduces to a differential system with constant coefficients.

(c) We assume that we are near the resonance: $|\omega - \omega_0| \ll \omega_1$, and that terms proportional to $(\omega - \omega_0)$ may be neglected in the previous equations. Check that, within this approximation, one has, for $t_0 \le t \le t_1$,

$$\beta_\pm(t) = \beta_\pm(t_0) \cos \theta - i e^{\mp i\omega t_0} \beta_\mp(t_0) \sin \theta, \qquad (18.2)$$

where $\theta = \omega_1(t - t_0)/2$.

(d) Show that the spin state at time $t_1$, when the neutron leaves the cavity, can be written as:

$$\begin{pmatrix} \alpha_+(t_1) \\ \alpha_-(t_1) \end{pmatrix} = U(t_0, t_1) \begin{pmatrix} \alpha_+(t_0) \\ \alpha_-(t_0) \end{pmatrix} \qquad (18.3)$$

where the matrix $U(t_0, t_1)$ is

$$U(t_0, t_1) = \begin{pmatrix} e^{-i\chi} \cos \phi & -ie^{-i\delta} \sin \phi \\ -ie^{i\delta} \sin \phi & e^{i\chi} \cos \phi \end{pmatrix}, \qquad (18.4)$$

with $\phi = \omega_1(t_1 - t_0)/2$, $\chi = \omega(t_1 - t_0)/2$ and $\delta = \omega(t_1 + t_0)/2$.

## 18.2   Ramsey Fringes

The neutrons are initially in the spin state $|n : -\rangle$. They successively cross two identical cavities of the type described above. This is called Ramsey's configuration and it is shown in Fig. 18.1. The same oscillating field $B_1(t)$, given by 16.1, is applied in both cavities. The modulus $B_1$ of this field is adjusted so as to satisfy the condition $\phi = \pi/4$. The constant field $B_0$ is applied throughout the experimental setup. At the end of this setup, one measures the number of outgoing neutrons which have flipped their spin and are in the final state $|n : +\rangle$. This is done for several values of $\omega$ in the vicinity of $\omega = \omega_0$.

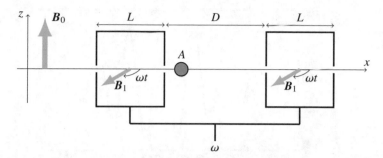

**Fig. 18.1** Ramsey's configuration; the role of the detecting atom $A$ is specified in parts 3 and 4

**18.2.1** At time $t_0$, a neutron enters the first cavity in the state $|n : -\rangle$. What is its spin state, and what is the probability to find it in the state $|n : +\rangle$, when it leaves the cavity?

**18.2.2** The same neutron enters the second cavity at time $t'_0 = t_1 + T$, with $T = D/v$ where $D$ is the distance between the two cavities. Between the two cavities the spin precesses freely around $\boldsymbol{B}_0$. What is the spin state of the neutron at time $t'_0$?

**18.2.3** Let $t'_1$ be the time when the neutron leaves the second cavity: $t'_1 - t'_0 = t_1 - t_0$. Express the quantity $\delta' = \omega(t'_1 + t'_0)/2$ in terms of $\omega$, $t_0$, $t_1$ and $T$. Write the transition matrix $U(t'_0, t'_1)$ in the second cavity.

**18.2.4** Calculate the probability $P_+$ to detect the neutron in the state $|n : +\rangle$ after the second cavity. Show that it is an oscillating function of $(\omega_0 - \omega)T$. Explain why this result can be interpreted as an interference process.

**18.2.5** In practice, the velocities of the neutrons have some dispersion around the mean value $v$. This results in a dispersion in the time $T$ to get from one cavity to the other. A typical experimental result giving the intensity of the outgoing beam in the state $|n : +\rangle$ as a function of the frequency $v = \omega/2\pi$ of the rotating field $\boldsymbol{B}_1$ is shown in Fig. 18.2.

(a) Explain the shape of this curve by averaging the previous result over the distribution

$$dp(T) = \frac{1}{\tau\sqrt{2\pi}} e^{-(T-T_0)^2/2\tau^2} \, dT. \tag{18.5}$$

We recall that $\int_{-\infty}^{\infty} \cos(\Omega T) \, dp(T) = e^{-\Omega^2 \tau^2/2} \cos(\Omega T_0)$.

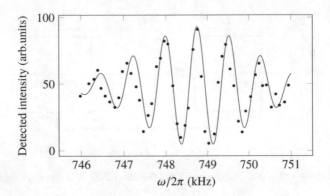

**Fig. 18.2** Intensity of the outgoing beam in the state $|n : +\rangle$ as a function of the frequency $\omega/2\pi$ for a neutron beam with some velocity dispersion

(b) In the above experiment, the value of the magnetic field was $B_0 = 2.57 \times 10^{-2}$ T and the distance $D = 1.6$ m. Calculate the magnetic moment of the neutron. Evaluate the average velocity $v_0 = D/T_0$ and the velocity dispersion $\delta v = v_0 \tau / T_0$ of the neutron beam.

(c) Which optical interference experiment is the result reminiscent of?

**18.2.6** Suppose one inserts between the two cavities of Fig. 18.1 a device which can measure the $z$ component of the neutron spin (the principle of such a detector is presented in the next section). Determine the probability $P_{+,+}$ to detect the neutron in the state $|n : +\rangle$ between the two cavities and in the state $|n : +\rangle$ when it leaves the second cavity, and the probability $P_{-,+}$ to detect the neutron in the state $|n : -\rangle$ between the cavities and in the state $|n : +\rangle$ when it leaves the second cavity. Check that one does *not* have $P_+ = P_{+,+} + P_{-,+}$ and comment on this fact.

## 18.3    Detection of the Neutron Spin State

In order to measure the spin of a neutron, one lets it interact during a time $\tau$ with a spin 1/2 atom at rest. The atom's spin operator is $\hat{S}_a$. Let $|a : \pm\rangle$ be the two eigenstates of the observable $\hat{S}_{az}$. After the interaction between the neutron and the atom, one measures the spin of the atom. Under certain conditions, as we shall see, one can deduce the spin state of the neutron after this measurement.

**18.3.1** Spin States of the Atom.

Let $|a : \pm x\rangle$ be the eigenstates of $\hat{S}_{ax}$ and $|a : \pm y\rangle$ those of $\hat{S}_{ay}$. Write $|a : \pm x\rangle$ and $|a : \pm y\rangle$ in the basis $\{|a : +\rangle, |a : -\rangle\}$. Express $|a : \pm y\rangle$ in terms of $|a : \pm x\rangle$.

**18.3.2** We assume that the neutron–atom interaction does not affect the neutron's trajectory. We represent the interaction between the neutron and the atom by a very simple model. This interaction is assumed to last a finite time $\tau$ during which the neutron-atom interaction Hamiltonian has the form

$$\hat{V} = \frac{2A}{\hbar}\hat{S}_{nz} \otimes \hat{S}_{ax},\qquad(18.6)$$

where $A$ is a constant. We neglect the action of any external field, including $\boldsymbol{B}_0$, during the time $\tau$.

Explain why $\hat{S}_{nz}$ and $\hat{V}$ commute. Give their common eigenstates and the corresponding eigenvalues.

**18.3.3** We hereafter assume that the interaction time $\tau$ is adjusted in such a way that

$$A\tau = \pi/2.\qquad(18.7)$$

Suppose the initial state of the system is

$$|\psi(0)\rangle = |n : +\rangle \otimes |a : +y\rangle.\qquad(18.8)$$

Calculate the final state of the system $|\psi(\tau)\rangle$. Answer the same question if the initial state is $|\psi(0)\rangle = |n : -\rangle \otimes |a : +y\rangle$.

**18.3.4** We now suppose that the initial spin state is

$$|\psi(0)\rangle = (\alpha_+|n : +\rangle + \alpha_-|n : -\rangle) \otimes |a : +y\rangle.\qquad(18.9)$$

After the neutron–atom interaction described above, one measures the $z$ component $S_{az}$ of the atom's spin.

(a) What results can one find, and with what probabilities?
(b) *After* this measurement, what prediction can one make about the value of the $z$ component of the neutron spin? Is it necessary to let the neutron interact with another measuring apparatus in order to know $S_{nz}$ once the value of $S_{az}$ is known?

## 18.4   A Quantum Eraser

We have seen above that if one measures the spin state of the neutron between the two cavities, the interference signal disappears. In this section, we will show that it is possible to recover an interference if the information left by the neutron on the detecting atom is "erased" by an appropriate measurement.

A neutron, initially in the spin state $|n : -\rangle$, is sent into the two-cavity system. Immediately after the first cavity, there is a detecting atom of the type discussed above, prepared in the spin state $|a : +y\rangle$. By assumption, the spin state of the atom evolves only during the time interval $\tau$ when it interacts with the neutron.

**18.4.1** Write the spin state of the neutron–atom system when the neutron is:

(a) just leaving the first cavity (time $t_1$), before interacting with the atom;
(b) just after the interaction with the atom (time $t_1 + \tau$);
(c) entering the second cavity (time $t_0'$);
(d) just leaving the second cavity (time $t_1'$).

**18.4.2** What is the probability to find the neutron in the state $|n : +\rangle$ at time $t_1'$? Does this probability reflect an interference phenomenon? Interpret the result.

**18.4.3** At time $t_1'$, Bob measures the $z$ component of the neutron spin and Alice measures the $y$ component of the atom's spin. Assume both measurements give $+\hbar/2$. Show that the corresponding probability reflects an interference phenomenon.

**18.4.4** Is this result compatible with the conclusion of question 4.2?

**18.4.5** In your opinion, which of the following three statements are appropriate, and for what reasons?

(a) When Alice performs a measurement on the atom, Bob sees at once an interference appear in the signal he is measuring on the neutron.
(b) Knowing the result obtained by Alice on each event, Bob can select a subsample of his own events which displays an interference phenomenon.
(c) The experiment corresponds to an interference between two quantum paths for the neutron spin. By restoring the initial state of the atom, the measurement done by Alice erases the information concerning which quantum path is followed by the neutron spin, and allows interferences to reappear.

**18.4.6** Alice now measures the component of the atom's spin along an arbitrary axis defined by the unit vector $\boldsymbol{w}$. Show that the contrast of the interferences varies proportionally to $|\sin \eta|$, where $\cos \eta = \boldsymbol{w}.\boldsymbol{u}_z$. Interpret the result.

## 18.5 Solutions

### 18.5.1 Magnetic Resonance

**18.5.1.1** The magnetic energy levels are: $E_\pm = \mp \gamma_n \hbar B_0/2 = \pm \hbar \omega_0/2$.

**18.5.1.2 (a)** The Hamiltonian is

$$H = \frac{\hbar}{2} \begin{pmatrix} \omega_0 & \omega_1 e^{-i\omega t} \\ \omega_1 e^{i\omega t} & -\omega_0 \end{pmatrix}. \tag{18.10}$$

Therefore, the evolution equations are

$$i\dot{\alpha}_+ = \frac{\omega_0}{2}\alpha_+ + \frac{\omega_1}{2}e^{-i\omega t}\alpha_-; \quad i\dot{\alpha}_- = -\frac{\omega_0}{2}\alpha_- + \frac{\omega_1}{2}e^{+i\omega t}\alpha_+ . \tag{18.11}$$

**(b)** With the variables $\beta_\pm(t) = \alpha_\pm(t) \exp[\pm i\omega(t - t_0)/2]$, we obtain

$$i\dot{\beta}_+ = \frac{\omega_0 - \omega}{2}\beta_+ + \frac{\omega_1}{2}e^{-i\omega t_0}\beta_-; \quad i\dot{\beta}_- = \frac{\omega - \omega_0}{2}\beta_- + \frac{\omega_1}{2}e^{i\omega t_0}\beta_+ . \tag{18.12}$$

**(c)** If $|\omega_0 - \omega| \ll \omega_1$, we have, to a good approximation, the differential system

$$i\dot{\beta}_+ = \frac{\omega_1}{2}e^{-i\omega t_0}\beta_-; \quad i\dot{\beta}_- = \frac{\omega_1}{2}e^{i\omega t_0}\beta_+, \tag{18.13}$$

whose solution is indeed

$$\beta_\pm(t) = \beta_\pm(t_0) \cos\frac{\omega_1(t - t_0)}{2} - i\, e^{\mp i\omega t_0}\beta_\mp(t_0) \sin\frac{\omega_1(t - t_0)}{2}. \tag{18.14}$$

**(d)** Defining $\phi = \omega_1(t_1 - t_0)/2$, $\chi = \omega(t_1 - t_0)/2$, $\delta = \omega(t_1 + t_0)/2$, we obtain

$$\alpha_+(t_1) = e^{-i\chi}\beta_+(t_1) = e^{-i\chi}\left[\alpha_+(t_0)\cos\phi - i\alpha_-(t_0)e^{-i\omega t_0}\sin\phi\right]$$

$$\alpha_-(t_1) = e^{i\chi}\beta_-(t_1) = e^{+i\chi}\left[\alpha_-(t_0)\cos\phi - i\alpha_+(t_0)e^{+i\omega t_0}\sin\phi\right], \tag{18.15}$$

and, therefore,

$$U = \begin{pmatrix} e^{-i\chi}\cos\phi & -i\, e^{i\delta}\sin\phi \\ -i\, e^{i\delta}\sin\phi & e^{i\chi}\cos\phi \end{pmatrix}. \tag{18.16}$$

## 18.5.2 Ramsey Fringes

**18.5.2.1** We assume $\phi = \pi/4$; the initial conditions are: $\alpha_+(t_0) = 0, \alpha_-(t_0) = 1$. At time $t_1$ the state is

$$|\psi(t_1)\rangle = \frac{1}{\sqrt{2}} \left( -i\, e^{-i\delta} |n : +\rangle + e^{i\chi} |n : -\rangle \right). \tag{18.17}$$

In other words $\alpha_+(t_1) = -i e^{-i\delta}/\sqrt{2}, \; \alpha_-(t_1) = e^{i\chi}/\sqrt{2}$, and $P\pm = 1/2$.

**18.5.2.2** We set $T = D/v$. The neutron spin precesses freely between the two cavities during time $T$, and we obtain

$$\begin{pmatrix} \alpha_+(t_0') \\ \alpha_-(t_0') \end{pmatrix} = \frac{1}{\sqrt{2}} \begin{pmatrix} -i e^{-i\delta} e^{-i\omega_0 T/2} \\ e^{i\chi} e^{+i\omega_0 T/2} \end{pmatrix}. \tag{18.18}$$

**18.5.2.3** By definition, $t_0' = t_1 + T$ and $t_1' = 2t_1 - t_0 + T$, therefore the transition matrix in the second cavity is

$$U' = \begin{pmatrix} e^{-i\chi'} \cos\phi' & -i e^{-i\delta'} \sin\phi' \\ -i e^{i\delta'} \sin\phi' & e^{i\chi'} \cos\phi' \end{pmatrix} \tag{18.19}$$

with $\phi' = \phi = \omega_1(t_1 - t_0)/2, \; \chi' = \chi = \omega(t_1 - t_0)/2$. Only the parameter $\delta$ is changed into

$$\delta' = \omega(t_1' + t_0')/2 = \omega(3t_1 + 2T - t_0)/2. \tag{18.20}$$

**18.5.2.4** The probability amplitude for detecting the neutron in state $|+\rangle$ after the second cavity is obtained by (i) applying the matrix $U'$ to the vector (18.18), (ii) calculating the scalar product of the result with $|n : +\rangle$. We obtain

$$\alpha_+(t_1') = \frac{1}{2} \left( -i e^{-i(\chi+\delta+\omega_0 T/2)} - i e^{-i(\delta'-\chi-\omega_0 T/2)} \right). \tag{18.21}$$

Since

$$\delta + \chi = \omega t_1 \quad \delta' - \chi = \frac{\omega}{2}(3t_1 + 2T - t_0 - t_1 + t_0) = \omega(t_1 + T), \tag{18.22}$$

we have

$$\alpha_+(t_1') = -\frac{i}{2}\, e^{-i\omega(t_1+T/2)} \left( e^{-i(\omega_0-\omega)T/2} + e^{i(\omega_0-\omega)T/2} \right). \tag{18.23}$$

Therefore, the probability that the neutron spin has flipped in the two-cavity system is

$$P_+ = |\alpha_+(t_1')|^2 = \cos^2 \frac{(\omega - \omega_0)T}{2}. \tag{18.24}$$

With the approximation of Sect. 18.5.1.2(c), the probability for a spin flip in a single cavity is independent of $\omega$, and is equal to 1/2. In contrast, the present result for two cavities exhibits a strong modulation of the spin flip probability, between 1 (e.g. for $\omega = \omega_0$) and 0 (e.g. for $(\omega - \omega_0)T = \pi$). This modulation results from an interference process of the two quantum paths corresponding respectively to:

- a spin flip in the first cavity, and no flip in the second one,
- no flip in the first cavity and a spin flip in the second one.

Each of these paths has a probability 1/2, so that the sum of the probability amplitudes (18.23) is fully modulated.

**18.5.2.5 (a)** Since $\cos^2 \phi/2 = (1 + \cos\phi)/2$, the averaged probability distribution is

$$\left\langle \cos^2 \frac{(\omega - \omega_0)T}{2} \right\rangle = \frac{1}{2} + \frac{1}{2} e^{-(\omega-\omega_0)^2\tau^2/2} \cos\left[(\omega - \omega_0)T_0\right]. \tag{18.25}$$

This form agrees with the observed variation in $\omega$ of the experimental signal. The central maximum, which is located at $\omega/2\pi = 748.8\,\text{kHz}$ corresponds to $\omega = \omega_0$. For that value, a constructive interference appears whatever the neutron velocity. The lateral maxima and minima are less peaked, however, since the position of a lateral peak is velocity dependent. The first two lateral maxima correspond to $(\omega - \omega_0)T_0 \simeq \pm 2\pi$. Their amplitude is reduced, compared to the central peak, by a factor $\exp(-2\pi^2\tau^2/T_0^2)$.

**(b)** The angular frequency $\omega_0$ is related to the magnetic moment of the neutron by $\hbar\omega_0 = 2\mu_n B_0$ which leads to $\mu_n = 9.65 \times 10^{-27}\,\text{JT}^{-1}$. The time $T_0$ can be deduced from the spacing between the central maximum and a lateral one. The first lateral maximum occurs at 0.77 kHz from the resonance, hence $T_0 = 1.3\,\text{ms}$ This corresponds to an average velocity $v_0 = 1230\,\text{m s}^{-1}$. The ratio of intensities between the second lateral maximum and the central one is roughly 0.55. This is approximately equal to $\exp(-8\pi^2\tau^2/T_0^2)$, and gives $\tau/T_0 \approx 0.087$, and $\delta v \approx 110\,\text{m s}^{-1}$.

**(c)** This experiment can be compared to a Young double slit interference experiment with polychromatic light. The central fringe (corresponding to the peak at $\omega = \omega_0$) remains bright, but the contrast of the interferences decreases rapidly as one departs from the center. In fact, the maxima for some frequencies correspond to minima for others.

**18.5.2.6** The probability $P_{++}$ is the product of the two probabilities: the probability to find the neutron in the state $|n : +\rangle$ when it leaves the first cavity ($p = 1/2$) and, knowing that it is in the state $|n : +\rangle$, the probability to find it in the same state when it leaves the second cavity ($p = 1/2$); this gives $P_{+,+} = 1/4$. Similarly $P_{-,+} = 1/4$. The sum $P_{+,+} + P_{-,+} = 1/2$ does not display any interference, since one has measured in which cavity the neutron spin has flipped. This is very similar to an electron double-slit interference experiment if one measures which slit the electron goes through.

### 18.5.3  Detection of the Neutron Spin State

**18.5.3.1** By definition:

$$|a : \pm x\rangle = \frac{1}{\sqrt{2}}(|a : +\rangle \pm |a : -\rangle), \qquad |a : \pm y\rangle = \frac{1}{\sqrt{2}}(|a : +\rangle \pm i|a : -\rangle),$$

$$(18.26)$$

and these states are related to one another by

$$|a : \pm y\rangle = \frac{1}{2}((1 \pm i)|a : +x\rangle + (1 \mp i)|a : -x\rangle). \tag{18.27}$$

**18.5.3.2** The operators $\hat{S}_{nz}$ and $\hat{S}_{ax}$ commute since they act in two different Hilbert spaces; therefore $[\hat{S}_{nz}, \hat{V}] = 0$.

The common eigenvectors of $\hat{S}_{nz}$ and $\hat{V}$, and the corresponding eigenvalues are

$$\begin{aligned}
|n : +\rangle \otimes |a : \pm x\rangle && S_{nz} = +\hbar/2 && V = \pm A\hbar/2, \\
|n : -\rangle \otimes |a : \pm x\rangle && S_{nz} = -\hbar/2 && V = \mp A\hbar/2.
\end{aligned} \tag{18.28}$$

The operators $\hat{S}_{nz}$ and $\hat{V}$ form a complete set of commuting operators as far as spin variables are concerned.

**18.5.3.3** Expanding in terms of the energy eigenstates, one obtains for $|\psi(0)\rangle = |n : +\rangle \otimes |a : +y\rangle$:

$$|\psi(\tau)\rangle = \frac{1}{2}|n : +\rangle \otimes \left((1 + i)e^{-iA\tau/2}|a : +x\rangle + (1 - i)e^{iA\tau/2}|a : -x\rangle\right),$$

$$(18.29)$$

i.e. for $A\tau/2 = \pi/4$:

$$|\psi(\tau)\rangle = \frac{1}{\sqrt{2}}|n : +\rangle \otimes (|a : +x\rangle + |a : -x\rangle) = |n : +\rangle \otimes |a : +\rangle. \quad (18.30)$$

Similarly, if $|\psi(0)\rangle = |n : -\rangle \otimes |a : +y\rangle$, then $|\psi(\tau)\rangle = i|n : -\rangle \otimes |a : -\rangle$.

Physically, this means that the neutron's spin state does not change since it is an eigenstate of $\hat{V}$, while the atom's spin precesses around the $x$ axis with angular frequency $A$. At time $\tau = \pi/(2A)$, it lies along the $z$ axis.

**18.5.3.4** If the initial state is $|\psi(0)\rangle = (\alpha_+|n : +\rangle + \alpha_-|n : -\rangle) \otimes |a : +y\rangle$, the state after the interaction is

$$|\psi(\tau)\rangle = \alpha_+|n : +\rangle \otimes |a : +\rangle + i\,\alpha_-|n : -\rangle \otimes |a : -\rangle. \quad (18.31)$$

The measurement of the $z$ component of the atom's spin gives $+\hbar/2$, with probability $|\alpha_+|^2$ and state $|n : +\rangle \otimes |a : +\rangle$ after the measurement, or $-\hbar/2$ with probability $|\alpha_-|^2$ and state $|n : -\rangle \otimes |a : -\rangle$ after the measurement.

In both cases, after measuring the $z$ component of the atom's spin, the neutron spin state is known: it is the same as that of the measured atom. It is *not* necessary to let the neutron interact with another measuring apparatus in order to know the value of $S_{nz}$.

## 18.5.4 A Quantum Eraser

**18.5.4.1** The successive states are:

**(a)** $\frac{1}{\sqrt{2}}\left(-ie^{-i\delta}|n : +\rangle \otimes |a : +y\rangle + e^{i\chi}|n : -\rangle \otimes |a : +y\rangle\right)$

**(b)** $\frac{1}{\sqrt{2}}\left(-ie^{-i\delta}|n : +\rangle \otimes |a : +\rangle + ie^{i\chi}|n : -\rangle \otimes |a : -\rangle\right)$

**(c)** $\frac{1}{\sqrt{2}}\left(-ie^{-i(\delta+\omega_0 T/2)}|n : +\rangle \otimes |a : +\rangle\right.$

$\left. + ie^{i(\chi+\omega_0 T/2)}|n : -\rangle \otimes |a : -\rangle\right).$

Finally, when the neutron leaves the second cavity (step d), the state of the system is:

$$|\psi_f\rangle = \frac{1}{2}\left(-ie^{-i(\delta+\omega_0 T/2)}\left(e^{-i\chi}|n : +\rangle - ie^{i\delta'}|n : -\rangle\right) \otimes |a : +\rangle\right.$$

$$\left. + ie^{i(\chi+\omega_0 T/2)}\left(-ie^{-i\delta'}|n : +\rangle + e^{i\chi}|n : -\rangle\right) \otimes |a : -\rangle\right). \quad (18.32)$$

**18.5.4.2** The probability to find the neutron in state $|+\rangle$ is the sum of the probabilities for finding:

(a) the neutron in state $+$ and the atom in state $+$, i.e. the square of the modulus of the coefficient of $|n : +\rangle \otimes |a : +\rangle$ (1/4 in the present case),

(b) the neutron in state $+$ and the atom in state $-$ (probability 1/4 again). One gets therefore $P_+ = 1/4 + 1/4 = 1/2$: There are no interferences since the quantum path leading in the end to a spin flip of the neutron can be determined from the state of the atom.

**18.5.4.3** One can expand the vectors $|a : \pm\rangle$ on $|a : \pm y\rangle$:

$$|\psi_f\rangle = \frac{1}{2\sqrt{2}} \left(-ie^{-i(\delta + \omega_0 T/2)} \left(e^{-i\chi}|n : +\rangle - ie^{i\delta'}|n : -\rangle\right)\right.$$
$$\otimes (|a : +y\rangle + |a : -y\rangle)$$
$$\left. + e^{i(\chi + \omega_0 T/2)} \left(-ie^{-i\delta'}|n : +\rangle + e^{i\chi}|n : -\rangle\right)\right. \tag{18.33}$$
$$\otimes (|a : +y\rangle - |a : -y\rangle))$$

The probability amplitude that Bob finds $+\hbar/2$ along the $z$ axis while Alice finds $+\hbar/2$ along the $y$ axis is the coefficient of the term $|n : +\rangle \otimes |a : +y\rangle$ in the above expansion. Equivalently, the probability is obtained by projecting the state onto $|n : +\rangle \otimes |a : +y\rangle$, and squaring. One obtains

$$P\left(S_{nz} = \frac{\hbar}{2}, S_{ay} = \frac{\hbar}{2}\right) = \frac{1}{8} \left|-ie^{-i(\delta + \chi + \omega_0 T/2)} - ie^{i(\chi - \delta' + \omega_0 T/2)}\right|^2$$
$$= \frac{1}{2}\cos^2 \frac{(\omega - \omega_0)T}{2}, \tag{18.34}$$

which clearly exhibits a modulation reflecting an interference phenomenon. Similarly, one finds that

$$P\left(S_{nz} = \frac{\hbar}{2}, S_{ay} = -\frac{\hbar}{2}\right) = \frac{1}{2}\sin^2 \frac{(\omega - \omega_0)T}{2}, \tag{18.35}$$

which is also modulated.

**18.5.4.4** This result is compatible with the result 4.2. Indeed the sum of the two probabilities calculated above is 1/2 as in 4.2. If Bob does not know the result found by Alice, or if Alice does not perform a measurement, which is equivalent from his point of view, Bob sees no interferences. The interferences only arise for the joint probability $P(S_{nz}, S_{ay})$.

**18.5.4.5** (a) The first statement is obviously wrong. As seen in question 4.2, if the atom $A$ is present, Bob no longer sees oscillations (in $\omega - \omega_0$) of the probability for detecting the neutron in the state $|+\rangle$. This probability is equal to 1/2 whatever Alice does. Notice that if the statement were correct, this would imply instantaneous transmission of information from Alice to Bob. By seeing interferences appear, Bob would know immediately that Alice is performing an experiment, even though she may be very far away.

(b) The second statement is correct. If Alice and Bob put together all their results, and if they select the subsample of events for which Alice finds $+\hbar/2$, then the number of events for which Bob also finds $+\hbar/2$ varies like $\cos^2((\omega - \omega_0)T/2)$; they recover interferences for this subset of events. In the complementary set, where Alice has found $-\hbar/2$, the number of Bob's results giving $+\hbar/2$ varies like $\sin^2((\omega - \omega_0)T/2)$. This search for correlations between events occurring in different detectors is a common procedure, in particle physics for instance.

(c) The third statement, although less precise but more picturesque than the previous one, is nevertheless acceptable. The $\cos^2((\omega - \omega_0)T/2)$ signal found in Sect. 18.2 can be interpreted as the interference of the amplitudes corresponding to two quantum paths for the neutron spin which is initially in the state $|n : -\rangle$; either its spin flips in the first cavity, or it flips in the second one. If there exists a possibility to determine which quantum path is followed by the system, interferences cannot appear. It is necessary to "erase" this information, which is carried by the atom, in order to observe "some" interferences. After Alice has measured the atom's spin along the $y$ axis, she has, in some sense "restored" the initial state of the system, and this enables Bob to see some interferences. It is questionable to say that information has been erased: one may feel that, on the contrary, extra information has been acquired. Notice that the statement in the text does not specify in which physical quantity the interferences appear. Notice also that the order of the measurements made by Alice and Bob has no importance, contrary to what this third statement seems to imply.

**18.5.4.6** Alice can measure along the axis $w = \sin \eta\, u_y + \cos \eta\, u_z$, in the $(y, z)$ plane, for instance. Projecting $|\psi_f\rangle$ onto the eigenstate of $\hat{S}_{aw}$ with eigenvalue $+\hbar/2$, i.e. $\cos(\eta/2)|a : +\rangle + i \sin(\eta/2)|a : -\rangle$, a calculation similar to 4.3 leads to a probability $[1 + \sin \eta \cos((\omega - \omega_0)T)]/2$. If $\eta = 0$ or $\pi$ (measurement along the $z$ axis) there are no interferences. For $\eta = \pi/2$ and $3\pi/2$ or, more generally, if Alice measures in the $(x, y)$ plane, the contrast of the interferences, $|\sin \eta|$, is maximum.

## 18.5.5 Comments

This chapter covers several interesting physical phenomena, which have all led to important developments:

- *Ramsey Fringes with Neutrons.* The experimental curve given in the text is taken from J.H. Smith et al., Phys. Rev. **108**, 120 (1957). Since then, the technique of Ramsey fringes has been considerably improved. Nowadays one proceeds differently. One stores neutrons in a "bottle" for a time of the order of 100 s and applies two radiofrequency pulses at the beginning and at the end of the storage. The elapsed time between the two pulses is 70 s, compared to 1.3 ms here. This improves enormously the accuracy of the frequency measurement. Such experiments are actually devised to measure the *electric* dipole moment of the neutron, of fundamental interest in relation to time-reversal invariance. They set a very small upper bound on this quantity (K.F. Smith et al., Phys. Lett. **234**, 191 (1990)).

- *Non Destructive Quantum Measurements.* The structure of the interaction Hamiltonian considered in the text has been chosen in order to provide a simple description of the quantum eraser effect. Realistic examples of nondestructive quantum measurements can be found in J. P. Poizat and P. Grangier, Phys. Rev. Lett. **70**, 271 (1993), and S.M. Barnett, Nature, Vol. 362, p. 113, March 1993.

- *Quantum eraser.* The notion of a quantum eraser was put forward by M. O. Scully and K. Drühl, *Quantum eraser: A proposed photon correlation experiment concerning observation and "delayed choice" in quantum mechanics*, Phys. Rev. A 25, 2208 (1982). As the title of this paper indicates, it is closely related to the idea of a "delayed choice", put forward by J. A. Wheeler in *Quantum Theory and Measurement*, edited by J.A. Wheeler and W. H. Zurek, Princeton Univ. Press (1983). For experimental implementations of this concept, see Y-H. Kim, R. Yu, S. P. Kulik, Y. Shih, and M. O. Scully, *Delayed Choice Quantum Eraser*, Phys. Rev. Lett. 84, 1 (2000), V. Jacques, E. Wu, F. Grosshans, F. Treussart, P. Grangier, A. Aspect, and J.-F. Roch, *Experimental Realization of Wheeler's Delayed-Choice Gedanken Experiment*, Science **315**, 966 (2007), and the discussion in M. Schirber, *Another Step Back for Wave-Particle Duality* Physics **4**, 102 (2011)

# A Quantum Thermometer

# 19

We study here the measurement of the cyclotron motion of an electron. The particle is confined in a Penning trap and it is coupled to the thermal radiation which causes quantum transitions of the system between various energy levels. In all the chapter, we neglect spin effects. The method and results come from an experiment performed at Harvard University in 1999.

We consider an electron of mass $M$ and charge $q$ ($q < 0$), confined in a Penning trap. In all the chapter, we neglect spin effects. The Penning trap consists in the superposition of a uniform magnetic field $\boldsymbol{B} = B\boldsymbol{e}_z$ ($B > 0$) and an electric field which derives from the potential $\Phi(\boldsymbol{r})$ whose power expansion near the origin is:

$$\Phi(\boldsymbol{r}) = \frac{M\omega_z^2}{4q}\left(2z^2 - x^2 - y^2\right). \tag{19.1}$$

The positive quantity $\omega_z$ has the dimension of an angular frequency. In all this chapter we set $\omega_c = |q|B/M$ ($\omega_c$ is called the cyclotron angular frequency) and we assume that $\omega_z \ll \omega_c$.

*Useful constants*: $M = 9.1 \ 10^{-31}$ kg ; $q = -1.6 \ 10^{-19}$ C; $h = 6.63 \ 10^{-34}$ J s; Boltzmann's constant $k_B = 1.38 \ 10^{-23}$ J K$^{-1}$.

## 19.1 The Penning Trap in Classical Mechanics

We recall that the Lorentz force acting on a charged particle moving in an electromagnetic field is $\mathbf{F} = q(\mathbf{E} + \boldsymbol{v} \times \mathbf{B})$.

**19.1.1** Check that $\Phi(\boldsymbol{r})$ satisfies the Laplace equation $\Delta\Phi = 0$. What is the shape of a surface of constant potential $\Phi(\boldsymbol{r}) = $ Const?

© Springer Nature Switzerland AG 2019
J.-L. Basdevant, J. Dalibard, *The Quantum Mechanics Solver*,
https://doi.org/10.1007/978-3-030-13724-3_19

**19.1.2** Show that the classical equation of motion of the electron in the trap is:

$$\ddot{x} + \omega_c \dot{y} - \frac{\omega_z^2}{2} x = 0 \qquad \ddot{y} - \omega_c \dot{x} - \frac{\omega_z^2}{2} y = 0 \qquad \ddot{z} + \omega_z^2 z = 0. \qquad (19.2)$$

**19.1.3** What is the type of motion along the $z$ axis?

**19.1.4** In order to study the component of the motion in the $xy$ plane, we set $\alpha = x + iy$.

(a) What is the differential equation satisfied by $\alpha(t)$?
(b) We seek a solution of this equation of the form $\alpha(t) = \alpha_0\, e^{i\omega t}$. Show that $\omega$ is a solution of the equation:

$$\omega^2 - \omega_c \omega + \frac{\omega_z^2}{2} = 0. \qquad (19.3)$$

(c) We note $\omega_r$ and $\omega_l$ the two roots of this equation with $\omega_r > \omega_l$. Show that:

$$\omega_r \simeq \omega_c \qquad \omega_l \simeq \frac{\omega_z^2}{2\omega_c}. \qquad (19.4)$$

**19.1.5** We consider the values $B = 5.3$ T and $\omega_z/(2\pi) = 64$ MHz.

(a) Show that the most general motion of the electron in the Penning trap is the superposition of three harmonic oscillator motions.
(b) Calculate the frequencies of these motions.
(c) Draw the projection on the $xy$ plane of the classical trajectory of the trapped electron, assuming that $\alpha_r \ll \alpha_l$ (the positive quantities $\alpha_r$ and $\alpha_l$ represent the amplitudes of the motions of angular frequencies $\omega_r$ and $\omega_l$).

## 19.2   The Penning Trap in Quantum Mechanics

We note $\hat{r}$ and $\hat{p}$ the position and momentum operators of the electron. The Hamiltonian of the electron in the Penning trap is, neglecting spin effects:

$$\hat{H} = \frac{1}{2M} (\hat{p} - q\,\mathbf{A}(\hat{r}))^2 + q\,\Phi(\hat{r}), \qquad (19.5)$$

where the electrostatic potential $\Phi(r)$ is given by (19.1). For the magnetic vector potential, we choose the form $\mathbf{A}(r) = \mathbf{B} \times r/2$.

**19.2.1** Expand the Hamiltonian and show that it can be written as $\hat{H} = \hat{H}_{xy} + \hat{H}_z$, where $\hat{H}_{xy}$ only involves the operators $\hat{x}$, $\hat{y}$, $\hat{p}_x$ and $\hat{p}_y$, while $\hat{H}_z$ only involves the operators $\hat{z}$ and $\hat{p}_z$.

Do $\hat{H}_{xy}$ and $\hat{H}_z$ possess a common eigenbasis?

**19.2.2** We are now interested in the motion along the $z$ axis. This is called the axial motion. Recall without giving any proof:

(a) the expression of the operators $\hat{a}_z$ and $\hat{a}_z^\dagger$ which allow to write $\hat{H}_z$ in the form $\hat{H}_z = \hbar\omega_z(\hat{N}_z + 1/2)$ with $\hat{N}_z = \hat{a}_z^\dagger\hat{a}_z$ and $[\hat{a}_z, \hat{a}_z^\dagger] = 1$;
(b) the eigenvalues of $\hat{N}_z$ and $\hat{H}_z$.

**19.2.3** We now consider the motion in the $xy$ plane under the effect of the Hamiltonian $\hat{H}_{xy}$. We set $\Omega = \sqrt{\omega_c^2 - 2\omega_z^2}/2$. We introduce the right and left annihilation operators $\hat{a}_r$ and $\hat{a}_l$:

$$\hat{a}_r = \sqrt{\frac{M\Omega}{4\hbar}}(\hat{x} - i\hat{y}) + \frac{i}{\sqrt{4\hbar M\Omega}}(\hat{p}_x - i\hat{p}_y), \tag{19.6}$$

$$\hat{a}_l = \sqrt{\frac{M\Omega}{4\hbar}}(\hat{x} + i\hat{y}) + \frac{i}{\sqrt{4\hbar M\Omega}}(\hat{p}_x + i\hat{p}_y). \tag{19.7}$$

(a) Show that $[\hat{a}_r, \hat{a}_r^\dagger] = [\hat{a}_l, \hat{a}_l^\dagger] = 1$.
(b) Show that any left operator commutes with any right operator, i.e.:

$$[\hat{a}_r, \hat{a}_l] = 0 \quad [\hat{a}_r, \hat{a}_l^\dagger] = 0 \quad [\hat{a}_r^\dagger, \hat{a}_l] = 0 \quad [\hat{a}_r^\dagger, \hat{a}_l^\dagger] = 0. \tag{19.8}$$

(c) Recall the eigenvalues of $\hat{n}_r = \hat{a}_r^\dagger\hat{a}_r$ and $\hat{n}_l = \hat{a}_l^\dagger\hat{a}_l$ (no proof is required). Do $\hat{n}_r$ and $\hat{n}_l$ possess a common eigenbasis?
(d) Show that the Hamiltonian $\hat{H}_{xy}$ can be written as:

$$\hat{H}_{xy} = \hbar\omega_r(\hat{n}_r + 1/2) - \hbar\omega_l(\hat{n}_l + 1/2), \tag{19.9}$$

where the angular frequencies $\omega_r$ and $\omega_l$ have been introduced in Sect. 19.1.
(e) Deduce from this the eigenvalues of the Hamiltonian $\hat{H}_{xy}$.

**19.2.4** We note $|\psi(t)\rangle$ the state of the system at time $t$ and we define $a_r(t) = \langle\psi(t)|\hat{a}_r|\psi(t)\rangle$ and $a_l(t) = \langle\psi(t)|\hat{a}_l|\psi(t)\rangle$. Using the Ehrenfest theorem, calculate $da_r/dt$ and $da_l/dt$.

Integrate these equations and calculate the expectation value of the electron's position $(\langle x\rangle(t), \langle y\rangle(t))$ in the $xy$ plane. We set $a_r(0) = \rho_r\,e^{-i\phi_r}$ and $a_l(0) = \rho_l\,e^{i\phi_l}$, where $\rho_r$ and $\rho_l$ are real and positive.

Show that the time evolution of the expectation value of the electron position $\langle r \rangle (t)$ is similar to the classical evolution found in Sect. 19.1.

**19.2.5** We note $|\phi_0\rangle$ the eigenstate of $\hat{H}$ corresponding to the eigenvalues 0 for each of the operators $\hat{n}_r$, $\hat{n}_l$ and $\hat{N}_z$.

(a) Determine the corresponding wave function $\phi_0(r)$ (it is not necessary to normalize the result).
(b) Using the same numerical values as in the previous section, evaluate the spatial extension of $\phi_0(r)$.

**19.2.6** The experiment is performed at temperatures $T$ ranging between 0.1 and 4 K. Compare the characteristic thermal energy $k_B T$ to each of the energy quanta of the "cyclotron", "axial" and "magnetron" motions (associated respectively with $\hat{n}_r$, $\hat{N}_z$ and $\hat{n}_l$). For which of these motions does the discrete nature of the energy spectrum play an important role?

---

## 19.3　Coupling of the Cyclotron and Axial Motions

We now study a method for detecting the cyclotron motion. This method uses a small coupling between this motion and the axial motion. The coupling is produced by an inhomogeneous magnetic field, and it can be described by the additional term in the Hamiltonian:

$$\hat{W} = \frac{\epsilon}{2} M \omega_z^2 \hat{n}_r \hat{z}^2. \tag{19.10}$$

The experimental conditions are chosen such that $\epsilon = 4 \times 10^{-7}$.

**19.3.1** Write the total Hamiltonian $\hat{H}_c = \hat{H} + \hat{W}$ using the operators $\hat{n}_r$, $\hat{n}_l$, $\hat{p}_z$ and $\hat{z}$.

**19.3.2** Show that the excitation numbers of the cyclotron motion ($\hat{n}_r$) and of the magnetron motion ($\hat{n}_l$) are constants of the motion.

**19.3.3** Consider the eigensubspace $\mathcal{E}_{n_r, n_l}$ of $\hat{n}_r$ and $\hat{n}_l$, corresponding to the eigenvalues $n_r$ and $n_l$.

(a) Write the form of $\hat{H}_c$ in this subspace.
(b) Show that the axial motion is harmonic if the system is prepared in a state belonging to $\mathcal{E}_{n_r, n_l}$. Give its frequency in terms of $n_r$ and $n_l$.
(c) Give the eigenvalues and eigenstates of $\hat{H}_c$ inside $\mathcal{E}_{n_r, n_l}$.

**19.3.4** Deduce from the previous question that the eigenstates of $\hat{H}_c$ can be labeled by 3 quantum numbers, $n_r, n_l, n_z$. We write these states as $|n_r, n_l, n_z\rangle$. Give the energy eigenvalues in terms of these quantum numbers and of $\omega_r$, $\omega_l$, $\omega_z$ and $\epsilon$.

**19.3.5** One measures the beat between a highly stable oscillator of frequency $\omega_z/(2\pi)$ (delivering a signal proportional to $\sin(\omega_z t)$) and the current induced in an electric circuit by the axial motion. This latter current is proportional to $\langle p_z\rangle(t)$.

(a) Calculate the time evolution of the expectation values of the position and momentum operators $\hat{z}$ and $\hat{p}_z$ assuming that the state of the electron is restricted to be in the subspace $E_{n_r,n_l}$. We choose the initial conditions $\langle \hat{z}\rangle(t=0) = z_0$ and $\langle \hat{p}_z\rangle(t=0) = 0$.
(b) To first order in $\epsilon$, what is the phase difference $\varphi$ between the detected current and the stable oscillator after a time $\tau$? Show that the measurement of this phase difference provides a measurement of the excitation number of the cyclotron motion.

**19.3.6** We now assume that the electron is in an arbitrary state

$$|\Psi\rangle = \sum_{n_r,n_l,n_z} c_{n_r,n_l,n_z}|n_r, n_l, n_z\rangle. \tag{19.11}$$

(a) We measure the phase difference $\varphi$ on a time interval ranging from $t=0$ to $t=\tau$. What are the possible results $\varphi_k$ of the measurement? Show that this provides a means to determine the excitation number of the cyclotron motion.
(b) What is the state of the electron after a measurement giving the result $\varphi_k$?
(c) We choose $\tau = 0.1$ s and we assume that the measurement of $\varphi$ is done with an accuracy of $\pi/10$. Using the previous values of the physical parameters, show that this accuracy leads to an unambiguous determination of the cyclotron excitation number.
(d) After a measurement giving the result $\varphi_k$, we let system evolve for a length of time $T$ under the action of the Hamiltonian $\hat{H}_c$. We then perform a new measurement of $\varphi$. What results do we expect?

## 19.4   A Quantum Thermometer

In practice, the cyclotron motion is in thermal equilibrium with a thermostat at temperature $T$. We recall that, in that situation, the thermal fluctuations can excite the system in an energy level $E_n$ with some probability $p_n$.

We perform successive measurements of the phase difference $\varphi$ in the time intervals $[0, \tau], [\tau, 2\tau], \ldots, [(N-1)\tau, N\tau]$. The total duration $N\tau$ of this series of measurements for a given temperature $T$ is $N\tau = 3000$ s, i.e. a total number of

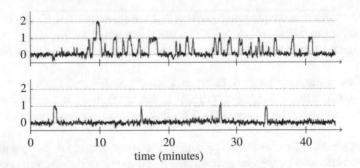

**Fig. 19.1** Time evolution of the quantum number $n_r$ corresponding to the cyclotron motion for two temperatures $T_a$ and $T_b$

results $N = 3 \times 10^4$ for $\tau = 0.1$ s. By this procedure, one can follow the variation of $n_r$ during the time interval $N\tau$, with a time resolution of $\tau$.

**19.4.1** Two recordings of this measurement are represented on Fig. 19.1 for two different temperatures. Comment on these results and explain in particular:

(a) to what phenomenon are associated the sudden changes of the signal;
(b) what is the fraction of the time during which the electron is in the levels $n_r = 0, n_r = 1, n_r = 2, \ldots$ (the accuracy of a usual graduated ruler is sufficient).

**19.4.2** The probability $p_n$ for a system to be in the energy level $E_n$ is given by the Boltzmann factor $p_n = \mathcal{N} \exp(-E_n/k_B T)$, where $\mathcal{N}$ is a normalization factor. Show that for a one-dimensional harmonic oscillator the ratio $p_{n+1}/p_n$ does not depend on $n$.

**19.4.3** Give an estimate of the two temperatures corresponding to the two recordings of Fig. 19.1.

**19.4.4** Determine the normalization factor $\mathcal{N}$ of the probability law $p_n$ for a one-dimensional harmonic oscillator of angular frequency $\omega$ in thermal equilibrium with a thermostat at temperature $T$, and calculate the average excitation number $\bar{n}$. It is convenient to set $\gamma = \hbar\omega/(k_B T)$.

**19.4.5** What is the order of magnitude of the lowest temperature one can measure with such a device?

**19.4.6** How can one improve the sensitivity of such a "quantum thermometer"?

## 19.5 Solutions

### 19.5.1 The Penning Trap in Classical Mechanics

**19.5.1.1** The electric field $E = -\nabla\Phi$ is:

$$E(r) = \frac{M\omega_z^2}{2q} \begin{pmatrix} x \\ y \\ -2z \end{pmatrix} \tag{19.12}$$

Therefore, one has $\Delta\Phi = -\nabla \cdot E = -\frac{M\omega_z^2}{2q}(1 + 1 - 2) = 0$. The potential satisfies Laplace's equation in the vacuum. The surfaces of constant potential are hyperboloids of axis $z$.

**19.5.1.2** Using the expression (19.12) for the electric field, the equation of motion is:

$$M \begin{pmatrix} \ddot{x} \\ \ddot{y} \\ \ddot{z} \end{pmatrix} = \frac{M\omega_z^2}{2} \begin{pmatrix} x \\ y \\ -2z \end{pmatrix} + q \begin{pmatrix} \dot{x} \\ \dot{y} \\ \dot{z} \end{pmatrix} \times \begin{pmatrix} 0 \\ 0 \\ B \end{pmatrix}, \tag{19.13}$$

or, by setting $\omega_c = -qB/M$:

$$\ddot{x} + \omega_c\dot{y} - \frac{\omega_z^2}{2}x = 0 \qquad \ddot{y} - \omega_c\dot{x} - \frac{\omega_z^2}{2}y = 0 \qquad \ddot{z} + \omega_z^2 z = 0. \tag{19.14}$$

**19.5.1.3** Along the $z$ axis, the motion is harmonic, of angular frequency $\omega_z$.

**19.5.1.4 (a)** The differential equation satisfied by $\alpha(t)$ is:

$$\ddot{\alpha} - i\omega_c\dot{\alpha} - \frac{\omega_z^2}{2}\alpha = 0 \tag{19.15}$$

**(b)** If we search a solution of the form $\alpha_0\, e^{i\omega t}$, we find that $\omega$ is given by the equation:

$$-\omega^2 + \omega_c\omega - \frac{\omega_z^2}{2} = 0 \tag{19.16}$$

**(c)** The roots of this equation are:

$$\omega_r = \frac{1}{2}\left(\omega_c + \sqrt{\omega_c^2 - 2\omega_z^2}\right) \qquad \omega_l = \frac{1}{2}\left(\omega_c - \sqrt{\omega_c^2 - 2\omega_z^2}\right) \tag{19.17}$$

We assume $\omega_z \ll \omega_c$, i.e. $\sqrt{\omega_c^2 - 2\omega_z^2} \simeq \omega_c \left(1 - \omega_z^2/\omega_c^2\right)$. The two roots $\omega_r$ and $\omega_l$ are given approximately by:

$$\omega_r \simeq \omega_c \qquad \omega_l \simeq \frac{\omega_z^2}{2\omega_c}. \tag{19.18}$$

**19.5.1.5 (a)** We have seen previously that the motion along $z$ is harmonic of angular frequency $\omega_z$. In order to obtain the motion in the $xy$ plane, we integrate the equation of motion of $\alpha$:

$$\alpha(t) = \alpha_l \, e^{i(\omega_l t + \phi_l)} + \alpha_r \, e^{i(\omega_r t + \phi_r)}, \tag{19.19}$$

where $\alpha_l$ and $\alpha_r$ are two positive real numbers, and $\phi_l$ and $\phi_r$ are two *a priori* arbitrary phases. This gives the forms of $x(t)$ and $y(t)$ by taking the real and imaginary parts of this expression:

$$x(t) = \alpha_l \cos(\omega_l t + \phi_l) + \alpha_r \cos(\omega_r t + \phi_r) \tag{19.20}$$

$$y(t) = \alpha_l \sin(\omega_l t + \phi_l) + \alpha_r \sin(\omega_r t + \phi_r). \tag{19.21}$$

The motion in the $xy$ plane is a superposition of two harmonic motions of angular frequencies $\omega_r$ and $\omega_l$.

**(b)** One finds $\omega_c/2\pi = 1.48 \times 10^{11}$ Hz. The frequencies of the three motions are therefore:

$$\omega_l/2\pi \simeq 14\,\text{kHz} \qquad \omega_z/2\pi = 64\,\text{MHz} \qquad \omega_r/2\pi \simeq 150\,\text{GHz} \tag{19.22}$$

**(c)** The motion in the $xy$ plane is the superposition of two circular motions, one is of radius $\alpha_r$ and has a high frequency ($\omega_r$), and the other is of radius $\alpha_l$ and has a much lower frequency ($\omega_l$). Assuming $\alpha_r \ll \alpha_l$, this results in a trajectory of the type represented on the following figure:

## 19.5.2 The Penning Trap in Quantum Mechanics

**19.5.2.1** The expansion of the Hamiltonian gives $\hat{H} = \hat{H}_{xy} + \hat{H}_z$ with:

$$\hat{H}_{xy} = \frac{\hat{p}_x^2}{2M} + \frac{\hat{p}_y^2}{2M} + \frac{M}{8}(\omega_c^2 - 2\omega_z^2)(\hat{x}^2 + \hat{y}^2) + \frac{\omega_c}{2}\hat{L}_z \quad \hat{H}_z = \frac{\hat{p}_z^2}{2M} + \frac{1}{2}M\omega_z^2\hat{z}^2,$$
$$(19.23)$$

where we have introduced the $z$ component of the angular momentum $\hat{L}_z = \hat{x}\hat{p}_y - \hat{y}\hat{p}_x$. Since $\hat{H}_{xy}$ involves only the operators $\hat{x}, \hat{y}$ and $\hat{p}_x, \hat{p}_y$, and since $\hat{H}_z$ only involves $\hat{z}, \hat{p}_z$, the two operators $\hat{H}_{xy}$ and $\hat{H}_z$ commute, and each of them commutes with the total Hamiltonian $\hat{H}$:

$$[\hat{H}_{xy}, \hat{H}_z] = 0 \qquad [\hat{H}_{xy}, \hat{H}] = 0 \qquad [\hat{H}_z, \hat{H}] = 0. \qquad (19.24)$$

We can therefore search for an eigenbasis of $\hat{H}$ in the form of a common eigenbasis of $\hat{H}_{xy}$ and $\hat{H}_z$.

**19.5.2.2 (a)** The Hamiltonian $\hat{H}_z$ corresponds to a harmonic motion of angular frequency $\omega_z$. Setting

$$\hat{a}_z = \sqrt{\frac{M\omega_z}{2\hbar}}\hat{x} + i\frac{\hat{p}}{\sqrt{2M\hbar\omega_z}} \quad \text{and} \quad \hat{N}_z = \hat{a}_z^\dagger\hat{a}_z, \qquad (19.25)$$

one easily finds its spectrum since the Hamiltonian can be written as $\hat{H}_z = \hbar\omega_z(\hat{N}_z + 1/2)$.

**(b)** From the commutation relation $[\hat{a}_z, \hat{a}_z^\dagger] = 1$, we deduce that the eigenvalues of $\hat{N}_z$ are the non-negative integers $n_z$. The eigenvalues of $\hat{H}_z$ are therefore of the form $\hbar\omega_z(n_z + 1/2)$.

**19.5.2.3** We first calculate the generic commutator:

$$C = \left[ \sqrt{\frac{M\Omega}{4\hbar}}(\hat{x} - i\eta\hat{y}) + \xi\frac{i}{\sqrt{4\hbar M\Omega}}(\hat{p}_x - i\eta\hat{p}_y), \right.$$

$$\left. \sqrt{\frac{M\Omega}{4\hbar}}(\hat{x} + i\eta'\hat{y}) - \xi'\frac{i}{\sqrt{4\hbar M\Omega}}(\hat{p}_x + i\eta'\hat{p}_y) \right] \qquad (19.26)$$

where the four numbers $\eta$, $\xi$, $\eta'$ and $\xi'$ are equal to $\pm 1$. Using $[\hat{x}, \hat{p}_x] = [\hat{y}, \hat{p}_y] = i\hbar$, one finds:

$$C = \frac{1}{4}(\xi + \xi')(1 + \eta\eta').$$  (19.27)

**(a)** The commutator $[\hat{a}_r, \hat{a}_r^\dagger]$ corresponds to $\eta = \eta' = +1$ and $\xi = \xi' = 1$, therefore $[\hat{a}_r, \hat{a}_r^\dagger] = 1$. Similarly, one obtains $[\hat{a}_l, \hat{a}_l^\dagger] = 1$ from $\eta = \eta' = -1$ and $\xi = \xi' = 1$.

**(b)** The commutator $[\hat{a}_r, \hat{a}_l]$ corresponds to $\xi = 1$ and $\xi' = -1$, hence $[\hat{a}_r, \hat{a}_l] = 0$. Similarly, $[\hat{a}_r, \hat{a}_l^\dagger]$ vanishes since it corresponds to $\eta = -\eta' = 1$. Therefore the other commutators ($[\hat{a}_r^\dagger, \hat{a}_l]$ and $[\hat{a}_r^\dagger, \hat{a}_l^\dagger]$) also vanish if we take the Hermitian conjugates of the previous commutators.

**(c)** The commutation relation $[\hat{a}_r, \hat{a}_r^\dagger] = 1$ results in the fact that the eigenvalues of $\hat{n}_r$ are the non-negative integers, and the same holds for $\hat{n}_l$.

**(d)** The operators $\hat{n}_r$ and $\hat{n}_l$ are expanded as:

$$\hat{n}_{r,l} = \frac{\hat{p}_x^2 + \hat{p}_y^2}{4\hbar M\Omega} + \frac{M\Omega}{4\hbar}(\hat{x}^2 + \hat{y}^2) - \frac{1}{2} \pm \frac{\hat{L}_z}{2\hbar}$$  (19.28)

where the $+$ (resp.$-$) sign corresponds to $\hat{n}_r$ (resp. $\hat{n}_l$). The sum and difference of the roots of the equation $\omega^2 - \omega_c\omega + \omega_z^2/2 = 0$ are:

$$\omega_r + \omega_l = \omega_c \qquad \omega_r - \omega_l = \sqrt{\omega_c^2 - 2\omega_z^2} = 2\Omega.$$  (19.29)

We obtain the anticipated result:

$$\hat{H}_{xy} = \hbar\omega_r(\hat{n}_r + 1/2) - \hbar\omega_l(\hat{n}_l + 1/2)$$  (19.30)

**(e)** The eigenvectors of $\hat{H}_{xy}$ can therefore be labeled by the two (non-negative) integer quantum numbers $n_r$ and $n_l$ corresponding to the eigenvalues of $\hat{n}_r$ and $\hat{n}_l$. The corresponding eigenstates are noted $|n_r, n_l\rangle$. The eigenvalue of $H_{xy}$ associated to the vector $|n_r, n_l\rangle$ is $\hbar\omega_r(n_r + 1/2) - \hbar\omega_l(n_l + 1/2)$.

**19.5.2.4**  We have:

$$[\hat{a}_r, \hat{H}] = \hbar\omega_r[\hat{a}_r, \hat{a}_r^\dagger\hat{a}_r] = \hbar\omega_r\hat{a}_r.$$  (19.31)

Owing to Ehrenfest theorem we therefore find: $\dot{a}_r = -i\omega_r a_r$ and $\dot{a}_l = +i\omega_l a_l$. The solutions of these two equations are:

$$a_r(t) = a_r(0)\, e^{-i\omega_r t} \quad a_l(t) = a_l(0)\, e^{+i\omega_l t} \tag{19.32}$$

The expectation value of the position in the $xy$ plane can be calculated by using:

$$a_r + a_l = \sqrt{\frac{M\Omega}{\hbar}}\langle x\rangle + \frac{i}{\sqrt{\hbar M\Omega}}\langle p_x\rangle \quad i(a_r - a_l) = \sqrt{\frac{M\Omega}{\hbar}}\langle y\rangle + \frac{i}{\sqrt{\hbar M\Omega}}\langle p_y\rangle. \tag{19.33}$$

In other words:

$$\langle x\rangle(t) = \sqrt{\frac{\hbar}{M\Omega}}\,\mathrm{Re}(a_r(t) + a_l(t)) \quad \langle y\rangle(t) = \sqrt{\frac{\hbar}{M\Omega}}\,\mathrm{Re}(ia_r(t) - ia_l(t)). \tag{19.34}$$

Setting $a_r(0) = \rho_r\, e^{-i\phi_r}$ and $a_l(0) = \rho_l\, e^{i\phi_l}$, we obtain:

$$\langle x\rangle(t) = \sqrt{\frac{\hbar}{M\Omega}}(\rho_r \cos(\omega_r t + \phi_r) + \rho_l \cos(\omega_l t + \phi_l)) \tag{19.35}$$

$$\langle y\rangle(t) = \sqrt{\frac{\hbar}{M\Omega}}(\rho_r \sin(\omega_r t + \phi_r) + \rho_l \sin(\omega_l t + \phi_l)) \tag{19.36}$$

As in the classical motion, the coordinates $\langle x\rangle$ and $\langle y\rangle$ are the sums of two sinusoidal functions of angular frequencies $\omega_r$ and $\omega_l$. The $x$ and $y$ components have equal amplitudes and are phase shifted by $\pi/2$ with respect to each other. The average motion in the $xy$ plane is therefore the superposition of two uniform circular motions of angular frequencies $\omega_r$ and $\omega_l$. The trajectory is the same as in question 19.1.5.

**19.5.2.5 (a)** The wave function $\phi_0(\mathbf{r})$ corresponding to $n_r = n_l = n_z = 0$ can be written as the product of three functions in the variables $x + iy$, $x - iy$ and $z$. This function must satisfy $\hat{a}_\mu \phi_0(\mathbf{r}) = 0$, with $\mu = r, l, z$. Therefore, setting $\eta = +$ and $\eta = -$ for $\hat{a}_r$ and $\hat{a}_l$ respectively:

$$\left(\frac{\partial}{\partial x} - i\eta\frac{\partial}{\partial y} + \frac{M\Omega}{\hbar}(x - i\eta y)\right)\phi_0(\mathbf{r}) = 0 \quad \left(\frac{\partial}{\partial z} + \frac{M\omega_z}{\hbar}z\right)\phi_0(\mathbf{r}) = 0. \tag{19.37}$$

By adding and subtracting the two equation for $\eta = \pm$, we obtain:

$$\left(\frac{\partial}{\partial x} + \frac{M\Omega}{\hbar}x\right)\phi_0(\mathbf{r}) = 0 \quad \left(\frac{\partial}{\partial y} + \frac{M\Omega}{\hbar}y\right)\phi_0(\mathbf{r}) = 0 \tag{19.38}$$

The function $\phi_0(\mathbf{r})$ is therefore a product of three gaussian functions in the variables $x, y, z$:

$$\phi_0(\mathbf{r}) \propto e^{-(x^2+y^2)/4r_0^2} \, e^{-z^2/4z_0^2} \tag{19.39}$$

with

$$r_0 = \sqrt{\frac{\hbar}{2M\Omega}} \quad \text{and} \quad z_0 = \sqrt{\frac{\hbar}{2M\omega_z}}. \tag{19.40}$$

**(b)** The probability distribution $|\phi_0(\mathbf{r})|^2$ is centered at $\mathbf{r} = 0$. Its extension along the $x$ and $y$ axes is $\Delta x = \Delta y = r_0 \simeq 11\,$nm. Along the $z$ axis, we have $\Delta z = z_0 \simeq 380\,$nm.

**19.5.2.6** For the range of temperatures of interest the ratio $k_B T / \hbar\omega_\mu$ for $\mu = r, l, z$ is

$$\frac{k_B T}{\hbar\omega_r} = 0.014 - 0.6 \quad \frac{k_B T}{\hbar\omega_l} = 1.5 \times 10^5 - 6 \times 10^6 \quad \frac{k_B T}{\hbar\omega_z} = 30 - 1300 \tag{19.41}$$

The discrete nature of the energy spectrum will play a decisive role only for the cyclotron motion (corresponding to $\hat{a}_r, \hat{a}_r^\dagger$). Only the first three levels of this motion $n_r = 0, 1, 2, 3$ will be occupied significantly in this very low temperature domain. For the other components of the motion, of much lower frequencies than the cyclotron motion, one expects that the thermal fluctuations will populate a large number of levels. The "quantum" character of the motions will be hidden under the thermal noise.

### 19.5.3  Coupling of the Cyclotron and Axial Motions

**19.5.3.1** In the presence of the *axial-cyclotron* coupling, the Hamiltonian $\hat{H}_c$ is:

$$\hat{H}_c = \hbar\omega_r(\hat{n}_r + 1/2) - \hbar\omega_l(\hat{n}_l + 1/2) + \frac{\hat{p}_z^2}{2M} + \frac{M\omega_z^2}{2}(1 + \epsilon\hat{n}_r)\,\hat{z}^2. \tag{19.42}$$

**19.5.3.2** It is straightforward to check that $\hat{n}_r$ and $\hat{n}_l$ commute with $\hat{H}_c$. The corresponding physical quantities (i.e. the excitation numbers of the cyclotron and magnetron motions) are therefore constants of the motion.

**19.5.3.3 (a)** Inside the subspace $\mathcal{E}_{n_r,n_l}$, the Hamiltonian $\hat{H}_c$ involves only the operators $\hat{z}$ and $\hat{p}_z$:

$$\hat{H}_c^{(n_r,n_l)} = \frac{\hat{p}_z^2}{2M} + \frac{M\omega_z^2}{2}(1 + \epsilon n_r)\hat{z}^2 + E_{n_r,n_l} \tag{19.43}$$

with $E_{n_r,n_l} = \hbar\omega_r(n_r + 1/2) - \hbar\omega_l(n_l + 1/2)$.

**(b)** If the system is prepared in a state belonging to the subspace $\mathcal{E}_{n_r,n_l}$, it will remain in this subspace since $n_r$ and $n_l$ are constants of the motion. Its motion is described by the Hamiltonian $\hat{H}_c^{(n_r,n_l)}$, corresponding to a harmonic oscillator along $z$, with frequency $\omega_z\sqrt{1 + \epsilon n_r}$.

**(c)** The eigenvalues of $\hat{H}_c^{(n_r,n_l)}$ are $\hbar\omega_z(n_z + 1/2)\sqrt{1 + \epsilon n_r} + E_{n_r,n_l}$. The corresponding eigenstates are the Hermite functions $\psi_n(Z)$, where:

$$Z = z\sqrt{M\omega_z(1 + \epsilon n_r)^{1/2}/\hbar}. \tag{19.44}$$

**19.5.3.4** We can perform the same operation as above inside each subspace of $\hat{n}_r$ and $\hat{n}_l$. We thus obtain a basis of eigenstates of $\hat{H}_c$ which we note $|n_r, n_l, n_z\rangle$. The eigenvalue corresponding to each eigenstate is:

$$E_{n_r,n_l,n_z} = \hbar\omega_r(n_r + 1/2) - \hbar\omega_l(n_l + 1/2) + \hbar\omega_z(n_z + 1/2)\sqrt{1 + \epsilon n_r} \tag{19.45}$$

Contrary to the result of Sect. 19.2, this basis no longer corresponds to factorized functions of the variables $x \pm iy$ and $z$. The coupling between the axial and cyclotron motions induces a correlation between the axial frequency and the state of the cyclotron motion.

**19.5.3.5 (a)** If the system is prepared in the subspace $\mathcal{E}_{n_r,n_l}$, the axial Hamiltonian corresponds to a harmonic oscillator of angular frequency $\omega_z\sqrt{1 + \epsilon n_r}$. The Ehrenfest theorem then gives:

$$\langle z\rangle(t) = z_0 \cos\left(\omega_z t\sqrt{1 + \epsilon n_r}\right) \tag{19.46}$$

$$\langle p_z\rangle(t) = -M\omega_z z_0\sqrt{1 + \epsilon n_r} \sin\left(\omega_z t\sqrt{1 + \epsilon n_r}\right) \tag{19.47}$$

since the evolution equations of the quantum expectation values coincide with the classical equations for a harmonic oscillator.

**(b)** The phase shift accumulated during a time interval $\tau$ between the detected current, proportional to $\langle p_z \rangle(t)$, and the external oscillator is:

$$\varphi = \omega_z \tau \sqrt{1 + \epsilon n_r} - \omega_z \tau \simeq \frac{\epsilon}{2} \omega_z \tau n_r. \tag{19.48}$$

Knowing the time $\tau$, the frequency $\omega_z$ and the coupling constant $\epsilon$, one can deduce the cyclotron excitation number $n_r$.

**19.5.3.6 (a)** The possible results of a measurement are the numbers

$$\varphi_k = \frac{\epsilon}{2} \omega_z \tau k, \tag{19.49}$$

where $k = n_r$ is a non-negative integer. A given experimental result determines unambiguously the excitation number of the cyclotron motion.

**(b)** The measurement postulate of quantum mechanics implies that the state $|\Psi'\rangle$ of the system after a measurement corresponds to the projection of the state vector before the measurement $|\Psi\rangle$ on the eigensubspace corresponding to the measured result:

$$|\Psi'\rangle \propto \sum_{n_l, n_z} c_{n_r^{(0)}, n_l, n_z} |n_r^{(0)}, n_l, n_z\rangle \tag{19.50}$$

where the integer $n_r^{(0)}$ corresponds to the result of the measurement of $\varphi$. One must further normalize the right hand side in order to obtain the state vector $|\Psi'\rangle$.

**(c)** For the values of the parameters given in the text, one finds $\phi_1 = \epsilon \omega_z \tau/2 \simeq 2\pi \times 1.28$. The accuracy of $\pi/10$ is much smaller than the difference between the phase shifts corresponding to $n_r$ and $n_r + 1$, and one can indeed measure unambiguously the excitation numbers $n_r = 0, 1, 2, \ldots$.

**(d)** A measurement of $n_r$ prepares the system in an eigensubspace of $\hat{n}_r$. Since $\hat{n}_r$ commutes with the Hamiltonian $\hat{H}_c$, $n_r$ is a constant of the motion. Any further measurement of the cyclotron excitation number will therefore give the same result $n_r$, corresponding to the same phase shift $\varphi_k$. Of course, this conclusion is no longer true if the system is not fully isolated and interacts with its environment. The coupling with the environment can cause transitions between different eigenstates of $\hat{H}_c$, as we shall see in the next section.

### 19.5.4 A Quantum Thermometer

**19.5.4.1 (a)** The sudden jumps of the signal are associated with a change of the cyclotron excitation number due to the coupling of the trapped electron with the thermostat. We recall that otherwise $n_r$ would be a constant of the motion.

**(b)** In the case of the first experimental curve of Fig. 19.1, the fractions of time spent in the $n_l = 0$, 1 and 2 levels are 80%, 19% and 1%, respectively. In the case of the second curve, the fractions are 97% and 3% for the time spent on $n_l = 0$ and $n_l = 1$, respectively.

**19.5.4.2** For a one-dimensional harmonic oscillator of frequency $\omega$, one finds:

$$\frac{p_{n+1}}{p_n} = \frac{e^{-(n+3/2)\hbar\omega/k_B T}}{e^{-(n+1/2)\hbar\omega/k_B T}} = e^{-\hbar\omega/k_B T}, \tag{19.51}$$

which is independent of $n$.

**19.5.4.3** For the curves of Fig. 19.1, we find:

- Top curve: $p_1/p_0 = 0.24$, i.e. $k_B T_a = \hbar\omega_c/|\ln(0.24)| \simeq 0.7\hbar\omega_c$. This corresponds to $T_a \simeq 5$ K. In principle, the determination of the temperature can also be made using $p_2/p_1$, but the accuracy is poor compared to that obtained with $p_1/p_0$.
- Bottom curve: $p_1/p_0 = 0.03$, i.e. $k_B T_b = \hbar\omega_c/|\ln(0.03)| \simeq 0.29\hbar\omega_c$. This corresponds to $T_b \simeq 2$ K.

**19.5.4.4** The normalization factor is determined from:

$$1 = \mathcal{N}\sum_{n=0}^{\infty} e^{-(n+1/2)\hbar\omega/k_B T}. \tag{19.52}$$

This is a geometric series in $e^{-\gamma}$:

$$1 = \mathcal{N}\frac{e^{-\gamma/2}}{1 - e^{-\gamma}} \quad \rightarrow \quad \mathcal{N} - 2\sinh(\gamma/2). \tag{19.53}$$

The mean excitation number is:

$$\bar{n} = \sum_n n P_n = \frac{\sum_n n e^{-n\gamma}}{\sum_n e^{-n\gamma}} = -\frac{d}{d\gamma}\ln\left(\sum_n e^{-n\gamma}\right) \tag{19.54}$$

or:

$$\bar{n} = \frac{1}{e^\gamma - 1}.$$ (19.55)

**19.5.4.5** One can see on the above expression that $\bar{n}$ is a rapidly increasing function of the temperature. If the temperature is such that $\gamma \sim 1$, i.e. $k_B T \sim \hbar \omega_c$ (or $T \sim 7.1$ K for this experiment), the mean excitation number is of the order of $(e-1)^{-1} \sim 0.6$. Below this temperature, the occupation of the level $n_l = 0$ becomes predominant, as can be seen on the curves of Fig. 19.1.

**19.5.4.6** In order to measure a temperature with such a device, one must use a statistical sample which is significantly populated in the level $n_l = 1$. It is experimentally difficult to go below a probability of $10^{-2}$ for the level $n_r = 1$, which corresponds to a temperature $T \sim 1.5$ K.

**19.5.4.7** In order to improve the sensitivity of this thermometer, one can:

- Increase significantly the total time of measurement in order to detect occupation probabilities of the level $n_r = 1$ significantly less than $10^{-2}$;
- Reduce the value of the magnetic field $B$, in order to reduce the cyclotron frequency $\omega_c$, and to increase (for a given temperature) the occupation probability of the level $n_l = 1$.

The data used in this chapter are adapted from the article of S. Peil and G. Gabrielse, *Observing the Quantum Limit of an Electron Cyclotron: QND Measurements of Quantum Jumps between Fock States*, Physical Review Letters **83**, p. 1287 (1999). The system presented here provides an additional example of a nondestructive quantum measurement, complementary to those studied in Chaps. 17 and 18.

# Laser Cooling and Trapping

# 20

By shining laser light onto an assembly of neutral atoms or ions, it is possible to cool and trap these particles. In this chapter we study a simple cooling mechanism, Doppler cooling, and we derive the corresponding equilibrium temperature. We then show that the cooled atoms can be confined in the potential well created by a focused laser beam.

We consider a "two state" atom, whose levels are denoted $|g\rangle$ (ground state) and $|e\rangle$ (excited state), with respective energies $0$ and $\hbar\omega_0$. This atom interacts with a classical electromagnetic wave of frequency $\omega_L/2\pi$. For an atom located at $r$, the Hamiltonian is

$$\hat{H} = \hbar\omega_0|e\rangle\langle e| - d \cdot (E(r,\ t)|e\rangle\langle g| + E^*(r,\ t)|g\rangle\langle e|), \qquad (20.1)$$

where $d$, which is assumed to be real, represents the matrix element of the atomic electric dipole operator between the states $|g\rangle$ and $|e\rangle$ (i.e. $d = \langle e|\hat{D}|g\rangle = \langle g|\hat{D}|e\rangle^*$). The quantity $E + E^*$ represents the electric field. We set

$$E(r,\ t) = E_0(r)\exp(-i\omega_L t). \qquad (20.2)$$

In all the chapter we assume that the detuning $\Delta = \omega_L - \omega_0$ is small compared with $\omega_L$ and $\omega_0$. We treat classically the motion $r(t)$ of the atomic center of mass.

## 20.1 Optical Bloch Equations for an Atom at Rest

**20.1.1** Write the evolution equations for the four components of the density operator of the atom $\rho_{gg}$, $\rho_{eg}$, $\rho_{ge}$ and $\rho_{ee}$ under the effect of the Hamiltonian $\hat{H}$.

**20.1.2** We take into account the coupling of the atom with the empty modes of the radiation field, which are in particular responsible for the spontaneous emission of

© Springer Nature Switzerland AG 2019
J.-L. Basdevant, J. Dalibard, *The Quantum Mechanics Solver*,
https://doi.org/10.1007/978-3-030-13724-3_20

the atom when it is in the excited state $|e\rangle$. We shall assume that this boils down to adding to the above evolution equations "relaxation" terms:

$$\left(\frac{d\rho_{ee}}{dt}\right)_{relax} = -\left(\frac{d\rho_{gg}}{dt}\right)_{relax} = -\Gamma\rho_{ee} \qquad (20.3)$$

and

$$\left(\frac{d\rho_{eg}}{dt}\right)_{relax} = -\frac{\Gamma}{2}\rho_{eg} \qquad \left(\frac{d\rho_{ge}}{dt}\right)_{relax} = -\frac{\Gamma}{2}\rho_{ge} \qquad (20.4)$$

where $\Gamma^{-1}$ is the radiative lifetime of the excited state. Justify qualitatively these terms.

**20.1.3** Check that for times much larger than $\Gamma^{-1}$, these equations have the following stationary solutions:

$$\rho_{ee} = \frac{s}{2(s+1)} \qquad \rho_{gg} = \frac{2+s}{2(s+1)}, \qquad (20.5)$$

and

$$\rho_{eg} = -\frac{\boldsymbol{d} \cdot \boldsymbol{E}(\boldsymbol{r},t)/\hbar}{\Delta + i\Gamma/2}\frac{1}{1+s} \qquad \rho_{ge} = -\frac{\boldsymbol{d} \cdot \boldsymbol{E}^*(\boldsymbol{r},t)/\hbar}{\Delta - i\Gamma/2}\frac{1}{1+s} \qquad (20.6)$$

where we have set

$$s = \frac{2|\boldsymbol{d} \cdot \boldsymbol{E}_0(\boldsymbol{r})|^2/\hbar^2}{\Delta^2 + \Gamma^2/4}. \qquad (20.7)$$

**20.1.4** Interpret physically the steady state value of the quantity $\Gamma\rho_{ee}$ in terms of spontaneous emission rate.

## 20.2   The Radiation Pressure Force

In this section, we limit ourselves to the case where the electromagnetic field is a progressive plane wave:

$$\boldsymbol{E}_0(\boldsymbol{r}) = \boldsymbol{E}_0 \exp(i\boldsymbol{k} \cdot \boldsymbol{r}). \qquad (20.8)$$

By analogy with the classical situation, we can define the radiative force operator at point $r$ as:

$$\hat{F}(r) = -\nabla_r \hat{H}. \tag{20.9}$$

**20.2.1**  Evaluate the expectation value of $\hat{F}(r)$ assuming that the atom is at rest in $r$ and that its internal dynamics is in steady state.

**20.2.2**  Interpret the result physically in terms of momentum exchanges between the atom and the radiation field. One can introduce the *recoil velocity* $v_{\text{rec}} = \hbar k / m$.

**20.2.3**  How does this force behave at high intensities? Give an order of magnitude of the possible acceleration for a sodium atom $^{23}$Na, with a resonance wavelength $\lambda = 0.589 \times 10^{-6}$ m and a lifetime of the excited state of $\Gamma^{-1} = 16 \times 10^{-9}$ s ($d = 2.1 \times 10^{-29}$ C m).

**20.2.4**  We now consider an atom in uniform motion: $r = r_0 + v_0 t \, (v_0 \ll c)$. Give the expression of the force acting on this atom.

**20.2.5**  The action of the force on the atom will modify its velocity. Under what condition is it legitimate to treat this velocity as a constant quantity for the calculation of the radiation pressure force, as done above? Is this condition valid for sodium atoms?

## 20.3  Doppler Cooling

The atom now moves in the field of two progressive plane waves of opposite directions ($+z$ and $-z$) and of same intensity (Fig. 20.1). We restrict ourselves to the motion along the direction of propagation of the two waves and we assume that for weak intensities ($s \ll 1$) one can add independently the forces exerted by the two waves.

**20.3.1**  Show that for sufficiently small velocities, the total force is linear in the velocity and can be cast in the form:

$$f = -\frac{m v}{\tau}. \tag{20.10}$$

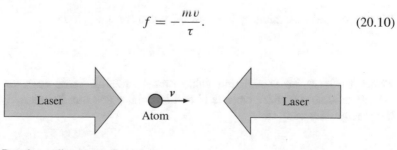

**Fig. 20.1**  Doppler cooling in one dimension

**20.3.2** What is the minimal (positive) value of $\tau_{min}$ for a fixed saturation parameter per wave $s_0$ for an atom at rest? Calculate $\tau_{min}$ for sodium atoms, assuming one fixes $s_0 = 0.1$.

**20.3.3** This cooling mechanism is limited by the heating due to the random nature of spontaneous emission. To evaluate the evolution of the velocity distribution $P(v, t)$ and find its steady state value, we shall proceed in the following way.

(a) Express $P(v, t + dt)$ in terms of $P(v, t)$. One can split the atoms into three classes:

- the atoms having undergone no photon scattering event between $t$ and $t + dt$,
- the atoms having scattered a photon from the $+z$ wave,
- the atoms having scattered a photon from the $-z$ wave.

We choose $dt$ short enough that the probability of the first option is dominant, and such that multiple scattering events are negligible. We also assume that the velocities contributing significantly to $P(v, t)$ are small enough for the linearization of the force performed above to be valid. For simplicity we will assume that spontaneously emitted photons propagate only along the $z$ axis, a spontaneous emission occurring with equal probabilities in the directions $+z$ and $-z$.

(b) Show that $P(v, t)$ obeys the Fokker–Planck equation

$$\frac{\partial P}{\partial t} = \alpha \frac{\partial}{\partial v}(vP) + \beta \frac{\partial^2 P}{\partial v^2} \tag{20.11}$$

and express of $\alpha$ and $\beta$ in terms of the physical parameters of the problem.

(c) Determine the steady state velocity distribution. Show that it corresponds to a Maxwell distribution and give the effective temperature.

(d) For which detuning is the effective temperature minimal? What is this minimal temperature for sodium atoms?

## 20.4    The Dipole Force

We now consider a stationary light wave (with a constant phase)

$$E_0(r) = E_0^*(r). \tag{20.12}$$

**20.4.1** Evaluate the expectation value of the radiative force operator $\hat{F}(r) = -\nabla_r \hat{H}$ assuming that the atom is at rest at point $r$ and that its internal dynamics has reached its steady state.

**20.4.2** Show that this force derives from a potential and evaluate the potential well depth that can be attained for sodium atoms with a laser beam of intensity $P = 1$ W, and wavelength $\lambda_L = 0.650\,\mu$m, focused on a circular spot of radius $10\,\mu$m.

## 20.5 Solutions

### 20.5.1 Optical Bloch Equations for an Atom at Rest

**20.5.1.1** The evolution of the density operator $\hat{\rho}$ is given by:

$$i\hbar\frac{d\hat{\rho}}{dt} = [\hat{H}, \hat{\rho}] \tag{20.13}$$

so that:

$$\begin{aligned}
\frac{d\rho_{ee}}{dt} &= i\frac{\mathbf{d}\cdot\mathbf{E}(\mathbf{r})\,e^{-i\omega_L t}}{\hbar}\rho_{ge} - i\frac{\mathbf{d}\cdot\mathbf{E}^*(\mathbf{r})\,e^{i\omega_L t}}{\hbar}\rho_{eg} \\
\frac{d\rho_{eg}}{dt} &= -i\omega_0\rho_{eg} + i\frac{\mathbf{d}\cdot\mathbf{E}(\mathbf{r})\,e^{-i\omega_L t}}{\hbar}(\rho_{gg} - \rho_{ee})
\end{aligned} \tag{20.14}$$

and

$$\frac{d\rho_{gg}}{dt} = -\frac{d\rho_{ee}}{dt} \qquad \frac{d\rho_{ge}}{dt} = \left(\frac{d\rho_{eg}}{dt}\right)^*. \tag{20.15}$$

**20.5.1.2** Assume that the atom-field system is placed at time $t = 0$ in the state

$$|\psi(0)\rangle = (\alpha|g\rangle + \beta|e\rangle) \otimes |0\rangle, \tag{20.16}$$

where $|0\rangle$ denotes the vacuum state of the electromagnetic field and neglect in a first step the action of the laser. At time $t$, the state of the system is derived from the Wigner–Weisskopf treatment of spontaneous emission:

$$|\psi(t)\rangle = (\alpha|g\rangle + \beta e^{-(i\omega_0 + \Gamma/2)t}|e\rangle) \otimes |0\rangle + |g\rangle \otimes |\phi\rangle, \tag{20.17}$$

where the state of the field $|\phi\rangle$ is a superposition of one-photon states for the various modes of the electromagnetic field. Consequently the evolution of the density matrix elements is $\rho_{ee}(t) = |\beta|^2 e^{-\Gamma t}$, $\rho_{eg}(t) = \alpha^*\beta e^{-(i\omega_0+\Gamma)t}$, or, in other words,

$$\left(\frac{d\rho_{ee}}{dt}\right)_{\text{relax}} = -\Gamma\rho_{ee} \qquad \left(\frac{d\rho_{eg}}{dt}\right)_{\text{relax}} = -\frac{\Gamma}{2}\rho_{eg}. \tag{20.18}$$

The two other relations originate from the conservation of the trace of the density operator ($\rho_{ee} + \rho_{gg} = 1$) and from its hermitian character $\rho_{eg} = \rho_{ge}^*$.

We assume in the following that the evolution of the atomic density operator is obtained by adding the action of the laser field and the spontaneous emission contribution. Since $\Gamma$ varies like $\omega_0^3$, this is valid as long as the shift of the atomic transition due to the laser irradiation remains small compared with $\omega_0$. This requires $dE \ll \hbar\omega_0$, which is satisfied for usual continuous laser sources.

**20.5.1.3** The evolution of the density operator components is given by

$$
\begin{aligned}
\frac{d\rho_{ee}}{dt} &= -\Gamma \,\rho_{ee} + i\frac{d\cdot E(r)\,e^{-i\omega_L t}}{\hbar}\rho_{ge} - i\frac{d\cdot E^*(r)\,e^{i\omega_L t}}{\hbar}\rho_{eg} \\
\frac{d\rho_{eg}}{dt} &= \left(-i\omega_0 - \frac{\Gamma}{2}\right)\rho_{eg} + i\frac{d\cdot E(r)\,e^{-i\omega_L t}}{\hbar}(\rho_{gg} - \rho_{ee}).
\end{aligned}
\tag{20.19}
$$

These equations are often called optical Bloch equations.

At steady-state, $\rho_{ee}$ and $\rho_{gg}$ tend to a constant value, while $\rho_{eg}$ and $\rho_{ge}$ oscillate respectively as $e^{-i\omega_L t}$ and $e^{i\omega_L t}$. This steady-state is reached after a characteristic time of the order of $\Gamma^{-1}$. From the second equation we extract the steady-state value of $\rho_{eg}$ as a function of $\rho_{gg} - \rho_{ee} = 1 - 2\rho_{ee}$:

$$
\rho_{eg} = i\frac{d \cdot E(r)e^{-i\omega_L t}/h}{i\Delta + \Gamma/2}(1 - 2\rho_{ee}).
\tag{20.20}
$$

We now insert this value in the evolution of $\rho_{ee}$ and we get:

$$
\rho_{ee} = \frac{s}{2(1+s)} \quad \text{with} \quad s(r) = \frac{2|d \cdot E(r)|^2/\hbar^2}{\Delta^2 + \Gamma^2/4}.
\tag{20.21}
$$

The three other values given in the text for $\rho_{gg}$, $\rho_{eg}$ and $\rho_{ge}$ follow immediately.

**20.5.1.4** The steady state value of $\rho_{ee}$ gives the average probability of finding the atom in the internal state $|e\rangle$. This value results from the competition between absorption processes, which tend to populate the level $|e\rangle$ and stimulated+spontaneous emission processes, which depopulate $|e\rangle$ in favor of $|g\rangle$.

The quantity $\Gamma\rho_{ee}$ represents the steady-state rate of spontaneous emission as the atom is irradiated by the laser wave. For a low saturation parameter $s$, this rate is proportional to the laser intensity $|E(r)|^2$. When the laser intensity increases, $s$ gets much larger than 1 and the steady state value of $\rho_{ee}$ is close to $1/2$. This means that the atom spends half of the time in level $|e\rangle$. In this case, the rate of spontaneous emission tends to $\Gamma/2$.

## 20.5.2 The Radiation Pressure Force

**20.5.2.1** For a plane laser wave the force operator is given by:

$$
\hat{F}(r) = ik\, d \cdot E_0(e^{i(k\cdot r - \omega_L t)}|e\rangle\langle g| - e^{-i(k\cdot r - \omega_L t)}|g\rangle\langle e|).
\tag{20.22}
$$

The expectation value in steady state is $\mathrm{Tr}(\hat{\rho}\hat{F})$ which gives:

$$f = \langle F \rangle = \mathrm{i}k\, \boldsymbol{d} \cdot \boldsymbol{E}_0\, \mathrm{e}^{\mathrm{i}(k \cdot r - \omega_\mathrm{L} t)} \rho_{ge} + c.c.$$

$$= \hbar k \frac{\Gamma}{2} \frac{s_0}{1 + s_0}$$

with

$$s_0 = \frac{2|\boldsymbol{d} \cdot \boldsymbol{E}_0|^2 / \hbar^2}{\Delta^2 + \Gamma^2/4}. \tag{20.23}$$

**20.5.2.2** The interpretation of this result is as follows. The atom undergoes absorption processes, where it goes from the ground internal state to the excited internal state, and gains the momentum $\hbar k$. From the excited state, it can return to the ground state by a stimulated or spontaneous emission process. In a stimulated emission the atom releases the momentum that it has gained during the absorption process, so that the net variation of momentum in a such a cycle is zero. In contrast, in a spontaneous emission process, the momentum change of the atom has a random direction and it averages to zero since the spontaneous emission process occurs with the same probability in two opposite directions. Therefore the net momentum gain for the atom in a cycle "absorption-spontaneous emission" is $\hbar k$ corresponding to a velocity change $v_\mathrm{rec}$. Since these cycles occur with a rate $(\Gamma/2)s_0/(1 + s_0)$ (as found at the end of Sect. 20.1), we recover the expression for the radiation force found above.

**20.5.2.3** For a large laser intensity, the force saturates to the value $\hbar k\Gamma/2$. This corresponds to an acceleration $a_\mathrm{max} = \hbar k\Gamma/(2m) = 9 \times 10^5\,\mathrm{ms}^{-2}$, which is 100,000 times larger than the acceleration due to gravity.

**20.5.2.4** In the rest frame of the atom, the laser field still corresponds to a plane wave with a modified frequency $\omega_\mathrm{L} - \boldsymbol{k} \cdot \boldsymbol{v}$ (first order Doppler effect). The change of momentum of the photon is negligible for non-relativistic atomic velocities. The previous result is then changed into:

$$f = \hbar k \frac{\Gamma}{2} \frac{s(\boldsymbol{v})}{1 + s(\boldsymbol{v})} \quad \text{with} \quad s(\boldsymbol{v}) = \frac{2\,|\boldsymbol{d} \cdot \boldsymbol{E}_0|^2 / \hbar^2}{(\Delta - \boldsymbol{k} \cdot \boldsymbol{v})^2 + \Gamma^2/4}. \tag{20.24}$$

**20.5.2.5** The notion of force derived above is valid if the elementary velocity change in a single absorption or emission process (the recoil velocity $v_\mathrm{rec} = \hbar k/m$) modifies only weakly the value of $f$. This is the case when the elementary change of Doppler shift $kv_\mathrm{rec} = \hbar k^2/m$ is very small compared with the width of the resonance:

$$\frac{\hbar k^2}{m} \ll \Gamma. \tag{20.25}$$

This is the so called *broad line condition*. This condition is well satisfied for sodium atoms since $\hbar k^2/(m\Gamma) = 5 \times 10^{-3}$ in this case.

### 20.5.3 Doppler Cooling

**20.5.3.1** The total force acting on the atom moving with velocity $v$ is

$$f(v) = \hbar k \, \Gamma \left( \frac{|\boldsymbol{d} \cdot \boldsymbol{E}_0|^2/\hbar^2}{(\Delta - kv)^2 + \Gamma^2/4} - \frac{|\boldsymbol{d} \cdot \boldsymbol{E}_0|^2/\hbar^2}{(\Delta + kv)^2 + \Gamma^2/4} \right), \qquad (20.26)$$

where we have used the fact that $s \ll 1$. For low velocities ($kv \ll \Gamma$) we get at first order in $v$

$$f(v) = -\frac{mv}{\tau} \quad \text{with} \quad \tau = \frac{m}{\hbar k^2 s_0} \frac{\Delta^2 + \Gamma^2/4}{2(-\Delta)\Gamma}. \qquad (20.27)$$

This corresponds to a damping force if the detuning $\Delta$ is negative. In this case the atom is cooled because of the Doppler effect. This is the so-called Doppler cooling: A moving atom feels a stronger radiation pressure force from the counterpropagating wave than from the copropagating wave. For an atom at rest the two radiation pressure forces are equal and opposite: the net force is zero.

**20.5.3.2** For a fixed saturation parameter $s_0$, the cooling time is minimal for $\Delta = -\Gamma/2$, which leads to

$$\tau_{\min} = \frac{m}{2\hbar k^2 s_0}. \qquad (20.28)$$

Note that this time is always much longer than the lifetime of the excited state $\Gamma^{-1}$ when the broad line condition is fullfilled. For sodium atoms this minimal cooling time is $16\,\mu s$ for $s_0 = 0.1$.

**20.5.3.3** **(a)** The probability that an atom moving with velocity $v$ scatters a photon from the $\pm z$ wave during the time $dt$ is

$$dP_\pm(v) = \frac{\Gamma s_0}{2} \left( 1 \pm \frac{2\Delta kv}{\Delta^2 + \Gamma^2/4} \right) dt. \qquad (20.29)$$

Since we assume that the spontaneously emitted photons also propagate along $z$, half of the scattering events do not change the velocity of the atom. This is the case when the spontaneously emitted photon propagates along the same direction as the absorbed photon. For the other half of the events, the change of the atomic velocity is $\pm 2v_{\text{rec}}$, corresponding to a spontaneously emitted photon propagating in the direction opposite to the absorbed photon. Consequently, the probability that the velocity of the atom does not change during the time $dt$ is $1 - (dP_+(v) + dP_-(v))/2$, and the probability that the atomic velocity changes by $\pm 2 v_{rec}$ is $dP \pm (v)/2$. Therefore one has:

$$P(v,\ t+dt) = \left(1 - \frac{dP_+(v) + dP_-(v)}{2}\right) P(v,\ t)$$

$$+ \frac{dP_+(v - 2v_{rec})}{2} P(v - 2v_{rec},\ t) + \frac{dP_-(v + 2v_{rec})}{2} P(v + 2v_{rec},\ t).$$

**(b)** Assuming that $P(v)$ varies smoothly over the recoil velocity scale (which will be checked in the end), we can transform the finite difference equation found above into a differential equation:

$$\frac{\partial P}{\partial t} = \alpha \frac{\partial}{\partial v}(vP) + \beta \frac{\partial^2 P}{\partial v^2}, \qquad (20.30)$$

with

$$\alpha = \frac{m}{\tau} \qquad \beta = \Gamma v_{rec}^2 s_0. \qquad (20.31)$$

The term proportional to $\alpha$ corresponds to the Doppler cooling. The term proportional to $\beta$ accounts for the heating due to the random nature of spontaneous emission processes. The coefficient $\beta$ is a diffusion constant in velocity space, proportional to the square of the elementary step of the random walk $v_{rec}$, and to its rate $\Gamma s_0$.

**(c)** The steady state for $P(v)$ corresponds to the solution of:

$$\alpha v P(v) + \beta \frac{dP}{dv} = 0, \qquad (20.32)$$

whose solution (for $\alpha > 0$, i.e. $\Delta < 0$) is a Maxwell distribution:

$$P(v) = P_0 \exp\left(-\frac{\alpha v^2}{2\beta}\right). \qquad (20.33)$$

The effective temperature is therefore

$$k_B T = \frac{m\beta}{\alpha} = \frac{\hbar}{2} \frac{\Delta^2 + \Gamma^2/4}{-\Delta}. \qquad (20.34)$$

**(d)** The minimal temperature is obtained for $\Delta = -\Gamma/2$:

$$k_B T_{min} = \frac{\hbar\Gamma}{2}. \qquad (20.35)$$

This is the *Doppler cooling limit*, which is independent of the laser intensity. Note that, when the broad line condition is fullfilled, the corresponding velocity scale $v_0$ is such that:

$$v_{rec} \ll v_0 = \sqrt{\hbar\Gamma}/m \ll \Gamma/k. \tag{20.36}$$

The two hypotheses at the basis of our calculation are therefore valid: (i) $P(v)$ varies smoothly over the scale $v_{rec}$ and (ii) the relevant velocities are small enough for the linearization of the scattering rates to be possible.

For sodium atoms, the minimal temperature is $T_{min} = 240\,\mu K$, corresponding to $v_0 = 40\,cm\,s^{-1}$.

### 20.5.4 The Dipole Force

**20.5.4.1**  For a real amplitude $E_0(r)$ of the electric field of the light wave (standing wave), the force operator $\hat{F}(r)$ is:

$$\hat{F}(r) = \left( \sum_{i=x,y,z} d_i \nabla E_{0i}(r) \right) \left( e^{-i\omega_L t}|e\rangle\langle g| + e^{-i\omega_L t}|g\rangle\langle e| \right). \tag{20.37}$$

Assuming that the internal dynamics of the atom has reached its steady-state value, we get for the expectation value of $\hat{F}$:

$$\begin{aligned} f(r) = \langle F \rangle &= -\nabla(d \cdot E_0(r))\,\frac{d \cdot E_0(r)}{1+s(r)}\,\frac{\Delta}{\Delta^2+\Gamma^2/4} \\ &= -\frac{\hbar\Delta}{2}\,\frac{\nabla s(r)}{1+s(r)} \end{aligned} \tag{20.38}$$

**20.5.4.2**  This force is called the dipole force. It derives from the dipole potential $U(r)$:

$$f(r) = -\nabla U(r) \qquad \text{with} \qquad U(r) = \frac{\hbar\Delta}{2}\log(1+s(r)). \tag{20.39}$$

For a laser field with intensity $P = 1\,W$, focused on a spot with radius $r = 10\,\mu m$, the electric field at the center is

$$E_0 = \sqrt{\frac{2P}{\pi\epsilon_0 cr^2}} = 1.6 \times 10^6\,V/m \tag{20.40}$$

We suppose here that the circular spot is uniformly illuminated. A more accurate treatment should take into account the transverse Gaussian profile of the laser beam, but this would not significantly change the following results. This value for $E_0$ leads to $dE_0/\hbar = 3.1 \times 10^{11}\,s^{-1}$ and the detuning $\Delta$ is equal to $3 \times 10^{14}\,s^{-1}$. The potential depth is then found to be equal to 2.4 mK, 10 times larger than the Doppler cooling limit. Due to the large detuning, the photon scattering rate is quite small: 70 photons/s.

### 20.5.5 Comments and References

The Physics Nobel Prize was awarded in 1997 to S. Chu, C. Cohen- Tannoudji and W. D. Phillips *for development of methods to cool and trap atoms with laser light*. Several important results that go beyond the simple modeling presented in the chapter can be found in the corresponding Nobel lectures published in Rev. Mod. Phys. **70**, p.685, 707 and 721 (1998).

Cooling and trapping of atoms by light is a tool that is now used in many laboratories worldwide, both for metrological purposes such as the building of atomic clocks (see Chap. 6) or for producing cold quantum gases like Bose–Einstein condensates (see Chap. 22). See e.g. H. J. Metcalf and P. Van der Straten, *Laser cooling and trapping* (Springer Science & Business Media, 2012).

The idea of trapping material particles by light can be extended to objects much bigger than isolated atoms: the 2018 Nobel Prize in Physics was awarded to Arthur Ashkin for the invention of the *optical tweezers* and their application to biological systems.

# Part III

# Complex Systems

# Exact Results for the Three-Body Problem  21

The three-body problem is a famous question of mechanics. Henri Poincaré was the first to prove exact properties, and this contributed to his celebrity. The purpose of this chapter is to derive some rigorous results for the three-body problem in quantum mechanics. Here we are interested in obtaining rigorous *lower bounds* on three-body ground state energies. Upper bounds are easier to obtain by variational calculations. We will see that our lower bounds are actually quite close to the exact answers, to which they provide useful approximations.

## 21.1 The Two-Body Problem

Consider a system of two particles with equal masses $m$ and momenta $p_1$ and $p_2$, interacting via a potential $V(r_{12})$ where $r_{12} = |r_1 - r_2|$.

**21.1.1** Write the Hamiltonian $\hat{H}$ of the system. Let $P = p_1 + p_2$ and $p = (p_1 - p_2)/2$ be the total and relative momentum. Separate the center of mass $\hat{H}_{cm}$ and the relative $\hat{H}_{12}$ Hamiltonians by writing $\hat{H}$ as:

$$\hat{H} = \hat{H}_{cm} + \hat{H}_{12}, \quad H_{cm} = \frac{\hat{P}^2}{2M}, \quad \hat{H}_{12} = \frac{\hat{p}^2}{2\mu} + V(\hat{r}_{12}), \tag{21.1}$$

where $M = 2m$ is the total mass of the system. Give the value of the reduced mass $\mu$ in terms of $m$.

**21.1.2** We denote by $E^{(2)}(\mu)$ the ground state energy of $\hat{H}_{12}$. Give the expression for $E^{(2)}(\mu)$ in the two cases $V(r) = -b^2/r$ and $V(r) = \kappa r^2/2$.

© Springer Nature Switzerland AG 2019
J.-L. Basdevant, J. Dalibard, *The Quantum Mechanics Solver*,
https://doi.org/10.1007/978-3-030-13724-3_21

## 21.2    The Variational Method

Let $\{|n\rangle\}$ be the orthonormal eigenstates of a Hamiltonian $\hat{H}$ and $\{E_n\}$ the ordered sequence of its corresponding eigenvalues: $E_0 < E_1 < E_2 < \cdots$.

**21.2.1**  Show that $\langle n|\hat{H}|n\rangle = E_n$.

**21.2.2**  Consider an arbitrary vector $|\psi\rangle$ of the Hilbert space of the system. By expanding $|\psi\rangle$ on the basis $\{|n\rangle\}$, prove the inequality

$$\forall\psi, \quad \langle\psi|\hat{H}|\psi\rangle \geq E_0\langle\psi|\psi\rangle. \tag{21.2}$$

**21.2.3**  Show that the previous result remains valid if $\hat{H}$ is the Hamiltonian of a two-body subsystem and $|\psi\rangle$ a three-body state. In order to do so, one can denote by $\hat{H}_{12}$ the Hamiltonian of the $(1, 2)$ subsystem in the three-body system of wave function $\psi(\boldsymbol{r}_1, \boldsymbol{r}_2, \boldsymbol{r}_3)$. One can first consider a given value of $\boldsymbol{r}_3$, and then integrate the result over this variable.

---

## 21.3    Relating the Three-Body and Two-Body Sectors

Consider a system of three-particles of equal masses $m$ with pairwise interactions:

$$V = V(r_{12}) + V(r_{13}) + V(r_{23}). \tag{21.3}$$

**21.3.1**  Check the identity

$$3(p_1^2+p_2^2+p_3^2) = (\boldsymbol{p}_1+\boldsymbol{p}_2+\boldsymbol{p}_3)^2+(\boldsymbol{p}_1-\boldsymbol{p}_2)^2+(\boldsymbol{p}_2-\boldsymbol{p}_3)^2+(\boldsymbol{p}_3-\boldsymbol{p}_1)^2 \tag{21.4}$$

and show that the three-body Hamiltonian $\hat{H}^{(3)}$ can be written as

$$\hat{H}^{(3)} = \hat{H}_{\mathrm{cm}} + \hat{H}_{\mathrm{rel}}^{(3)}, \quad \hat{H}_{\mathrm{cm}} = \frac{\hat{\boldsymbol{P}}^2}{6m},$$

where $\hat{\boldsymbol{P}} = \hat{\boldsymbol{p}}_1 + \hat{\boldsymbol{p}}_2 + \hat{\boldsymbol{p}}_3$ is the total three-body momentum, and where the relative Hamiltonian $\hat{H}_{\mathrm{rel}}^{(3)}$ is a sum of two-particle Hamiltonians of the type defined in (21.1),

$$\hat{H}_{\mathrm{rel}}^{(3)} = \hat{H}_{12} + \hat{H}_{23} + \hat{H}_{31} \tag{21.5}$$

with a *new* value $\mu'$ of the reduced mass. Express $\mu'$ in terms of $m$.

**21.3.2**  Do the two-body Hamiltonians $\hat{H}_{ij}$ commute in general? What would be the result if they did?

**21.3.3**  We call $|\Omega\rangle$ the normalized ground state of $\hat{H}^{(3)}_{\text{rel}}$, and $E^{(3)}$ the corresponding energy. Show that the three-body ground state energy is related to the ground state energy of each two-body subsystem by the inequality:

$$E^{(3)} \geq 3E^{(2)}\left(\mu'\right). \tag{21.6}$$

**21.3.4**  Which lower bounds on the three-body ground-state energy $E^{(3)}$ does one obtain in the two cases $V(r) = -b^2/r$ and $V(r) = \kappa r^2/2$?

In the first case, the exact result, which can be obtained numerically, is $E^{(3)} \simeq -1.067\, mb^4/\hbar^2$. How does this compare with the bound (21.6)?

## 21.4   The Three-Body Harmonic Oscillator

The three-body problem can be solved exactly in the case of harmonic interactions $V(r) = \kappa r^2/2$. In order to do this, we introduce the Jacobi variables:

$$\hat{R}_1 = (\hat{r}_1 - \hat{r}_2)/\sqrt{2}, \quad \hat{R}_2 = (2\hat{r}_3 - \hat{r}_1 - \hat{r}_2)/\sqrt{6}, \quad \hat{R}_3 = (\hat{r}_1 + \hat{r}_2 + \hat{r}_3)/\sqrt{3}, \tag{21.7}$$

$$\hat{Q}_1 = (\hat{p}_1 - \hat{p}_2)/\sqrt{2}, \quad \hat{Q}_2 = (2\hat{p}_3 - \hat{p}_1 - \hat{p}_2)/\sqrt{6}, \quad \hat{Q}_3 = (\hat{p}_1 + \hat{p}_2 + \hat{p}_3)/\sqrt{3}. \tag{21.8}$$

**21.4.1**  What are the commutation relations between the components $\hat{R}^{\alpha}_j$ and $\hat{Q}^{\beta}_k$ of $\hat{R}_j$ and $\hat{Q}_k$, ($\alpha = 1, 2, 3,\ and\ \beta = 1, 2, 3$)?

**21.4.2**  Check that one has $Q_1^2 + Q_2^2 + Q_3^2 = p_1^2 + p_2^2 + p_3^2$, and:

$$3(R_1^2 + R_2^2) = (r_1 - r_2)^2 + (r_2 - r_3)^2 + (r_3 - r_1)^2. \tag{21.9}$$

**21.4.3**  Rewrite the three-body Hamiltonian in terms of these variables for a harmonic two-body interaction $V(r) = \kappa r^2/2$. Derive the three-body ground state energy from the result. Show that the inequality (21.6) is saturated, i.e. the bound (21.6) coincides with the exact result in that case.

Do you think that the bound (21.6), which is valid for any potential, can be improved without further specifying the potential?

## 21.5   From Mesons to Baryons in the Quark Model

In elementary particle physics, the previous results are of particular interest since *mesons* are bound states of two quarks, whereas *baryons*, such as the proton, are bound states of three quarks. Furthermore, it is an empirical observation that the spectroscopy of mesons and baryons is very well accounted for by non-relativistic potential models for systems of quarks.

The $\phi$ meson, for instance, is a bound state of a strange quark $s$ and its antiquark $\bar{s}$, both of same mass $m_s$. The mass $m_\phi$ is given by $m_\phi = 2m_s + E^{(2)}(\mu)/c^2$ where $\mu = m_s/2, c$ is the velocity of light, and $E^{(2)}$ is the ground state energy of the $s\bar{s}$ system which is bound by a potential $V_{q\bar{q}}(r)$. The $\Omega^-$ baryon is made of three strange quarks. Its mass is given by $M_\Omega = 3m_s + E^{(3)}/c^2$, where $E^{(3)}$ is the ground state energy of the three $s$ quarks, which interact pairwise through a two-body potential $V_{qq}(r)$.

These potentials are related very simply to each other by

$$V_{qq}(r) = \frac{1}{2}V_{q\bar{q}}(r). \tag{21.10}$$

It is a remarkable property, called flavor independence, that these potentials are the same for all types of quarks.

**21.5.1** Following a procedure similar to that of Sect. 21.3, show that $E^{(3)} \geq (3/2)E^{(2)}(\mu')$; express $\mu'$ in terms of $\mu = m_s/2$.

**21.5.2** Consider the potential $V_{q\bar{q}}(r) = g \ln(r/r_0)$, and the two-body Hamiltonians $\hat{H}^{(2)}(\mu)$ and $\hat{H}^{(2)}(\tilde{\mu})$ corresponding to the same potential but different reduced masses $\mu$ and $\tilde{\mu}$. By rescaling $r$, transform $\hat{H}^{(2)}(\tilde{\mu})$ into $\hat{H}^{(2)}(\mu) + C$, where $C$ is a constant.

Calculate the value of $C$ and show that the eigenvalues $E_n^{(2)}(\mu)$ of $\hat{H}^{(2)}(\mu)$ and $E_n^{(2)}(\tilde{\mu})$ of $\hat{H}^{(2)}(\tilde{\mu})$ are related by the simple formula

$$E_n^{(2)}(\tilde{\mu}) = E_n^{(2)}(\mu) + \frac{g}{2}\ln\frac{\mu}{\tilde{\mu}}. \tag{21.11}$$

**21.5.3** A striking characteristic of the level spacings in quark-antiquark systems is that these spacings are approximately independent of the nature of the quarks under consideration, therefore independent of the quark masses. Why does this justify the form of the above potential $V_{q\bar{q}}(r) = g \ln(r/r_0)$?

**21.5.4** Show that the following relation holds between the $\Omega^-$ and $\phi$ masses $M_\Omega$ and $m_\phi$:

$$M_\Omega \geq \frac{3}{2}m_\phi + a \tag{21.12}$$

and express the constant $a$ in terms of the coupling constant $g$.

**21.5.5** The observed masses are $m_\phi = 1019\,\text{MeV}/c^2$ and $M_\Omega = 1672\,\text{MeV}/c^2$. The coupling constant is $g = 650\,\text{MeV}$. Test the inequality with these data.

## 21.6 Solutions

### 21.6.1 The Two-Body Problem

**21.6.1.1** The two-body Hamiltonian is

$$\hat{H} = \frac{\hat{p}_1^2}{2m} + \frac{\hat{p}_2^2}{2m} + \hat{V}(r_{12}). \tag{21.13}$$

The center of mass motion can be separated as usual:

$$\hat{H} = \frac{\hat{P}^2}{2M} + \frac{\hat{p}^2}{2\mu} + \hat{V}(r_{12}), \tag{21.14}$$

where $M = 2m$ and $\mu = m/2$ are respectively the total mass and the reduced mass of the system.

**21.6.1.2** For a Coulomb-type interaction $V(r) = -b^2/r$, we get

$$E^{(2)}(\mu) = -\frac{\mu b^4}{2\hbar^2}. \tag{21.15}$$

For a harmonic interaction $V(r) = \kappa r^2/2$, we get

$$E^{(2)}(\mu) = \frac{3}{2}\hbar\sqrt{\frac{\kappa}{\mu}}. \tag{21.16}$$

### 21.6.2 The Variational Method

**21.6.2.1** By definition, $\langle n|\hat{H}|n\rangle = E_n\langle n|n\rangle = E_n$.

**21.6.2.2** Since $\{|n\rangle\}$ is a basis of the Hilbert space, $|\psi\rangle$ can be expanded as $|\psi\rangle = \sum c_n|n\rangle$, and the square of its norm is $\langle\psi|\psi\rangle = \sum |c_n|^2$. We therefore have $\langle\psi|\hat{H}|\psi\rangle = \sum E_n|c_n|^2$.

If we simply write

$$\langle\psi|\hat{H}|\psi\rangle - E_0\langle\psi|\psi\rangle = \sum(E_n - E_0)|c_n|^2, \tag{21.17}$$

we obtain, since $E_n \geq E_0$ and $|c_n|^2 \geq 0$:

$$\langle\psi|\hat{H}|\psi\rangle \geq E_0\langle\psi|\psi\rangle. \tag{21.18}$$

**21.6.2.3** If $\hat{H} = \hat{H}_{12}$, for fixed $r_3$, $\psi(r_1, r_2, r_3)$ can be considered as a non-normalized two-body wave function. Therefore

$$\int \psi^*(r_1, r_2, r_3)\hat{H}_{12}\psi(r_1, r_2, r_3)d^3r_1d^3r_2 \geq E_0 \int |\psi(r_1, r_2, r_3)|^2 \, d^3r_1d^3r_2.$$
(21.19)

By integrating this inequality over $r_3$, one obtains the desired result.

### 21.6.3 Relating the Three-Body and Two-Body Sectors

**21.6.3.1** The identity is obvious, since the crossed terms vanish on the right-hand side. Therefore $\hat{H} = \hat{P}^2/(6m) + \hat{H}_{12} + \hat{H}_{23} + \hat{H}_{31}$, with

$$\hat{H}_{ij} = \frac{(\hat{p}_i - \hat{p}_j)^2}{6m} + \hat{V}(r_{ij}) = \frac{[(\hat{p}_i - \hat{p}_j)/2]^2}{2\mu'} + \hat{V}(r_{ij})$$
(21.20)

with a reduced mass $\mu' = 3m/4$.

**21.6.3.2** Obviously, $\hat{H}_{12}$ and $\hat{H}_{23}$ do not commute; for instance $\hat{p}_1 - \hat{p}_2$ does not commute with $\hat{V}(r_{23})$. If they did, the three-body energies would just be the sum of two-body energies as calculated with a reduced mass $\mu' = 3m/4$, and the solution of the three-body problem would be simple.

**21.6.3.3** By definition, $E^{(3)} = \langle\Omega|\hat{H}_{rel}^{(3)}|\Omega\rangle = \sum\langle\Omega|\hat{H}_{ij}|\Omega\rangle$. However, owing to the results of questions 21.2.2 and 21.2.3, we have $\langle\Omega|\hat{H}_{ij}|\Omega\rangle \geq E^{(2)}(\mu')$, so that

$$E^{(3)} \geq 3E^{(2)}(\mu') \quad \text{with} \quad \mu' = 3m/4.$$
(21.21)

**21.6.3.4** For a Coulomb-type potential, we obtain

$$E^{(3)} \geq -\frac{3}{2}\frac{\mu'b^4}{\hbar^2} = -\frac{9}{8}\frac{mb^4}{\hbar^2},$$
(21.22)

which deviates by only 6% from the exact answer $-1.067 \, mb^4/\hbar^2$.

In the harmonic case, we obtain:

$$E^{(3)} \geq 3\frac{3}{2}\hbar\sqrt{\frac{\kappa}{\mu'}} = 3\sqrt{3}\hbar\sqrt{\frac{\kappa}{m}}.$$
(21.23)

### 21.6.4 The Three-Body Harmonic Oscillator

**21.6.4.1** One easily verifies that Jacobi variables satisfy canonical commutation relations:

$$[\hat{R}_j^\alpha, \hat{Q}_k^\beta] = i\hbar \, \delta_{jk} \, \delta_{\alpha\beta}. \tag{21.24}$$

**21.6.4.2** These relations are a simple algebraic exercise.

**21.6.4.3** We find

$$\hat{H} = \frac{\hat{Q}_1^2}{2m} + \frac{3}{2}\kappa\,\hat{R}_1^2 + \frac{\hat{Q}_2^2}{2m} + \frac{3}{2}\kappa\,\hat{R}_2^2 + \frac{\hat{Q}_3^2}{2m} = \hat{H}_1 + \hat{H}_2 + \hat{H}_{\text{cm}}, \tag{21.25}$$

where $\hat{H}_{\text{cm}} = \hat{Q}_3^2/(2m) = \hat{P}^2/(6m)$ is the center of mass Hamiltonian. The three Hamiltonians $\hat{H}_1$, $\hat{H}_2$, and $\hat{H}_{\text{cm}}$ *commute*. The ground state energy (in the center of mass frame) is therefore

$$E^{(3)} = 2\frac{3}{2}\hbar\sqrt{\frac{3\kappa}{m}} = 3\sqrt{3}\hbar\sqrt{\frac{\kappa}{m}}, \tag{21.26}$$

which coincides with the lower bound obtained in question 21.3.4. The bound is therefore saturated if the interaction is harmonic.

In order to improve the bound, one must further specify the interaction. Actually, the bound is saturated if and only if the interaction potential is harmonic. Indeed the variational inequality we use becomes an equality if and only if the wave function coincides with the exact ground state wave function. Owing to the particular symmetry of quadratic forms, the Jacobi variables guarantee that this happens in the harmonic case. The property ceases to be true for any other potential.

### 21.6.5 From Mesons to Baryons in the Quark Model

**21.6.5.1** The $s\bar{s}$ relative Hamiltonian is

$$\hat{H}^{(2)} = \frac{\hat{p}^2}{m_s} + V_{q\bar{q}}(\hat{r}). \tag{21.27}$$

The $sss$ relative Hamiltonian is (cf. Sect. 21.3):

$$\hat{H}^{(3)} = \sum_{i<j}\left(\frac{(\hat{p}_i - \hat{p}_j)^2}{6m_s} + \frac{1}{2}V_{q\bar{q}}(\hat{r}_{ij})\right) \tag{21.28}$$

$$= \frac{1}{2}\sum_{i<j}\left(\frac{(\hat{p}_i - \hat{p}_j)^2}{3m_s} + V_{q\bar{q}}(\hat{r}_{ij})\right). \tag{21.29}$$

Therefore,

$$2\hat{H}^{(3)} = \sum_{i<j} \hat{H}_{ij} \quad \text{with} \quad \hat{H}_{ij} = \frac{((\hat{p}_i - \hat{p}_j)/2)^2}{2\mu'} + V_{q\bar{q}}(\hat{r}_{ij}) \tag{21.30}$$

with $\mu' = 3m_s/8 = 3\mu/4$. From this relation we deduce the inequality:

$$2 E^{(3)} \geq 3 E^{(2)}(\mu') \quad \text{with} \quad \mu' = 3\mu/4. \tag{21.31}$$

**21.6.5.2** With the rescaling $r \to \alpha r$, one obtains:

$$\hat{H}^{(2)}(\tilde{\mu}) = \frac{\hat{p}^2}{2\alpha^2\tilde{\mu}} + g \ln \frac{r}{r_0} + g \ln \alpha. \tag{21.32}$$

The choice $\alpha = \sqrt{\mu/\tilde{\mu}}$ leads to $\hat{H}^{(2)}(\tilde{\mu}) = \hat{H}^{(2)}(\mu) + g \ln \alpha$ so that

$$E_n^{(2)}(\tilde{\mu}) = E_n^{(2)}(\mu) + \frac{g}{2} \ln \frac{\mu}{\tilde{\mu}}. \tag{21.33}$$

**21.6.5.3** In a logarithmic potential, the level spacing is independent of the mass. This is a remarkable feature of the observed spectra, at least for heavy quarks, and justifies the investigation of the logarithmic potential. Amazingly enough, this empirical prescription works quite well for light quarks, although one might expect that a relativistic treatment is necessary.

**21.6.5.4** The binding energies satisfy

$$E^{(3)} \geq \frac{3}{2} \left( E^{(2)} + \frac{g}{2} \ln \frac{4}{3} \right) \tag{21.34}$$

with

$$M_\Omega = 3m_s + \frac{E^{(3)}}{c^2} \quad m_\phi = 2m_s + \frac{E^{(2)}}{c^2}. \tag{21.35}$$

We therefore obtain

$$M_\Omega \geq \frac{3}{2}m_\phi + \frac{3g}{4c^2} \ln \frac{4}{3}. \tag{21.36}$$

**21.6.5.5** For $g = 650$ MeV and $a = 140$ MeV/$c^2$, we obtain

$$M_\Omega c^2 = 1672 \text{ MeV} \geq 1669 \text{ MeV}, \tag{21.37}$$

which is remarkably accurate.

Actually, the quark-quark potential is only logarithmic at distances smaller than 1 fm, which corresponds to the $\phi$ mean square radius. At larger distances, it grows more rapidly (linearly). Such inequalities are quite useful in practice for deciding what choice to make for the potential and for its domain of validity. The generalization of such inequalities can be found in the literature quoted below. They are useful in a variety of physical problems.

## 21.6.6 References

The approach presented in this chapter is discussed in J.-L. Basdevant, J.-M. Richard, and A. Martin, Nuclear Physics **B343**, 60, 69 (1990); J-L. Basdevant, J.-M. Richard, A. Martin, and Tai Tsun Wu, Nuclear Physics **B393**, 111 (1993); J-L. Basdevant and A. Martin, Journal of Mathematical Physics **37**, 5916 (1996).

The most recent and spectacular result in this field comes from the discovery of the doubly charmed baryon (ucc), whose mass is in remarkable agreement with theoretical predictions: $mc^2 = 3621$ MeV compared to $mc^2 = 3627$ MeV and $mc^2 = 3613$ MeV. See e.g. S. Fleck and J.-M. Richard, Prog. Theor. Phys. **82**, 760 (1989) and further references therein; M. Karliner and J.L. Rosner, Phys. Rev. D **90**, 094007 (2014); Phys. Rev. D 96, 033004 (2017); R. Aaij et al. (LHCb Collaboration, CERN), *Observation of the Doubly Charmed Baryon* $\Xi_{cc}^{++}$, Phys. Rev. Lett. **119**, 112001 (2017).

# Properties of a Bose–Einstein Condensate  22

By cooling down a gas of integer spin atoms to a very low temperature (typically less than one micro-Kelvin), one can observe the phenomenon of Bose–Einstein condensation. This results in a situation where a large fraction of the atoms accumulate in the same quantum state. We study here the ground state of such an $N$ particle system, hereafter called a condensate. We show that the nature of the system depends crucially on whether the two-body interactions between the atoms are attractive or repulsive.

## 22.1 Particle in a Harmonic Trap

We consider a particle of mass $m$ placed in a harmonic potential with a frequency $\omega/2\pi$. The Hamiltonian of the system is

$$\hat{H} = \frac{\hat{p}^2}{2m} + \frac{1}{2}m\omega^2\hat{r}^2, \tag{22.1}$$

where $\hat{r} = (\hat{x}, \hat{y}, \hat{z})$ and $\hat{p} = (\hat{p}_x, \hat{p}_y, \hat{p}_z)$ are respectively the position and momentum operators of the particle. We set $a_0 = \sqrt{\hbar/(m\omega)}$.

**22.1.1** Recall the energy levels of this system and its ground state wave function $\phi_0(r)$.

**22.1.2** We wish to obtain an upper bound on this ground state energy by the variational method. We use a Gaussian trial wave function:

$$\psi_\sigma(r) = \frac{1}{(\sigma^2\pi)^{3/4}} \exp(-r^2/(2\sigma^2)) \quad \text{with} \quad \sigma > 0, \tag{22.2}$$

© Springer Nature Switzerland AG 2019
J.-L. Basdevant, J. Dalibard, *The Quantum Mechanics Solver*,
https://doi.org/10.1007/978-3-030-13724-3_22

where $\sigma$ is a variational parameter. Using the set of integrals given below, find an upper bound on the ground state energy. Compare the bound with the exact value, and comment on the result.

$$\int |\psi_\sigma(\boldsymbol{r})|^2 \, dx \, dy \, dz = 1, \qquad \int |\psi_\sigma(\boldsymbol{r})|^4 \, dx \, dy \, dz = \frac{1}{(2\pi)^{3/2}} \frac{1}{\sigma^3},$$

$$\int x^2 |\psi_\sigma(\boldsymbol{r})|^2 \, dx \, dy \, dz = \frac{\sigma^2}{2}, \qquad \int \left| \frac{\partial \psi_\sigma(\boldsymbol{r})}{\partial x} \right|^2 \, dx \, dy \, dz = \frac{1}{2\sigma^2}.$$

## 22.2    Interactions Between Two Confined Particles

We now consider two particles of equal masses $m$, both placed in the same harmonic potential. We denote the position and momentum operators of the two particles by $\hat{\boldsymbol{r}}_1, \hat{\boldsymbol{r}}_2$ and $\hat{\boldsymbol{p}}_1, \hat{\boldsymbol{p}}_2$.

**22.2.1**  In the absence of interactions between the particles, the Hamiltonian of the system is

$$\hat{H} = \frac{\hat{\boldsymbol{p}}_1^2}{2m} + \frac{\hat{\boldsymbol{p}}_2^2}{2m} + \frac{1}{2} m\omega^2 \hat{\boldsymbol{r}}_1^2 + \frac{1}{2} m\omega^2 \hat{\boldsymbol{r}}_2^2. \tag{22.3}$$

(a)  What are the energy levels of this Hamiltonian?
(b)  What is the ground state wave function $\Phi_0(\boldsymbol{r}_1, \boldsymbol{r}_2)$?

**22.2.2**  We now suppose that the two particles interact via a potential $v(\boldsymbol{r}_1 - \boldsymbol{r}_2)$. We assume that, on the scale of $a_0$, this potential is of very short range and that it is peaked around the origin. Therefore, for two functions $f(\boldsymbol{r})$ and $g(\boldsymbol{r})$ which vary appreciably only over domains larger than $a_0$, one has

$$\iint f(\boldsymbol{r}_1) \, g(\boldsymbol{r}_2) \, v(\boldsymbol{r}_1 - \boldsymbol{r}_2) \, d^3r_1 \, d^3r_2 \simeq \frac{4\pi\hbar^2 a}{m} \int f(\boldsymbol{r}) \, g(\boldsymbol{r}) \, d^3r. \tag{22.4}$$

The quantity $a$, which is called the scattering length, is a characteristic of the atomic species under consideration. It can be positive (repulsive interaction) or negative (attractive interaction). One can measure for instance that for sodium atoms (isotope $^{23}$Na) $a = 3.4$ nm, whereas $a = -1.5$ nm for lithium atoms (isotope $^7$Li).

(a)  Using perturbation theory, calculate to first order in $a$ the shift of the ground state energy of $\hat{H}$ caused by the interaction between the two atoms. Comment on the sign of this energy shift.
(b)  Under what condition on $a$ and $a_0$ is this perturbative approach expected to hold?

## 22.3  Energy of a Bose–Einstein Condensate

We now consider $N$ particles confined in the same harmonic trap of angular frequency $\omega$. The particles have pairwise interactions through the potential $v(r)$ defined by (22.4). The Hamiltonian of the system is

$$\hat{H} = \sum_{i=1}^{N} \left( \frac{\hat{p}_i^2}{2m} + \frac{1}{2}m\omega^2 \hat{r}_i^2 \right) + \sum_{i<j} v(\hat{r}_i - \hat{r}_j). \tag{22.5}$$

In order to find an (upper) estimate of the ground state energy of the system, we use the variational method with factorized trial wave functions of the type:

$$\Psi_\sigma(r_1, r_2, \ldots, r_N) = \psi_\sigma(r_1)\, \psi_\sigma(r_2) \, \ldots \, \psi_\sigma(r_N), \tag{22.6}$$

where $\psi_\sigma(r)$ is defined in (22.2).

**22.3.1**  Calculate the expectation values of the kinetic energy, of the potential energy and of the interaction energy, if the $N$ particle system is in the state $|\Psi_\sigma\rangle$:

$$E_k(\sigma) = \langle \Psi_\sigma | \sum_{i=1}^{N} \frac{\hat{p}_i^2}{2m} |\Psi_\sigma\rangle \quad E_p(\sigma) = \langle \Psi_\sigma | \sum_{i=1}^{N} \frac{1}{2}m\omega^2 \hat{r}_i^2 |\Psi_\sigma\rangle$$

$$E_{int}(\sigma) = \langle \Psi_\sigma | \sum_{i<j} v(\hat{r}_i - \hat{r}_j)|\Psi_\sigma\rangle$$

We set $E(\sigma) = \langle \Psi_\sigma | \hat{H} |\Psi_\sigma\rangle$.

**22.3.2**  We introduce the dimensionless quantities $\tilde{E}(\sigma) = E(\sigma)/(N\hbar\omega)$ and $\tilde{\sigma} = \sigma/a_0$. Give the expression of $\tilde{E}$ in terms of $\tilde{\sigma}$. Cast the result in the form

$$\tilde{E}(\sigma) = \frac{3}{4}\left( \frac{1}{\tilde{\sigma}^2} + \tilde{\sigma}^2 \right) + \frac{\eta}{\tilde{\sigma}^3} \tag{22.7}$$

and express the quantity $\eta$ as a function of $N$, $a$ and $a_0$. In all what follows, we shall assume that $N \gg 1$.

**22.3.3**  For $a = 0$, recall the ground state energy of $\hat{H}$.

---

## 22.4  Condensates with Repulsive Interactions

In this part, we assume that the two-body interaction between the atoms is repulsive, i.e. $a > 0$.

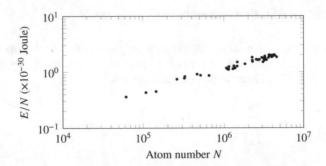

**Fig. 22.1** Energy per atom $E/N$ in a sodium condensate, as a function of the number of atoms $N$ in the condensate

**22.4.1** Draw qualitatively the value of $\tilde{E}$ as a function of $\tilde{\sigma}$. Discuss the variation with $\eta$ of the position of its minimum $\tilde{E}_{\min}$.

**22.4.2** We consider the case $\eta \gg 1$. Show that the contribution of the kinetic energy to $\tilde{E}$ is negligible. In that approximation, calculate an approximate value of $\tilde{E}_{\min}$.

**22.4.3** In this variational calculation, how does the energy of the condensate vary with the number of atoms $N$? Compare the prediction with the experimental result shown in Fig. 22.1.

**22.4.4** Figure 22.1 has been obtained with a sodium condensate (mass $m = 3.8 \times 10^{-26}$ kg) in a harmonic trap of frequency $\omega/(2\pi) = 142$ Hz.

(a) Calculate $a_0$ and $\hbar\omega$ for this potential.
(b) Above which value of $N$ does the approximation $\eta \gg 1$ hold?
(c) Within the previous model, calculate the value of the sodium atom scattering length that can be inferred from the data of Fig. 22.1. Compare the result with the value obtained in scattering experiments $a = 3.4$ nm. Is it possible a priori to improve the accuracy of the variational method?

## 22.5   Condensates with Attractive Interactions

We now suppose that the scattering length $a$ is negative.

**22.5.1** Draw qualitatively $\tilde{E}$ as a function of $\tilde{\sigma}$.

**22.5.2** Comment on the approximation (22.4) in the region $\sigma \to 0$.

**22.5.3** Show that there exists a critical value $\eta_c$ of $|\eta|$ above which $\tilde{E}$ no longer has a local minimum for a value $\tilde{\sigma} \neq 0$. Calculate the corresponding size $\sigma_c$ as a function of $a_0$.

**22.5.4** In an experiment performed with lithium atoms ($m = 1.17 \times 10^{-26}$ kg), it has been noticed that the number of atoms in the condensate never exceeds 1200 for a trap of frequency $\omega/(2\pi) = 145$ Hz. How can this result be explained?

## 22.6 Solutions

### 22.6.1 Particle in a Harmonic Trap

**22.6.1.1** The Hamiltonian of a three-dimensional harmonic oscillator can be written

$$\hat{H} = \hat{H}_x + \hat{H}_y + \hat{H}_z, \tag{22.8}$$

where $\hat{H}_i$ represents a one dimensional harmonic oscillator of same frequency along the axis $i = x, y, z$. We therefore use a basis of eigenfunctions of $\hat{H}$ of the form $\phi(x, y, z) = \chi_{n_x}(x)\chi_{n_y}(y)\chi_{n_z}(z)$, i.e. products of eigenfunctions of $\hat{H}_x, \hat{H}_y, \hat{H}_z$, where $\chi_n(x)$ is the $n$th Hermite function. The eigenvalues of $\hat{H}$ can be written as $E_n = (n + 3/2)\hbar\omega$, where $n = n_x + n_y + n_z$ is a non-negative integer.

The ground state wave function, of energy $(3/2)\hbar\omega$, corresponds to $n_x = n_y = n_z = 0$, i.e.

$$\phi_0(\boldsymbol{r}) = \frac{1}{(a_0^2\pi)^{3/4}} \exp[-r^2/(2a_0^2)]. \tag{22.9}$$

**22.6.1.2** The trial wave functions $\psi_\sigma$ are normalized. In order to obtain an upper bound for the ground-state energy of $\hat{H}$, we must calculate $E(\sigma) = \langle \psi_\sigma | \hat{H} | \psi_\sigma \rangle$ and minimize the result with respect to $\sigma$. Using the formulas given in the text, one obtains

$$\left\langle \psi_\sigma \left| \frac{\hat{p}^2}{2m} \right| \psi_\sigma \right\rangle = 3\frac{\hbar^2}{2m}\frac{1}{2\sigma^2} \qquad \left\langle \psi_\sigma \left| \frac{1}{2}m\omega^2 r^2 \right| \psi_\sigma \right\rangle = 3\frac{m\omega^2}{2}\frac{\sigma^2}{2} \tag{22.10}$$

and

$$E(\sigma) = \frac{3}{4}\hbar\omega \left( \frac{a_0^2}{\sigma^2} + \frac{\sigma^2}{a_0^2} \right). \tag{22.11}$$

This quantity is minimum for $\sigma = a_0$, and we find $E_{\min}(\sigma) = (3/2)\hbar\omega$. In this particular case, the upper bound coincides with the exact result. This is due to the fact that the set of trial wave functions contains the ground state wave function of $\hat{H}$.

## 22.6.2 Interactions Between Two Confined Particles

**22.6.2.1 (a)** The Hamiltonian $\hat{H}$ can be written as $\hat{H} = \hat{H}_1 + \hat{H}_2$, where $\hat{H}_1$ and $\hat{H}_2$ are respectively the Hamiltonians of particle 1 and particle 2. A basis of eigenfunctions of $\hat{H}$ is formed by considering products of eigenfunctions of $\hat{H}_1$ (functions of the variable $r_1$) and eigenfunctions of $\hat{H}_2$ (functions of the variable $r_2$). The energy eigenvalues are $E_n = (n+3)\hbar\omega$, where $n$ is a non-negative integer.

**(b)** The ground state of $\hat{H}$ is:

$$\Phi_0(r_1, r_2) = \phi_0(r_1)\phi_0(r_2) = \frac{1}{a_0^3 \pi^{3/2}} \exp[-(r_1^2 + r_2^2)/(2a_0^2)]. \tag{22.12}$$

**22.6.2.2 (a)** Since the ground state of $\hat{H}$ is non-degenerate, its shift to first order in $a$ can be written as

$$\Delta E = \langle \Phi_0 | \tilde{v} | \Phi_0 \rangle = \iint |\Phi_0(r_1, r_2)|^2 \, v(r_1 - r_2) \, \mathrm{d}^3 r_1 \mathrm{d}^3 r_2$$

$$\simeq \frac{4\pi \hbar^2 a}{m} \int |\phi_0(r)|^4 \, \mathrm{d}^3 r = \frac{4\pi \hbar^2 a}{m} \frac{1}{(2\pi)^{3/2}} \frac{1}{a_0^3} \tag{22.13}$$

therefore

$$\frac{\Delta E}{\hbar\omega} = \sqrt{\frac{2}{\pi}} \frac{a}{a_0}. \tag{22.14}$$

For a repulsive interaction ($a > 0$), there is an increase in the energy of the system. Conversely, in the case of an attractive interaction ($a < 0$), the ground state energy is lowered.

**(b)** The perturbative approach yields a good approximation provided the energy shift $\Delta E$ is small compared to the level spacing $\hbar\omega$ of $\hat{H}$. Therefore, one must have $|a| \ll a_0$, i.e. the scattering length must be small compared to the spreading of the ground state wave function.

## 22.6.3 Energy of a Bose–Einstein Condensate

**22.6.3.1** Using the formulas provided in the text, one obtains:

$$E_k(\sigma) = N \frac{3}{4} \frac{\hbar^2}{m\sigma^2} \qquad E_p(\sigma) = N \frac{3}{4} m\omega^2 \sigma^2$$

$$E_{\text{int}}(\sigma) = \frac{N(N-1)}{2} \sqrt{\frac{2}{\pi}} \hbar\omega \frac{a a_0^2}{\sigma^3}.$$

Indeed, there are $N$ kinetic energy and potential energy terms, and $N(N-1)/2$ pairs which contribute to the interaction energy.

**22.6.3.2** With the change of variables introduced in the text, one finds

$$\tilde{E}(\sigma) = \frac{3}{4}\left(\frac{1}{\tilde{\sigma}^2} + \tilde{\sigma}^2\right) + \frac{N-1}{\sqrt{2\pi}}\frac{a}{a_0}\frac{1}{\tilde{\sigma}^3} \tag{22.15}$$

so that

$$\eta = \frac{N-1}{\sqrt{2\pi}}\frac{a}{a_0}. \tag{22.16}$$

**22.6.3.3** If the scattering length is zero, there is no interaction between the particles. The ground state of the system is the product of the $N$ functions $\phi_0(r_i)$ and the ground state energy is $E = (3/2)N\hbar\omega$.

### 22.6.4 Condensates with Repulsive Interactions

**22.6.4.1** Figure 22.2 gives the variation of $\tilde{E}(\tilde{\sigma})$ as a function of $\tilde{\sigma}$ for increasing values of $\eta$. The value of the function for a given value of $\tilde{\sigma}$ increases as $\eta$ increases. For large $\tilde{\sigma}$, the behavior of $\tilde{E}$ does not depend on $\eta$. It is dominated by the potential energy term $3\tilde{\sigma}^2/4$.

The minimum $\tilde{E}_{\min}$ increases as $\eta$ increases. This minimum corresponds the point where the potential energy term, which tends to favor small values of $\sigma$, matches the kinetic and interaction energy terms which, on the contrary, favor large sizes $\sigma$. Since the interactions are repulsive, the size of the system is larger than in the absence of interactions, and the corresponding energy is also increased.

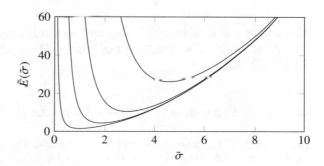

**Fig. 22.2** Variation of $\tilde{E}(\tilde{\sigma})$ with $\tilde{\sigma}$ for $\eta = 0, 10, 100, 1000$ (*from bottom to top*)

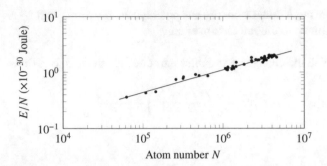

**Fig. 22.3** Fit of the experimental data with an $N^{2/5}$ law

**22.6.4.2** Let us assume $\eta$ is much larger than 1 and let us neglect a priori the kinetic energy term $1/\tilde{\sigma}^2$. The function $(3/4)\tilde{\sigma}^2 + \eta/\tilde{\sigma}^3$ is minimum for $\tilde{\sigma}_{\min} = (2\eta)^{1/5}$ where its value is

$$\tilde{E}_{\min} = \frac{5}{4}(2\eta)^{2/5}. \tag{22.17}$$

One can check a posteriori that it is legitimate to neglect the kinetic energy term $1/\tilde{\sigma}^2$. In fact it is always smaller than one of the two other contributions to $\tilde{E}$:

- For $\tilde{\sigma} < \tilde{\sigma}_{\min}$, one has $1/\tilde{\sigma}^2 \ll \eta/\tilde{\sigma}^3$.
- For $\tilde{\sigma} > \tilde{\sigma}_{\min}$, one has $1/\tilde{\sigma}^2 \ll \tilde{\sigma}^2$.

**22.6.4.3** For a number of atoms $N \gg 1$, the energy of the system as calculated by the variational method is

$$\frac{E}{N} = \frac{5}{4}\hbar\omega \left( \sqrt{\frac{2}{\pi}} N \frac{a}{a_0} \right)^{2/5}. \tag{22.18}$$

This variation of $E/N$ as $N^{2/5}$ is very well reproduced by the data. In Fig. 22.3 we have plotted a fit of the data with this law. One finds $E/N \simeq \alpha N^{2/5}$ with $\alpha = 8.2 \times 10^{-33}$ Joule.

**22.6.4.4** **(a)** One finds $a_0 = 1.76\mu\text{m}$ and $\hbar\omega = 9.4 \ 10^{-32}$ Joule.

**(b)** Consider the value $a = 3.4$ nm given in the text. The approximation $\eta \gg 1$ will hold if $N \gg 1300$. This is clearly the case for the data of Fig. 22.1.

**(c)** The coefficient $\alpha = 8.2 \times 10^{-33}$ Joule found by fitting the data leads to $a = 2.8$ nm. This value is somewhat lower than the expected value $a = 3.4$ nm. This

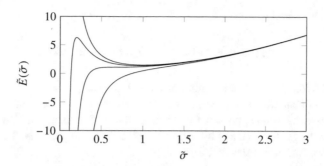

**Fig. 22.4** Plot of $\tilde{E}(\tilde{\sigma})$ for $\eta = 0$; $\eta = -0.1$; $\eta = -0.27$; $\eta = -1$ (*from top to bottom*)

is due to the fact that the result (22.3), $E/(N\hbar\omega) \simeq 1.142(Na/a_0)^{2/5}$, obtained in a variational calculation using simple Gaussian trial functions, does not yield a sufficiently accurate value of the ground state energy. With more appropriate trial wave functions, one can obtain, in the mean field approximation and in the limit $\eta \gg 1 : E_{gs}/(N\hbar\omega) \simeq 1.055(Na/a_0)^{2/5}$. The fit to the data is then in agreement with the experimental value of the scattering length.

### 22.6.5 Condensates with Attractive Interactions

**22.6.5.1** The function $\tilde{E}(\tilde{\sigma})$ is represented in Fig. 22.4. We notice that it has a local minimum only for small enough values of $\eta$. For $\eta < 0$, there is always a minimum at 0, where the function tends to $-\infty$.

**22.6.5.2** The absolute minimum at $\sigma = 0$ corresponds to a highly compressed atomic cloud. For such small sizes, approximation (22.4) for a "short range" potential loses its meaning. Physically, one must take into account the formation of molecules and/or atomic aggregates which have not been considered here.

**22.6.5.3** The local minimum at $\tilde{\sigma} \neq 0$ disappears when $\tilde{E}(\tilde{\sigma})$ has an inflexion point where the derivative vanishes. This happens for a critical value of $\eta$ determined by the two conditions:

$$\frac{d\tilde{E}}{d\tilde{\sigma}} = 0 \qquad \frac{d^2\tilde{E}}{d\tilde{\sigma}^2} = 0. \tag{22.19}$$

This leads to the system

$$0 = -\frac{1}{\tilde{\sigma}^4} + 1 - \frac{2\eta}{\tilde{\sigma}^5}, \qquad 0 = \frac{3}{\tilde{\sigma}^4} + 1 + \frac{8\eta}{\tilde{\sigma}^5} \tag{22.20}$$

from which we obtain

$$|\eta_c| = \frac{2}{5^{5/4}} \simeq 0.27 \qquad \tilde{\sigma}_c = \frac{1}{5^{1/4}} \simeq 0.67 \qquad (22.21)$$

or $\sigma_c \simeq 0.67 a_0$. If the local minimum exists, i.e. for $|\eta| < |\eta_c|$, one can hope to obtain a metastable condensate, whose size will be of the order of the minimum found in this variational approach. On the other hand, if one starts with a value of $|\eta|$ which is too large, for instance by trying to gather too many atoms, the condensate will collapse, and molecules will form.

**22.6.5.4** For the given experimental data one finds $a_0 = 3.1 \, \mu\text{m}$, and a critical number of atoms:

$$N_c = \sqrt{2\pi} \, \eta_c \, \frac{a_0}{|a|} \sim 1400, \qquad (22.22)$$

in good agreement with experimental observations.

## 22.6.6 Comments and References

The first Bose–Einstein condensate of a dilute gas of rubidium atoms was observed in Boulder (USA) in 1995: M.H. Anderson, J.R. Ensher, M.R. Matthews, C.E. Wieman, and E.A. Cornell, *Observation of Bose-Einstein Condensation in a Dilute Atomic Vapor*, Science **269**, 198 (1995).

The experimental data shown in this chapter for a sodium condensate are adapted from the results reported by M.-O. Mewes, M.R. Andrews, N.J. van Druten, D.M. Kurn, D.S. Durfee, and W. Ketterle, *Bose-Einstein Condensation in a Tightly Confining dc Magnetic Trap*, Phys. Rev. Lett. **77**, 416 (1996). The measurement of the energy $E/N$ is done by suddenly switching off the confining potential and by measuring the resulting ballistic expansion. The motion of the atoms in this expansion essentially originates from the conversion of the potential energy of the atoms in the trap into kinetic energy.

Experimental results on the attractive case (lithium) have first been reported by C. Bradley, C.A. Sackett, and R.G. Hulet, *Bose-Einstein Condensation of Lithium: Observation of Limited Condensate Number*, Phys. Rev. Lett. **78**, 985 (1997). For a recent study of the dynamics of the collapse of the gas in this attractive case, see C. Eigen, A. L. Gaunt, A. Suleymanzade, N. Navon, Z. Hadzibabic, R. P. Smith, *Observation of Weak Collapse in a Bose-Einstein Condensate*, Phys. Rev. X **6**, 041058 (2016).

A general overview on the subject can be found in the books by C. J. Pethick and H. Smith, *Bose-Einstein condensation in dilute gases* (Cambridge university press, 2002) and by L. Pitaevskii and S. Stringari, *Bose–Einstein condensation and Superfluidity* (second edition, Oxford University Press, 2016). During the last two decades, ultra-cold gases of atoms (bosons or fermions) have been extensively used to study many-body physics. For a review, see e.g. I. Bloch, J. Dalibard, and W. Zwerger, *Many-Body Physics with Ultracold Gases*, Rev. Mod. Phys. **80**, 885 (2008). These systems are now considered as good candidates for the realization of a quantum simulator, see the special issue of Nature Physics, April 2012.

The Physics Nobel prize was awarded in 2001 to E. Cornell, W. Ketterle, and C. Wieman for the achievement of Bose-Einstein condensation in dilute gases of alkali atoms, and for early fundamental studies of the properties of the condensates.

# Quantized Vortices

**23**

The rotation of a quantum fluid, a Bose–Einstein condensate for instance, is very different from that of a classical gas. When a Bose–Einstein condensate is stirred at a frequency $\Omega$, one first notes that below a critical frequency $\Omega_c$, the condensate stays at rest, which is a manifestation of its superfluid character. Then for $\Omega > \Omega_c$, the condensate is set in motion in a very specific manner, with the nucleation of one or several quantized vortices (see Fig. 23.1). These vortices are points (in two dimensions) or lines (in three dimensions) where the atomic density vanishes, and around which the phase of the wavefunction describing the atomic assembly varies by $2n\pi$, where $n$ is nonzero integer.

In this chapter, the atomic motion is restricted to the $xy$ plane and we use the polar coordinates $(r, \varphi)$ in this plane, with $\boldsymbol{u}_r$ and $\boldsymbol{u}_\varphi$ the radial and orthoradial unit vectors.

Reminder: The eigenenergies of a one-dimension harmonic oscillator of mass $m$ and frequency $\omega$ are $E_n = (n + 1/2)\hbar\omega$, with $n$ positive or zero integer. The corresponding eigenfunctions are the Hermite functions $\phi_n(x) = c_n\, P_n(x/a)\, e^{-x^2/(2a^2)}$, where $a = \sqrt{\hbar/(m\omega)}$, $P_n$ is a $n$-th degree polynomial and $c_n$ is a normalisation coefficient:

$$P_0(u) = 1 \qquad P_1(u) = 2u \qquad P_2(u) = 4u^2 - 2 \qquad c_n = \frac{\pi^{-1/4}}{\sqrt{2^n\, n!\, a}}.$$

## 23.1  Magnetic Trapping

A neutral particle of spin $S = 1/2$ is placed in the magnetic field

$$\boldsymbol{B}(\boldsymbol{r}) = b'(x\boldsymbol{u}_x - y\boldsymbol{u}_y) + B_0\boldsymbol{u}_z, \qquad (23.1)$$

© Springer Nature Switzerland AG 2019
J.-L. Basdevant, J. Dalibard, *The Quantum Mechanics Solver*,
https://doi.org/10.1007/978-3-030-13724-3_23

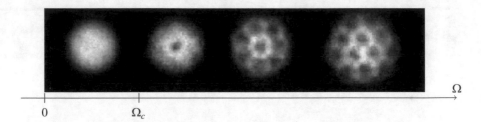

Fig. 23.1 Nucleation of quantized vortices in a rotating Bose–Einstein condensate. The rotation axis $z$ is perpendicular to the figure. The number of vortices depends on the rotation frequency $\Omega$. (Figure from Laboratoire Kastler Brossel)

where $\boldsymbol{u}_j$ denotes a unit vector along the axis $j$. We denote $\hat{\boldsymbol{\mu}} = \gamma \hat{\boldsymbol{S}}$ the operator associated with the magnetic moment of the particle and $|\pm\rangle_z$ the eigenstates of the spin projection along the $z$ axis. In this section, we treat the position of the particle $\boldsymbol{r} = x\boldsymbol{u}_x + y\boldsymbol{u}_y$ as a classical variable, which is thus a parameter of the Hamiltonian.

**23.1.1** Using the Pauli matrices, write the $2 \times 2$ matrix describing the coupling between the spin and the magnetic field in the basis $\{|+\rangle_z, |-\rangle_z\}$.

**23.1.2** Give the magnetic energy levels $E_\pm(\boldsymbol{r})$ of the particle.

**23.1.3** Assuming that the distance $r = (x^2 + y^2)^{1/2}$ is small compared to $B_0/b'$, expand the expression of the energy levels to second order in $r$ and write them as:

$$E_\pm(\boldsymbol{r}) = \pm \left( E_0 + \frac{1}{2}m\omega^2 r^2 \right) . \tag{23.2}$$

Give the expression of the oscillation frequency $\omega$ in terms of the magnetic field parameters.

**23.1.4** Calculate the oscillation frequency for the magnetic field configuration $B_0 = 10^{-4}$ T, $b' = 1$ T/m. Use the mass and magnetic moment of a rubidium atom: $m = 1.4 \times 10^{-25}$ kg, $\mu = \hbar\gamma/2 = 9.3 \times 10^{-24}$ J/T.

## 23.2   The Two-Dimensional Harmonic Oscillator

We assume that the atom is prepared in the spin state corresponding to the energy $E_+(\boldsymbol{r})$ in (23.2). In the following this energy plays the role of a potential for the motion of the atomic center-of-mass. We now treat this motion using the quantum formalism with the Hamiltonian:

$$\hat{H} = \frac{\hat{p}^2}{2m} + \frac{1}{2}m\omega^2\hat{r}^2 \quad \text{with} \quad \hat{p}^2 = \hat{p}_x^2 + \hat{p}_y^2 . \tag{23.3}$$

The constant $E_0$ is omitted to simplify the algebra. In what follows, the spin degree of freedom will not appear anymore since its sole role in the present problem was to give rise to the trapping potential $m\omega^2 r^2/2$.

**23.2.1** Write the Hamiltonian $\hat{H}$ as $\hat{H} = \hat{H}_x + \hat{H}_y$. Deduce from this expression the energy levels of $\hat{H}$ and their degeneracy.

**23.2.2** We introduce the $z$ component of the orbital angular momentum $\hat{L}_z = \hat{x}\hat{p}_y - \hat{y}\hat{p}_x$. Explain without explicit calculation why $\hat{H}$ and $\hat{L}_z$ commute.

**23.2.3** The expression of $\hat{L}_z$ in polar coordinates is $\hat{L}_z = \dfrac{\hbar}{i}\dfrac{\partial}{\partial\varphi}$. Give its eigenvalues and its eigenfunctions $\psi(r, \varphi)$.

**23.2.4** Give the expression of the ground state $\psi_0(r)$ and the value of its angular momentum along the $z$ axis.

**23.2.5** We are now interested in the energy level $E = 2\hbar\omega$.

(a) Show that one can choose the functions

$$\psi_{\pm 1}(r) = \sqrt{2}\,\frac{c_0 c_1}{a}\,(x \pm iy)\,e^{-r^2/2a^2}\,. \tag{23.4}$$

as a normalized basis of the corresponding subspace.
(b) Show that the functions $\psi_{\pm 1}(r)$ have a well-defined angular momentum $L_z$.
(c) Calculate the probability density per unit area $\rho(r)$ corresponding to $\psi_{\pm 1}(r)$.
(d) One can associate to any wavefunction $\psi(r)$ the probability current

$$J(r) = \frac{\hbar}{m}\mathrm{Im}\left[\psi^*(r)\,\nabla(\psi(r))\right], \tag{23.5}$$

and the corresponding velocity field is $v(r) = J(r)/\rho(r)$. Using the expression of the gradient operator in polar coordinates

$$\nabla\psi = \frac{\partial\psi}{\partial r}\,\mathbf{u}_r + \frac{1}{r}\frac{\partial\psi}{\partial\varphi}\,\mathbf{u}_\varphi, \tag{23.6}$$

calculate the velocity field associated with $\psi_{\pm 1}(r)$. Does it correspond to what one would expect for a vortex?

## 23.3  Quantum Physics in a Rotating Frame

We now set the magnetic trap in rotation, by superposing a small anisotropic perturbation to the isotropic potential $m\omega^2 r^2/2$ previously studied. The axes of the small perturbation rotate at frequency $\Omega$. We will not study the details of this perturbation since its only purpose is to impose a privileged frame, i.e. the frame rotating at frequency $\Omega$. Indeed this is the only frame in which the Hamiltonian is time-independent. Within Hamilton's formalism, the change from the laboratory reference frame to the rotating frame is performed by replacing $\hat{H}$ given in (23.3) by:

$$\hat{H}' = \hat{H} - \Omega \hat{L}_z . \tag{23.7}$$

**23.3.1** Using Ehrenfest's theorem, calculate the components of the average velocity $\langle v_x \rangle = d\langle x \rangle / dt$ and $\langle v_y \rangle = d\langle y \rangle / dt$. Does one recover the usual result $\langle v_i \rangle = \langle p_i \rangle / m$, $i = x, y$? Do you know another system for which a similar result holds?

**23.3.2** Still using Ehrenfest's theorem, calculate $d\langle p_x \rangle / dt$ and $d\langle p_y \rangle / dt$.

**23.3.3** Write the preceding results as

$$m \frac{d^2 \langle r \rangle}{dt^2} = -m\omega^2 \langle r \rangle + F(\langle r \rangle, \langle v \rangle) . \tag{23.8}$$

**23.3.4** We recall that in classical physics, two forces appear when working in a rotating frame: the centrifugal force $F_{\text{cen.}}$ and the Coriolis force $F_{\text{Cor.}}$ :

$$F_{\text{cen.}}(r) = m\Omega^2 r \qquad F_{\text{Cor.}}(v) = 2mv \times \Omega \qquad (\Omega = \Omega u_z) . \tag{23.9}$$

Relate the results derived above to the classical case.

**23.3.5** When the rotation frequency $\Omega$ exceeds the trapping frequency $\omega$, one observes experimentally that the atomic cloud is expelled from the trap. Explain this phenomenon using the classical equations of motion (one can neglect the Coriolis force).

## 23.4   Nucleation of Quantized Vortices

We consider an ensemble of $N$ identical atoms, all prepared in the same spin state. These atoms are confined in a trap of frequency $\omega$, which rotates at frequency $\Omega$. Each atom is labeled by the index $j$, $j = 1, \ldots, N$, and the Hamiltonian in the rotating frame reads:

$$\hat{H}_N = \sum_{j=1}^{N} \frac{\hat{p}_j^2}{2m} + \frac{1}{2} m\omega^2 \hat{r}_j^2 - \Omega \hat{L}_{j,z} , \tag{23.10}$$

where we neglected the interactions between particles. We suppose that the expression of the trapping potential $m\omega^2 r^2 / 2$ remains valid even if the spin of the atoms is larger than $1/2$.

**23.4.1** The atoms used in the experiment have a spin equal to 1. Explain why it is legitimate to look for the orbital part of the ground state of the system with the following form:

$$\Phi(r_1, r_2, \ldots, r_N) = \phi(r_1)\,\phi(r_2)\ldots\phi(r_N). \tag{23.11}$$

**23.4.2** Explain why the choice $\phi(r) = \psi_0(r)$ and the choice $\phi(r) = \psi_{\pm1}(r)$ both lead to states $\Phi$ that are eigenstates of $\hat{H}_N$. Calculate the corresponding energy in both cases.

**23.4.3** Plot on a same graph the variation of the energy with $\Omega$ for the states $\Phi$ corresponding to $\phi(r) = \psi_0(r)$ and $\phi(r) = \psi_{+1}(r)$.

Show that there exists a rotation frequency $\Omega_c$ above which the state $\Phi$ corresponding to $\phi(r) = \psi_0(r)$ cannot be the ground state of the system. Considering the result obtained in 23.3.5, do you think that this phenomenon can be observed in practice?

**23.4.4** We now take into account the interactions between atoms in a perturbative way. These interactions are modelled by a potential composed of $N(N-1)/2$ binary terms:

$$V = g\sum_{i<j} G(r_i - r_j), \quad g > 0, \tag{23.12}$$

where $G(r)$ is a positive function, centered in $r = 0$, tending to 0 at infinity and with an integral equal to 1. The width of $G$ is supposed to be much smaller than any other relevant length scale of the problem and one can conveniently replace $G$ by a Dirac distribution: $G(r) = \delta(r)$.

(a) Using perturbation theory in the non-degenerate case, evaluate at first order in $g$ the displacement of the energy levels corresponding to $\phi(r) = \psi_0(r)$ and to $\phi(r) = \psi_{+1}(r)$. We give

$$\iint e^{-2r^2/a^2}\, d^2r = \frac{\pi a^2}{2} \qquad \iint r^4 e^{-2r^2/a^2}\, d^2r = \frac{\pi a^6}{4}.$$

(b) Determine the new value for the critical angular frequency $\Omega_c$ above which the state $\Phi$ corresponding to $\phi(r) = \psi_0(r)$ cannot be the ground state of the system.
(c) Can this phenomenon be observed, taking into account the constraint found in question 23.3.5?
(d) For rubidium atoms confined in the trap described above, one finds $g \approx 10^{-3}\hbar^2/m$. What is the value of the relative shift $|\Omega_c - \omega|/\omega$ for $N = 1000$ atoms?
(e) To which picture(s) of Fig. 23.1 does the present phenomenon correspond?

**23.4.5** In order to study the rotation range $\Omega \simeq \omega$, one can enrich the previous model and assume that the ground state of the system now corresponds to a function $\phi(\mathbf{r})$ such that

$$\phi(\mathbf{r}) = \sum_{n=1}^{N} C_n \, \psi_{+n}(\mathbf{r}), \qquad \psi_{\pm n}(\mathbf{r}) = d_n \, (x \pm iy)^n \, e^{-r^2/2a^2}, \tag{23.13}$$

where $n$ is a positive integer and $d_n$ a normalization coefficient. The coefficients $C_n$ are determined from a minimization of the energy, the upper bound $N$ being an adjustable parameter.

(a) Show that the states $\psi_{\pm n}(\mathbf{r})$ have a well defined angular momentum.
(b) Show that the states $\psi_{\pm n}(\mathbf{r})$ are eigenstates of the Hamiltonian $\hat{H}$ given in (23.3) with the energy $(n+1)\hbar\omega$.
(c) Justify the choice of $\phi(\mathbf{r})$ in (23.13) as a good approximation of the ground state $\Phi$ in the rotation regime $\Omega \simeq \omega$.
(d) What is the expected density profile for such states?
(e) Do the experimental results of Fig. 23.1 correspond to this model?

## 23.5  Solutions

### 23.5.1  Magnetic Trapping

**23.5.1.1** The Hamiltonian reads

$$\hat{H} = -\hat{\boldsymbol{\mu}} \cdot \mathbf{B} = -\frac{\hbar\gamma}{2} \begin{pmatrix} B_0 & b'(x+iy) \\ b'(x-iy) & -B_0 \end{pmatrix}.$$

**23.5.1.2** Its eigenvalues are

$$E_{\pm}(\mathbf{r}) = \pm \frac{\hbar\gamma}{2} \left( B_0^2 + b'^2 r^2 \right)^{1/2}.$$

**23.5.1.3** For $b'r \ll B_0$ we find

$$E_{\pm}(\mathbf{r}) \simeq \pm \frac{\hbar\gamma B_0}{2} \left( 1 + \frac{b'^2 r^2}{2B_0^2} \right) = \pm \left( E_0 + \frac{1}{2} m\omega^2 r^2 \right)$$

with

$$E_0 = \frac{\hbar\gamma B_0}{2} \qquad \omega = \left( \frac{\hbar\gamma b'^2}{2m B_0} \right)^{1/2}.$$

**23.5.1.4** We take $\mu = \hbar\gamma/2$ equal to the Bohr magneton and we get

$$\nu = \frac{\omega}{2\pi} = \frac{b'}{2\pi}\sqrt{\frac{\mu}{mB_0}} = 130 \text{ Hz}.$$

## 23.5.2 The Two-Dimensional Harmonic Oscillator

**23.5.2.1** The Hamiltonian $\hat{H}$ can be written $\hat{H} = \hat{H}_x + \hat{H}_y$ with

$$\hat{H}_x = \frac{\hat{p}_x^2}{2m} + \frac{1}{2}m\omega^2\hat{x}^2, \qquad \hat{H}_y = \frac{\hat{p}_y^2}{2m} + \frac{1}{2}m\omega^2\hat{y}^2.$$

Each of these two Hamiltonians corresponds to a one-dimensional oscillator and they commute with each other. Their common eigenbasis is the set of functions

$$\phi_{n_x,n_y}(\boldsymbol{r}) = \phi_{n_x}(x)\,\phi_{n_y}(y)\,.$$

These functions are eigenstates of $\hat{H}$ with the eigenvalues $E_{n_x,n_y} = (n_x+1/2)\hbar\omega + (n_y + 1/2)\,\hbar\omega$. Therefore the eigenvalues of $\hat{H}$ can be written $E_n = (n + 1)\hbar\omega$, where $n$ is a nonnegative integer. The subspace associated to $E_n$ is generated by the functions $\phi_{n_x,n_y}(\boldsymbol{r})$ with $n_x + n_y = n$ and its dimension is equal to $n + 1$.

**23.5.2.2** An operator $\hat{O}$ commutes with $\hat{L}_z$ if it is invariant under a rotation around the $z$-axis. The kinetic energy term $\hat{p}^2/(2m)$ is indeed rotationally invariant. Since we consider here an isotropic oscillator, with the same oscillation frequency along $x$ and $y$, the potential energy term $m\omega^2\hat{r}^2/2$ is also invariant in a rotation around the $z$ axis and commutes with $\hat{L}_z$. This entails that $\hat{H}$ and $\hat{L}_z$ have a common eigenbasis.

**23.5.2.3** The eigenvalue equation for $\hat{L}_z$ reads in polar coordinates

$$\frac{\hbar}{i}\frac{\partial\psi(r,\varphi)}{\partial\varphi} = A\,\psi(r,\varphi)\,, \quad A \text{ real}\,,$$

and leads to $\psi(r,\varphi) = F(r)\,e^{iA\varphi/\hbar}$, where $F(r)$ is an arbitrary function. The condition $\psi(r,\varphi) = \psi(r,\varphi + 2\pi)$ then imposes that $A/\hbar$ is an integer. Denoting this integer by $m$, we can write the eigenfunction as $\psi(r,\varphi) = F(r)\,e^{im\varphi}$, and the corresponding eigenvalue is $m\hbar$.

**23.5.2.4** The ground level of the two-dimensional harmonic oscillator corresponds to the eigenvalue $\hbar\omega$. This eigenvalue is not degenerate and the eigenstate is

$$\psi_0(\boldsymbol{r}) = c_0^2\,e^{-r^2/(2a^2)} \qquad \text{with} \quad c_0^2 = \frac{1}{\sqrt{\pi}}\,.$$

The expression of this state in polar coordinates is independent of the angle $\varphi$. $\psi_0(\mathbf{r})$ is thus an eigenstate of $\hat{L}_z$ with zero angular momentum.

**23.5.2.5  (a)** The subspace associated with the energy $2\hbar\omega$ has dimension 2 and it is spanned by the two functions

$$\phi_{1,0}(\mathbf{r}) = \phi_1(x)\phi_0(y) = \frac{2c_0c_1}{a} x\, e^{-r^2/(2a^2)} \tag{23.14}$$

$$\phi_{0,1}(\mathbf{r}) = \phi_0(x)\phi_1(y) = \frac{2c_0c_1}{a} y\, e^{-r^2/(2a^2)} . \tag{23.15}$$

We can also choose the two functions suggested in question 23.2.5 as a basis of this energy level:

$$\psi_+(\mathbf{r}) = \frac{1}{\sqrt{2}} \left(\phi_{1,0}(\mathbf{r}) + i\phi_{0,1}(\mathbf{r})\right), \qquad \psi_+(\mathbf{r}) = \frac{1}{\sqrt{2}} \left(\phi_{1,0}(\mathbf{r}) - i\phi_{0,1}(\mathbf{r})\right).$$

**(b)**  These two functions read in polar coordinate

$$\psi_\pm(\mathbf{r}) = \frac{\sqrt{2}\, c_0c_1}{a} r\, e^{-r^2/(2a^2)}\, e^{\pm i\varphi},$$

which shows that they are eigenstates of $\hat{L}_z$ with eigenvalue $\pm\hbar$.

**(c)**  The probability density $\rho(\mathbf{r})$ is the same for both states $\psi_\pm(\mathbf{r})$:

$$\rho(\mathbf{r}) = |\psi_\pm(\mathbf{r})|^2 = \frac{2c_0^2c_1^2}{a^2} r^2 e^{-r^2/a^2} = \frac{r^2\, e^{-r^2/a^2}}{\pi a^4} .$$

It is zero in $r = 0$, as expected for a state with a nonzero angular momentum.

**(d)**  The probability current is orthoradial and the velocity field reads

$$v(\mathbf{r}) = \frac{\hbar}{mr}\, \mathbf{u}_\varphi .$$

It corresponds indeed to what is expected for a vortex, with a velocity that increases when the distance to the center decreases. This is the opposite of rigid body rotation, where the velocity is proportional to the distance to center. Note that the divergence of the velocity in $r = 0$ does not lead to a divergence of the kinetic energy, because the probability density tends to 0 at the origin.

### 23.5.3 Quantum Physics in a Rotating Frame

**23.5.3.1** Ehrenfest's theorem leads to

$$\langle v_x \rangle = \frac{\mathrm{d}\langle x \rangle}{\mathrm{d}t} = \frac{1}{i\hbar}\langle [x, \hat{H}'] \rangle \qquad \langle v_y \rangle = \frac{\mathrm{d}\langle y \rangle}{\mathrm{d}t} = \frac{1}{i\hbar}\langle [y, \hat{H}'] \rangle.$$

Let us now calculate the commutators $[\hat{x}, \hat{H}']$ et $[\hat{y}, \hat{H}']$. One can readily show that

$$[\hat{x}, \hat{H}] = i\hbar \hat{p}_x/m \qquad\qquad [\hat{y}, \hat{H}] = i\hbar \hat{p}_y/m$$

and that

$$[\hat{x}, \Omega L_z] = -i\hbar\Omega\hat{y} \qquad\qquad [\hat{y}, \Omega L_z] = i\hbar\Omega\hat{x}.$$

This leads to

$$\frac{\mathrm{d}\langle x \rangle}{\mathrm{d}t} = \frac{\langle p_x \rangle}{m} + \Omega\langle y \rangle \qquad \frac{\mathrm{d}\langle y \rangle}{\mathrm{d}t} = \frac{\langle p_y \rangle}{m} - \Omega\langle x \rangle .$$

Consequently, the relation $\langle p \rangle = m\langle v \rangle$ does not hold in the rotating frame. This a situation analogous to that of a charged particle in a magnetic field. The correspondance between the two cases can be made more explicit by writing the Hamiltonian $\hat{H}'$ as

$$\hat{H}' = \frac{(\hat{p} - q A(\hat{r}))^2}{2m} + \frac{1}{2}m(\omega^2 - \Omega^2)r^2$$

with

$$q A(r) = m\boldsymbol{\Omega} \times r , \quad \boldsymbol{\Omega} = \Omega u_z .$$

In a uniform magnetic field $\boldsymbol{B} = B\boldsymbol{u}_z$, a possible gauge choice for the vector potential is $A(r) = B \times \boldsymbol{u}_z/2$. The motion of a neutral particle in the frame rotating at frequency $\Omega$ and in the trapping potential $m\omega^2 r^2/2$ is thus formally equivalent to the motion of a charged particle in the magnetic field $\boldsymbol{B} = 2m\boldsymbol{\Omega}/q$ with the trapping potential $m(\omega^2 - \Omega^2)r^2/2$.

**23.5.3.2** Ehrenfest's theorem for $\hat{p}_x$ and $\hat{p}_y$ leads to

$$\frac{\mathrm{d}\langle p_x \rangle}{\mathrm{d}t} = \frac{1}{i\hbar}\langle [p_x, \hat{H}'] \rangle = -m\omega^2\langle x \rangle + \Omega\langle p_y \rangle$$

$$\frac{\mathrm{d}\langle p_y \rangle}{\mathrm{d}t} = \frac{1}{i\hbar}\langle [p_y, \hat{H}'] \rangle = -m\omega^2\langle y \rangle - \Omega\langle p_x \rangle .$$

**23.5.3.3** Combining the answers to the two preceding questions we get

$$m\frac{\mathrm{d}^2\langle x\rangle}{\mathrm{d}t^2} = \frac{\mathrm{d}\langle p_x\rangle}{\mathrm{d}t} + m\Omega\frac{\mathrm{d}\langle y\rangle}{\mathrm{d}t} = -m\omega^2\langle x\rangle + \Omega\langle p_y\rangle + m\Omega\langle v_y\rangle$$

$$= -m\omega^2\langle x\rangle + 2m\Omega\langle v_y\rangle + m\Omega^2\langle x\rangle$$

and

$$m\frac{\mathrm{d}^2\langle y\rangle}{\mathrm{d}t^2} = -m\omega^2\langle y\rangle - 2m\Omega\langle v_x\rangle + m\Omega^2\langle y\rangle,$$

which has indeed the structure proposed in question 23.3.3:

$$\boldsymbol{F}(\langle \boldsymbol{r}\rangle, \langle \boldsymbol{v}\rangle) = 2m\langle \boldsymbol{v}\rangle \times \boldsymbol{\Omega} + m\Omega^2\langle \boldsymbol{r}\rangle.$$

**23.5.3.4** The equation of motion for the center of the wave packet contains the same three forces that enter in the classical description of the problem:

- The trapping force $-m\omega^2\langle \boldsymbol{r}\rangle$;
- The Coriolis force $2m\langle \boldsymbol{v}\rangle \times \boldsymbol{\Omega}$, with a structure similar to the Lorentz force $q\langle \boldsymbol{v}\rangle \times \boldsymbol{B}$ for a charged particle moving in a magnetic field;
- The centrifugal force $+m\Omega^2\langle \boldsymbol{r}\rangle$. Since the particle is confined in a harmonic trap, the centrifugal force simply reduces the spring constant of the trap, which shifts from $m\omega^2$ to $m(\omega^2 - \Omega^2)$.

**23.5.3.5** If one neglects the Coriolis force, the classical equation of motion for a particle in the rotating trap is

$$m\frac{d^2\boldsymbol{r}}{dt^2} = -m(\omega^2 - \Omega^2)\boldsymbol{r}.$$

When the rotation frequency $\Omega$ exceeds the trap frequency $\omega$, the solution of this equation diverges with time. This instability originates from the fact that the restoring force from the trap $-m\omega^2\boldsymbol{r}$ is not large enough to balance the expelling centrifugal force $+m\Omega^2\boldsymbol{r}$.

## 23.5.4 Nucleation of Quantized Vortices

**23.5.4.1** The atoms have an integer spin and are bosons. Therefore, the state vector describing the $N$ particles must be symmetric under the exchange of any pair of particles. Since we assume here that all particles are prepared in the same spin state, the orbital wave function must also be symmetric by exchange of the two particles. This is indeed the case of the proposed wave function, for which all particles are prepared in the same state $\phi(\boldsymbol{r})$. Such a state is called a *Hartree* wave function.

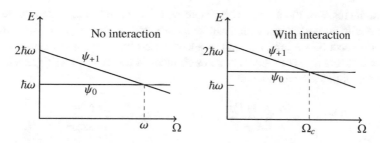

**Fig. 23.2** Variation of the energy per particle vs. $\Omega$, without (left) and with (right) repulsive interactions between the particles

**23.5.4.2** The wave functions $\psi_0(r)$ and $\psi_{\pm1}(r)$ are eigenstates of the Hamiltonian $\hat{H}$ (eigenvalues $\hbar\omega$ and $2\hbar\omega$) and of $\hat{L}_z$ (eigenvalues 0 and $\pm\hbar$). Therefore, they are eigenstates of the Hamiltonian in the rotating frame $\hat{H}'$ with eigenvalues $\hbar\omega$ for $\psi_0$, and $2\hbar\omega \mp \hbar\Omega$ for $\psi_{\pm1}$.

The $N$-body wave functions constructed from $\psi_0$ and $\psi_{\pm1}$ are thus eigenfunctions of $\hat{H}_N$, which is simply the sum of the Hamiltonians $\hat{H}'_j$ for $j = 1 \ldots N$. The corresponding energies are $N\hbar\omega$ for $\psi_0$ and $N\hbar(2\omega \mp \Omega)$ for $\psi_{\pm1}$.

**23.5.4.3** The variation of these energies with $\Omega$ is plotted in Fig. 23.2 (left). We see that for $\Omega > \Omega_c = \omega$, the energy of the state corresponding to $\psi_0$ is larger than that corresponding to $\psi_{+1}$. Therefore, for $\Omega > \omega$, the $N$-body ground state cannot be $\psi_0(r_1) \ldots \psi_0(r_N)$. The state with one vortex $\psi_{+1}(r_1) \ldots \psi_{+1}(r_N)$ could be eligible as the ground state. Unfortunately for such rotation frequencies (larger than the trap frequency $\omega$) we saw in question 23.3.5 that the trap was unstable. We cannot explain (yet) the existence of vortices.

**23.5.4.4 (a)** The interaction term displaces the energy of the $N$-body state associated to $\psi_0$ by a different amount from the state associated to $\psi_{\pm1}$. At first order of perturbation theory, the displacement for $\psi_0$ is

$$\Delta E_0 = g \frac{N(N-1)}{2} \int |\psi_0(r_1)|^2\, |\psi_0(r_2)|^2\, G(r_1 - r_2)\, d^2r_1\, d^2r_2$$

$$\simeq g \frac{N(N-1)}{2} \int |\psi_0(r)|^4\, d^2r.$$

We used here the fact that the $N(N-1)/2$ terms entering in the interaction energy have the same contribution, and we approximated the function $G$ by the Dirac distribution. A similar procedure holds for $\psi_{\pm1}$ and we obtain after a short calculation

$$\Delta E_0 = \frac{N(N-1)g}{4\pi a^2} \qquad \Delta E_{\pm1} = \frac{N(N-1)g}{8\pi a^2} = \frac{\Delta E_0}{2}.$$

**(b)** The increase of the energy of the state with one vortex $\psi_{+1}$ is smaller than for the state with no vortex $\psi_0$, which can be understood by noticing that the state with one vortex occupies a larger region of space. Therefore the rotation frequency $\Omega_c$ for which the energies of these two states coincide is down-shifted (Fig. 23.2, right). The two energies are equal when

$$N\hbar\omega + \frac{N(N-1)g}{4\pi a^2} = N\hbar(2\omega - \Omega_c) + \frac{N(N-1)g}{8\pi a^2}$$

corresponding to the critical rotation frequency

$$\Omega_c = \omega - \frac{(N-1)g}{8\pi a^2 \hbar} = \omega\left(1 - \frac{(N-1)gm}{8\pi\hbar^2}\right).$$

**(c)** The critical frequency $\Omega_c$ is now strictly lower than the trap frequency $\omega$. Therefore, somewhere in the frequency range $(\Omega_c, \omega)$, one can hope that the state with one vortex will be at least a good approximation of the true many-body ground state.

**(d)** For $g = 10^{-3}\hbar^2/m$ and $N = 1000$ atoms, we find the relative shift

$$\frac{\omega - \Omega_c}{\omega} \simeq \frac{1}{8\pi} \simeq 0.04.$$

**(e)** This state corresponds qualitatively to the second picture of Fig. 23.1. However in practice, interactions are stronger than what we considered here and the critical frequency is even lower than what we just calculated. Note also that for such strong interactions, the perturbation theory is not valid and one has to resort to other types of approximations.

**23.5.4.5 (a)** The wave function $\psi_{\pm n}$ varies as $e^{\pm in\varphi}$, which corresponds to the angular momentum $\pm n\hbar$.

**(b)** One calculates the energies of the states $\psi_{\pm n}$ by acting on them with $\hat{H}$. In order to evaluate the contribution of the kinetic energy, one can write $\partial_x^2 + \partial_y^2 = (\partial_x - i\partial_y)(\partial_x + i\partial_y)$ and notice that $(\partial_x + i\partial_y)$ acting on $(x + iy)^n$ gives a null result.

**(c)** In the absence of interaction, the energy of the $N$-body state corresponding to $\psi_{+n}$ is $N[(n + 1)\hbar\omega - n\hbar\Omega]$. For $\Omega = \omega$, this energy is $N\hbar\omega$ for all $n$. All states $\psi_{+n}$ are thus degenerate for $\Omega = \omega$. In the preceding study, we found the degeneracy of $\psi_0$ and $\psi_{+1}$, but this was actually a particular case of a much wider class of degenerate states.

In order to describe properly the ground state of the interacting gas for $\Omega$ slightly smaller than $\omega$, it is reasonable to approximate this state by a wave function that is a linear combination of states with an energy close to the ground state energy in the absence of interactions.

**(d)** The proposed state $\phi$ is equal to the gaussian function $e^{-r^2/(2a^2)}$, multiplied by a polynomial of degree $N$ for the complex variable $u = x + iy$. This complex polynomial generally has $N$ distinct roots in the complex plane $(x, y)$. For each root the wave function is zero and its phase winding evaluated on a small contour encircling the root is equal to $2\pi$. Each root of the polynomial thus corresponds to a vortex.

**(e)** The two pictures on the right of Fig. 23.1 correspond indeed to states with several vortices, which arrange in a regular manner. One can explain this ordering by searching the $C_n$ coefficients that minimize the interaction energy and by studying the position of the roots of the corresponding polynomial in the complex plane. However, such a study is out of the scope of this chapter.

### 23.5.5 References

The data shown in this problem are adapted from K.W. Madison, F. Chevy, W. Wohlleben, and J. Dalibard, *Vortex formation in a stirred Bose-Einstein condensate.*, Phys. Rev. Lett. **84**, 806 (2000) [see also Jour. Mod. Optics **47**, 2715 (2000)]. Observing quantized vortices in a rotating fluid provides a direct signature of its superfluid character and of its macroscopic coherence. The present problem focuses on the existence of vortices in a dilute, Bose-condensed atomic gas. Vortices have been detected in several different quantum fluids, such as liquid helium, superconductors, and cavity polaritons. For a detailed discussion, see e.g. A. J. Leggett, *Quantum liquids* (Oxford University Press, 2006), L. Pitaevskii and S. Stringari, *Bose–Einstein condensation and Superfluidity* (second edition, Oxford University Press, 2016).

# Motion in a Periodic Potential and Bloch Oscillations

**24**

The study of the motion of a particle in a spatially periodic potential is an important problem in quantum physics. A key example is the modeling of electrical conductivity of metals, which originates from the motion of electrons in the periodic potential created by a crystal. Here we study a simplified version of this problem, where a single particle moves in a one-dimensional chain of regularly spaced sites. In particular we encounter a surprising phenomenon called *Bloch oscillation*: When one adds a constant force to the one created by the periodic lattice, the particle oscillates in space. The resulting dynamics is thus very different from the uniformly accelerated motion that occurs in the absence of a lattice.

## 24.1  The Two-Site Problem

In a first step we consider the case where the particle can occupy only two sites separated by the distance $a$ (Fig. 24.1a). We denote $|0\rangle$ and $|1\rangle$ the states corresponding to the particle localized on the left and on the right site, respectively. The two states $\{|0\rangle, |1\rangle\}$ are supposed to be orthogonal to each other and we assume that the particle has the same energy $E_0$ on each of them. The tunnel effect across the potential barrier separating the two sites allows the particle to jump from one site to the other, and the corresponding matrix element is denoted $-J$. The Hamiltonian of the particle for this discretized two-site problem is therefore

$$\hat{H} = E_0 \left( |0\rangle\langle 0| + |1\rangle\langle 1| \right) - J \left( |0\rangle\langle 1| + |1\rangle\langle 0| \right) .$$

**24.1.1** Write this Hamiltonian as a $2 \times 2$ matrix in the basis $\{|0\rangle, |1\rangle\}$.

**24.1.2** Give the eigenstates of $\hat{H}$ and the corresponding energies.

© Springer Nature Switzerland AG 2019
J.-L. Basdevant, J. Dalibard, *The Quantum Mechanics Solver*,
https://doi.org/10.1007/978-3-030-13724-3_24

(a)                                            (b)

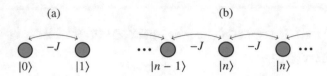

**Fig. 24.1** (**a**) Two-site model studied in Sect. 24.1. (**b**) Infinite chain studied in Sects. 24.2, 24.3 and 24.4

**24.1.3** The particle is prepared in state $|0\rangle$ at time $t = 0$. What is the probability to find the particle in state $|1\rangle$ at time $t$?

## 24.2    The Infinite Periodic Chain

We consider now the case where the particle can move along a one-dimensional chain of sites (Fig. 24.1b). We denote by $|n\rangle$ the state where the particle is localized on the $n$-th site ($n$ integer). The states $|n\rangle$ form a normalized, orthogonal basis set: $\langle n|n'\rangle = \delta_{n,n'}$. They all have the same energy $E_0$.

Tunnel effect allows the particle to jump from site $n$ to the neighbouring sites $n \pm 1$. As in the previous section, we denote the corresponding matrix element as $-J$. The Hamiltonian of the particle is thus

$$\hat{H} = \sum_n E_0|n\rangle\langle n| - J\left(|n\rangle\langle n+1| + |n+1\rangle\langle n|\right) .$$

**24.2.1** We want to find the eigenstates $|\phi\rangle$ of $\hat{H}$. Expand $|\phi\rangle$ as $|\phi\rangle = \sum_n D_n|n\rangle$ and write a recursion relation for the coefficients $D_n$.

**24.2.2** We look for a solution of the recursion relation with the form $D_n = D_0 \, e^{iqna}$, where $q$ is a real number which has the dimension of a wavenumber (inverse of a length). Show that one can obtain a solution for all $q$'s and that the corresponding energy $E(q)$ satisfies the equation

$$E(q) = E_0 - 2J\cos(qa) . \tag{24.1}$$

We denote $|\phi_q\rangle$ the energy eigenstate obtained in this way. We will not try to normalise it and we will set $D_0 = 1$ by convention.

**24.2.3** Among all energy eigenstates $|\phi_q\rangle$ found above, show that states corresponding to $q$ and $q + 2\pi/a$ are identical. In the following, we will thus restrict the values of $q$ to the interval $-\pi/a < q \le \pi/a$.

### 24.2.4 Some particular energy eigenstates

(a) Give the expression of the ground state and its energy.
(b) Give the expression of the eigenstate with the largest energy.
(c) Compare these two states with those found in Sect. 24.1.

### 24.2.5 The momentum operator for this discretized problem. We define the momentum operator by

$$\hat{P} = \frac{\hbar}{i} \frac{1}{2a} \sum_n |n\rangle\langle n+1| - |n+1\rangle\langle n| \,. \tag{24.2}$$

(a) Consider an arbitrary state vector $|\psi\rangle = \sum_n \alpha_n |n\rangle$ and define $|\chi\rangle = \hat{P}|\psi\rangle = \sum_n \beta_n |n\rangle$. Give the expression of the coefficients $\beta_n$ in terms of the coefficients $\alpha_n$. Consider the limit $a \to 0$ and connect the result to the standard definition of the momentum operator in (continuous) quantum mechanics.
(b) Show that the states $|\phi_q\rangle$ are eigenstates of $\hat{P}$ with a momentum $p(q)$ that can be expressed in terms of $\hbar$, $a$ and $q$. How does this relation simplify for $|q| \ll 1/a$?

---

## 24.3 Dynamics Along the Infinite Chain

Consider the state vector $|\psi_0\rangle = \sum_n \alpha_n |n\rangle$ with $\sum_n |\alpha_n|^2 = 1$, which is the analog of a wave packet for this discretized problem. In order to ensure the convergence of all mathematical expressions, we assume that $|\alpha_n|$ tends to 0 faster than any power of $n$ when $|n|$ tends to $\infty$.

### 24.3.1 Show that one can write $|\psi_0\rangle$ as

$$|\psi_0\rangle = \int_{-\pi/a}^{\pi/a} C(q) |\phi_q\rangle \, dq \,, \tag{24.3}$$

where $C(q)$ is a periodic function of period $2\pi/a$. We recall that any (sufficiently regular) periodic function can be expanded as a Fourier series:

$$C(q) = \sum_n \beta_n \, e^{-iqna} \quad \text{with} \quad \beta_n = \frac{a}{2\pi} \int_{-\pi/a}^{\pi/a} C(q) \, e^{inqa} \, dq \,.$$

Give the expression of $C(q)$ in terms of the coefficients $\alpha_n$.

### 24.3.2 The particle is prepared at initial time in the state $|\psi(0)\rangle = |\psi_0\rangle$ given in (24.3). Give the expression of the state of the particle $|\psi(t)\rangle$ at time $t$.

**24.3.3** We assume that the coefficients $C(q)$ entering into (24.3) are nonzero only in a region $|q| \ll \pi/a$. Simplify the preceding result using an expansion of $E(q)$ at the lowest significant order in $q$.

**24.3.4 Evolution of a free particle.** We wish to connect the previous results with the case of the quantum one-dimensional motion of a free particle with mass $M$. We denote $\psi(x, 0)$ the wave function of the free particle at time $t = 0$. Recall

(a) The expression of the Fourier transform $\varphi(p)$ of $\psi(x, 0)$;
(b) The expression of $\psi(x, t)$ as an integral over $p$ involving $\varphi(p)$.

**24.3.5** Draw an analogy between the results 24.3.3 and 24.3.4, and explain why the motion of the particle along the discrete chain is similar to the motion of a free particle in continuous space with an effective mass $M_{\text{eff}}$. How does $M_{\text{eff}}$ vary with the tunnel coefficient $J$ when $J \to 0$?

## 24.4　Bloch Oscillations

We assume now that the particle moving along the chain is in addition submitted to the constant force $F$, corresponding to the linear potential $-Fx$. In the discrete problem considered here, this potential reads

$$\hat{V} = -F \sum_n na \, |n\rangle\langle n| \, .$$

The total Hamiltonian is thus $\hat{H}' = \hat{H} + \hat{V}$. We will not calculate the exact eigenstates of $\hat{H}'$, but rather characterize the dynamics of the particle for a given initial condition.

**24.4.1** Suppose that the particle is prepared at time $t = 0$ in one of the states $|\phi_q\rangle$ discussed above:

$$|\psi(0)\rangle = |\phi_{q0}\rangle = \sum_n e^{inq_0 a} |n\rangle.$$

We want to prove that the state of the particle will remain at any time equal to one of the states $|\phi_q\rangle$, up to a global phase. In order to show this result, we tentatively write the state at time $t$ as

$$|\psi(t)\rangle = \sum_n e^{i[nq(t)a - \alpha(t)]}|n\rangle, \qquad (24.4)$$

where $q(t)$ and $\alpha(t)$ are real functions of time. Insert (24.4) into the Schrödinger equation and show that one can indeed obtain a solution with this mathematical structure. Write the two differential equations that $q(t)$ and $\alpha(t)$ must satisfy so that this result holds.

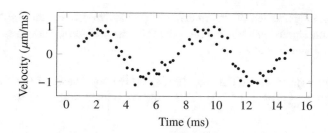

**Fig. 24.2** Measurement of the average velocity of cesium atoms placed in a periodic potential and submitted to an additional uniform force $F$. The line is a guide to the eye. (Figure extracted from M. Ben Dahan's PhD thesis)

**24.4.2** Show that the solutions of these equations are:

$$q(t) = q_0 + Ft/\hbar , \qquad \alpha(t) = \frac{1}{\hbar} \int_0^t E(q(t')) \, dt'. \qquad (24.5)$$

**24.4.3** What is the state of the particle at time $T = 2\pi\hbar/(Fa)$? Show that $\alpha(T)$ does not depend on $q_0$ and give its expression.

**24.4.4** Consider now the initial state $|\psi_0\rangle$ given in (24.3), where the function $C(q)$ is arbitrary. Give the state of the system at time $T = 2\pi\hbar/(Fa)$ and compare the physical properties of the system at time 0 and time $T$.

**24.4.5** Cesium atoms ($M = 2.2 \times 10^{-25}$ kg) that have been previously cooled by laser light, are placed in the periodic potential created by a laser standing wave of spatial period $a = 425$ nm. An additional force $F$ is applied to the atoms. Using a time-of-flight technique, one can measure at various times the mean velocity $\bar{v} = \langle \hat{P} \rangle / M$, where $\hat{P}$ is the momentum operator defined in question 24.2.5. The result of the measurement is shown in Fig. 24.2. Does this result, in particular the amplitude of the oscillations of $\bar{v}$, confirm the predictions made above?

**24.4.6** Calculate the applied force $F$ and the corresponding acceleration $F/M$.

## 24.5 Solutions

### 24.5.1 The Two-Site Problem

**24.5.1.1** The $2 \times 2$ matrix representing the Hamiltonian is:

$$\hat{H} = \begin{pmatrix} E_0 & -J \\ -J & E_0 \end{pmatrix} .$$

**24.5.1.2** The eigenstates of this matrix are (i) the symmetric state $|\psi_s\rangle = (|0\rangle + |1\rangle)/\sqrt{2}$ with eigenvalue $E_0 - J$ and (ii) the antisymmetric state $|\psi_a\rangle = (|0\rangle - |1\rangle)/\sqrt{2}$ with eigenvalue $E_0 + J$.

**24.5.1.3** We expand the initial state $|\psi(0)\rangle$ on the eigenbasis of the Hamiltonian: $|\psi(0)\rangle = |0\rangle = \frac{1}{\sqrt{2}} (|\psi_s\rangle + |\psi_a\rangle)$. Therefore the state at time $t$ reads:

$$|\psi(t)\rangle = \frac{1}{\sqrt{2}} \left( e^{-i(E_0-J)t/\hbar}|\psi_s\rangle + e^{-i(E_0+J)t/\hbar}|\psi_a\rangle \right)$$

and the probability to find the particle in state $|1\rangle$ is

$$P_1(t) = |\langle 1|\psi(t)\rangle|^2 = \frac{1}{4} \left| e^{-i(E_0-J)t/\hbar} - e^{-i(E_0+J)t/\hbar} \right|^2 = \sin^2(Jt/\hbar).$$

We recover the sinusoidal oscillation characteristic of the dynamics of a two-state system.

### 24.5.2 The Infinite Periodic Chain

**24.5.2.1** We multiply the eigenvalue equation $\hat{H}|\phi\rangle = E|\phi\rangle$ by the bra $\langle n|$ and find the recursion relation

$$E_0 D_n - J(D_{n+1} + D_{n-1}) = E D_n.$$

**24.5.2.2** We insert the proposed ansatz and obtain

$$e^{iqna} \left( E - E_0 + 2J\cos(qa) \right) D_0 = 0.$$

Since the coefficient $D_0$ is nonzero (otherwise $|\phi\rangle$ would be null), we conclude that $E(q)$ is given by

$$E(q) = E_0 - 2J\cos(qa) \qquad |\phi_q\rangle = \sum_n e^{iqna}|n\rangle.$$

The state $|\phi_q\rangle$ is called a *Bloch state*.

**24.5.2.3** If we change $q$ in $q + 2\pi/a$, the expression of $|\phi_q\rangle$ is unchanged. Indeed for each coefficient $D_n = e^{iqna}$ entering in the expansion of $|\phi\rangle$, we find $e^{i(q+2\pi/a)na} = e^{iqna}$ since $e^{i2\pi n} = 1$ for all $n$. To avoid multiple counting, we will therefore restrict the domain of relevant values of $q$ to an interval of length $2\pi/a$, for instance $-\pi/a < q \leq \pi/a$. This particular choice is called the *first Brillouin zone* in condensed matter physics.

### 24.5.2.4 Some particular energy eigenstates

(a) The ground state of the system is obtained for $q = 0$ and its energy is $E_0 - 2J$.

(b) In this discretized model, the state with the largest energy is obtained for $q = \pi/a$ and its energy is $E_0 + 2J$.

(c) The state with the lowest energy is $\sum_n |n\rangle$, which can be viewed as the generalization of the symmetric state $|0\rangle + |1\rangle$. The state with the highest energy is $\sum_n (-1)^n |n\rangle$. These alternating $+$ and $-$ amplitudes on the various sites can be viewed as the generalization of the antisymmetric state $|0\rangle - |1\rangle$.

### 24.5.2.5 The momentum operator for this discretized problem

(a) The action of $\hat{P}$ on $|\psi\rangle$ gives

$$\hat{P}|\psi\rangle = \frac{\hbar}{i} \frac{1}{2a} \sum_n \alpha_{n+1}|n\rangle - \alpha_n|n+1\rangle = \frac{\hbar}{i} \frac{1}{2a} \sum_n (\alpha_{n+1} - \alpha_{n-1})|n\rangle$$

so that

$$\beta_n = \frac{\hbar}{i} \frac{(\alpha_{n+1} - \alpha_{n-1})}{2a}. \tag{24.6}$$

This result is directly connected to the definition of the momentum operator in wave mechanics

$$\hat{P}\psi(x) = \chi(x) \quad \text{with} \quad \chi(x) = \frac{\hbar}{i} \frac{d\psi}{dx}.$$

The expression (24.6) for the chain is a discretized version of the usual momentum operator. Indeed if the function $\psi(x)$ varies slowly on the scale of $a$, one has

$$\frac{d\psi}{dx} \simeq \frac{\psi(x+a) - \psi(x-a)}{2a}.$$

(b) When inserting $\alpha_n = e^{iqna}$ in (24.6), we find $\beta_n = (\hbar/a)\sin(qa)\alpha_n$. This shows that $|\phi_q\rangle$ is an eigenstate of $\hat{P}$ with the momentum

$$p(q) = \frac{\hbar}{a} \sin(qa). \tag{24.7}$$

When $q$ is close to 0, this gives $p(q) \simeq \hbar q$, which is the usual relation between the momentum and the wave vector of a particle.

### 24.5.3 Dynamics Along the Infinite Chain

**24.5.3.1** Suppose that the expression suggested in the text,

$$|\psi_0\rangle = \int_{-\pi/a}^{\pi/a} C(q) \, |\phi_q\rangle \, dq \, , \tag{24.8}$$

indeed exists and insert the expansion $|\phi_q\rangle = \sum_n e^{iqna} |n\rangle$ in this relation. We find

$$|\psi_0\rangle = \sum_n \int_{-\pi/a}^{\pi/a} C(q) e^{iqna} \, dq \, |n\rangle = \sum_n \frac{2\pi}{a} \beta_n |n\rangle \, ,$$

where we introduced the coefficients $\beta_n$ of the expansion of $C(q)$ as a Fourier series. This expression coincides with the definition $|\psi_0\rangle = \sum_n \alpha_n |n\rangle$ if we choose $\beta_n = \alpha_n a/(2\pi)$ or equivalently $C(q)$ :

$$C(q) = \frac{a}{2\pi} \sum_n \alpha_n e^{-iqna} \, . \tag{24.9}$$

Note that one can check the validity of this result by injecting it in (24.8), which gives:

$$\int_{-\pi/a}^{\pi/a} C(q) \, |\phi_q\rangle \, dq = \frac{a}{2\pi} \int_{-\pi/a}^{\pi/a} \left( \sum_n \alpha_n e^{-iqna} \right) \left( \sum_{n'} e^{iqn'a} |n\rangle \right) dq \, .$$

The integral over $q$ is readily calculated:

$$\int_{-\pi/a}^{\pi/a} e^{iq(n'-n)a} \, dq = \frac{2\pi}{a} \delta_{n,n'}$$

and we recover $\sum_n \alpha_n |n\rangle = |\psi_0\rangle$.

**24.5.3.2** The initial state is written as a linear combination of the eigenstates of the Hamiltonian. The expression of the state at time $t$ follows immediately:

$$|\psi(t)\rangle = \int_{-\pi/a}^{\pi/a} C(q) \, e^{-iE(q)t/\hbar} \, |\phi_q\rangle \, dq \, .$$

**24.5.3.3** If $|q| \ll \pi/a$, we can use the approximation $E(q) \approx E_0 - 2J + Jq^2a^2$, which leads to

$$|\psi(t)\rangle \approx e^{-i(E_0-2J)t/\hbar} \int_{-\pi/a}^{\pi/a} C(q) \, e^{-iJq^2a^2t/\hbar} \, |\phi_q\rangle \, dq \, . \tag{24.10}$$

### 24.5.3.4 Evolution of a free particle

**(a)** The Fourier transform $\varphi(p)$ is defined as

$$\varphi(p) = \frac{1}{\sqrt{2\pi\hbar}} \int \psi(x)\, e^{-ixp/\hbar}\, dx\ . \tag{24.11}$$

**(b)** At time $t$, $\psi(x, t)$ can be written

$$\psi(x, t) = \frac{1}{\sqrt{2\pi\hbar}} \int \varphi(p)\, e^{-ip^2t/(2M\hbar)}\, e^{ixp/\hbar}\, dp\ . \tag{24.12}$$

**24.5.3.5** There is a strong analogy between the two sets of Eqs. (24.9)–(24.10) and (24.11)–(24.12):

- For the chain considered in this chapter, the position of the particle can only take discrete values and the probability amplitude to find the particle on site $n$ is $\alpha_n$. For a free particle, the position $x$ is continuous and the probability amplitude to find the particle in $x$ is $\psi(x)$.
- For the chain, the states $|\phi_q\rangle$ have a well-defined momentum and the coefficients $C(q)$ entering in the expansion of $|\psi_0\rangle$ are obtained from the discrete sum (24.9). For the free particle, the coefficients $\varphi(p)$ in the expansion of $\psi(x)$ on the plane wave basis $e^{ipx/\hbar}$ (which are also states with a well defined momentum) are obtained from the integral (24.11).
- The Eqs. (24.10) and (24.12) give the time evolution of the position distribution, with in both cases a contribution of the square of the momentum.

When we identify $e^{-iJq^2a^2t/\hbar}$ and $e^{-ip^2t/(2M\hbar)}$, we are led to define an effective mass for the particle moving along the chain:

$$M_{\text{eff}} = \frac{\hbar^2}{2Ja^2}\ .$$

This mass is inversely proportional to the tunnel matrix element. This confirms the intuitive idea that a weak tunneling corresponds to a strong inertia of the particle, with a small probability to jump from site to the next in a given time interval.

Note that the analogy between the free particle case and the motion along the chain is valid only for $|q| \ll \pi/a$. For larger values of $|q|$, the dispersion relation which relates $q$ and $E(q)$ is not quadratic and there is no simple correspondance between the two problems.

### 24.5.4 Bloch Oscillations

**24.5.4.1** We insert the proposed state vector in the Schrödinger equation and project the resulting result on $|n\rangle$ :

$$i\hbar(in\dot{q}a - i\dot{\alpha}) = E_0 - 2J\cos(q(t)a) - nFa .$$

Since this relation must be verified for all $n$, we require that

$$\hbar\dot{\alpha} = E(q(t)) \qquad\qquad \hbar\dot{q} = F .$$

**24.5.4.2** The integration of the equation on $q$ gives $\hbar q(t) = \hbar q_0 + Ft$. Using this result for $q(t)$, we can solve the equation of motion for $\alpha$ :

$$\hbar\alpha(t) = \int_0^t E(q(t'))\, dt' .$$

The evolution of $q(t)$ is reminiscent of the result for a particle with a uniformly accelerated motion, for which the momentum varies as $p(t) = p_0 + Ft$, hence its wave vector as $k(t) = k_0 + Ft/\hbar$. However, this analogy is only partial since, as we saw above, the Bloch states $|\phi_q\rangle$ are periodic with respect to $q$, which is not the case of the momentum eigenstates of a free particle.

**24.5.4.3** When we choose $T = 2\pi\hbar/(Fa)$, we find $q(T) = q_0 + 2\pi/a$. Now, we have seen in Sect. 24.2 that the state vectors $|\phi_q\rangle$ and $|\phi_{q+2\pi/a}\rangle$ coincide for any $q$. The state of the particle at time $T = 2\pi\hbar/(Fa)$ is thus the same as the state at time 0, up to the phase factor $e^{-i\alpha(T)}$. Note that the phase $\alpha(T)$ does not depend on $q_0$ since

$$\hbar\alpha(T) = \int_0^T [E_0 - 2J\cos(q(t)a)]\, dt = E_0 T.$$

The last equality holds because the $\cos(q(t)a)$ term has a vanishing contribution when integrated over a full period.

**24.5.4.4** Because of the linearity of the Schrödinger equation, the state of the particle at time $T$ is equal to the state at time 0 for any initial condition, up to the global phase factor $e^{-iE_0T/\hbar}$. The expectation values at time 0 and time $T$ of any physical observable are thus identical: the evolution is periodic in time.

**24.5.4.5** For the experimental data plotted in the text, we observe a periodic evolution of the average atomic velocity with the time period $\approx 8$ ms and an oscillation amplitude slightly smaller than 1 mm/s. If the initial state were exactly a Bloch state $|\phi_{q_0}\rangle$, we would expect that the state of the particle at any time $t$ would be again a Bloch state $|\phi_{q(t)}\rangle$, with $q(t) = q_0 + Ft/\hbar$. Using (24.7), we find that this would correspond to the velocity

$$v(t) = \frac{p(t)}{M} = \frac{\hbar}{Ma} \sin[q(t)a], \tag{24.13}$$

i.e., a sinusoidal oscillation with amplitude $\hbar/(Ma) = 1.1$ mm/s. The measured amplitude is slightly smaller than this prediction, which is due to the fact that the initial state is a wave packet formed by a continuous superposition of states $|\phi_q\rangle$. We also notice that in contrast with prediction (24.13), the measured $v(t)$ is not exactly a sinusoidal function: parts with a positive velocity have a slightly longer duration than parts with a negative velocity. This asymmetry can be explained by a theoretical treatment that goes beyond the discretized model used here.

**24.5.4.6** The measured period $T \approx 8$ ms corresponds to the force $F = 2\pi\hbar/(aT) = 2 \times 10^{-25}$ N and the acceleration $A = F/M = 2\pi\hbar/(MaT) = 0.9$ m/s$^2$, i.e. 1/10 of the acceleration of gravity.

## 24.5.5 Discussion and References

The Bloch oscillation phenomenon is a paradoxical result, where a constant force $F$ leads to an oscillatory motion instead of a constant acceleration. It shows that an ideal crystal cannot be a good conductor: when one applies a difference of electric potential between two edges of the crystal, the electrons of the conduction band feel a constant force in addition to the periodic potential created by the crystal. Therefore they should oscillate instead of being accelerated towards the positive edge. The conduction phenomenon results actually from the defects present in real metals.

One can interpret Bloch oscillations using an analogy with the Bragg diffraction phenomenon. The particle initially at rest is accelerated by the force $F$ and its velocity increases. The motion of the particle is then modified by the periodic lattice, which can cause a reflection (from $v > 0$ to $v < 0$) of the wave associated to the particle. This reflection occurs with a large efficiency when Bragg condition is satisfied. It corresponds to the case where the phase of the wave function has an opposite sign on adjacent sites of the lattice, i.e. when the Bloch wavevector $q$ is on the edge of the Brillouin zone ($q = \pi/a$). Once reflected, the particle moves backward, it is again accelerated, and so on.

The experimental data shown in this chapter were adapted from the PhD thesis of M. Ben Dahan (Université Pierre et Marie Curie, Paris, 1997), and from M. Ben Dahan, E. Peik, J. Reichel, Y. Castin, and C. Salomon, *Bloch Oscillations of Atoms in an Optical Potential*, Phys. Rev. Lett. **76**, 4508 (1996), E. Peik, M. Ben Dahan, I. Bouchoule, Y. Castin, and C. Salomon, *Bloch oscillations of atoms, adiabatic rapid passage, and monokinetic atomic beams*, Phys. Rev. A **55**, 2989 (1997). These early experiments were reviewed in M. Raizen, C. Salomon, and Q. Niu, *New Light on Quantum Transport*, Physics Today, July 1997, p. 30.

During the last two decades, Bloch oscillations have been widely used in atomic physics for metrological applications, for example for the measurement of gravity (see e.g. N. Poli, , F.-Y. Wang, M. G. Tarallo, A. Alberti, M. Prevedelli and G. M. Tino, *Precision Measurement of Gravity with Cold Atoms in an Optical Lattice and Comparison with a Classical Gravimeter*, Phys. Rev. Lett. **106**, 038501 (2011)) and the determination of fundamental constants (P. Cladé, E. de Mirandes, M. Cadoret, S. Guellati-Khélifa, C. Schwob, F. Nez, L. Julien and F. Biraben (2006), *Determination of the Fine Structure Constant Based on Bloch Oscillations of Ultracold Atoms in a Vertical Optical Lattice*, Phys. Rev. Lett. **96**, 033001 (2006)).

Bloch oscillations also constitute a very useful tool for probing some characteristics of topological matter, see e.g. Marcos Atala, Monika Aidelsburger, Julio T. Barreiro, Dmitry Abanin, Takuya Kitagawa, Eugene Demler, and Immanuel Bloch, *Direct measurement of the Zak phase in topological Bloch bands*, Nature Physics **9**, 795 (2013) and G. Jotzu, M. Messer, R. Desbuquois, M. Lebrat, T. Uehlinger, D. Greif, and T. Esslinger, *Experimental realization of the topological Haldane model with ultracold fermions*, Nature **515**, 237 (2014).

# Magnetic Excitons

# 25

Quantum field theory deals with systems that possess a large number of degrees of freedom. This chapter presents a simple model, where we study the magnetic excitations of a long chain of coupled spins. We show that one can associate the excited states of the system with quasi-particles that propagate along the chain.

We recall that, for any integer $k$:

$$\sum_{n=1}^{N} e^{2i\pi kn/N} = N \text{ if } k = pN, \text{ with } p \text{ integer,}$$

$$= 0 \text{ otherwise.}$$

## 25.1 The Molecule CsFeBr$_3$

Consider a system with angular momentum equal to 1, i.e. $j = 1$ in the basis $|j, m\rangle$ common to $\hat{J}^2$ and $\hat{J}_z$.

**25.1.1** What are the eigenvalues of $\hat{J}^2$ and $\hat{J}_z$?

**25.1.2** For simplicity, we shall write $|j, m\rangle = |\sigma\rangle$, where $\sigma = m = 1, 0, -1$. Write the action of the operators $\hat{J}_\pm = \hat{J}_x \pm i\hat{J}_y$ on the states $|\sigma\rangle$.

**25.1.3** In the molecule Cs Fe Br$_3$, the ion Fe$^{2+}$ has an intrinsic angular momentum, or *spin*, equal to 1. We write the corresponding observable $\hat{\boldsymbol{J}}$, and we note $|\sigma\rangle$ the eigenstates of $\hat{J}_z$. The molecule has a plane of symmetry, and the magnetic interaction Hamiltonian of the ion Fe$^{2+}$ with the rest of the molecule is

$$\hat{H}_r = \frac{D}{\hbar^2} \hat{J}_z^2 \quad D > 0. \tag{25.1}$$

© Springer Nature Switzerland AG 2019
J.-L. Basdevant, J. Dalibard, *The Quantum Mechanics Solver*,
https://doi.org/10.1007/978-3-030-13724-3_25

What are the eigenstates of $\hat{H}_r$ and the corresponding energy values? Are there degeneracies?

## 25.2  Spin–Spin Interactions in a Chain of Molecules

We consider a one-dimensional closed chain made up with an *even number N* of Cs Fe Br$_3$ molecules. We are only interested in the magnetic energy states of the chain, due to the magnetic interactions of the $N$ Fe$^{2+}$ ions, each with spin 1.

We take $\{|\sigma_1, \sigma_2, \cdots, \sigma_N\rangle\}, \sigma_n = 1, 0, -1$, to be the orthonormal basis of the states of the system; it is an eigenbasis of the operators $\{\hat{J}_z^n\}$ where $\hat{\boldsymbol{J}}^n$ is the spin operator of the $n$-th ion ($n = 1, \cdots, N$).

The magnetic Hamiltonian of the system is the sum of two terms $\hat{H} = \hat{H}_0 + \hat{H}_1$ where

$$\hat{H}_0 = \frac{D}{\hbar^2} \sum_{n=1}^{N} (\hat{J}_z^n)^2 \qquad (25.2)$$

has been introduced in (25.1), and $\hat{H}_1$ is a nearest-neighbor spin–spin interaction term

$$\hat{H}_1 = \frac{A}{\hbar^2} \sum_{n=1}^{N} \hat{\boldsymbol{J}}^n \cdot \hat{\boldsymbol{J}}^{n+1} \qquad A > 0. \qquad (25.3)$$

To simplify the notation of $\hat{H}_1$, we define $\hat{\boldsymbol{J}}^{N+1} \equiv \hat{\boldsymbol{J}}^1$.

We assume that $\hat{H}_1$ is a small perturbation compared to $\hat{H}_0 (A \ll D)$, and we shall treat it in first order perturbation theory.

**25.2.1**  Show that $|\sigma_1, \sigma_2, \cdots, \sigma_N\rangle$ is an eigenstate of $\hat{H}_0$, and give the corresponding energy value.

**25.2.2**  What is the ground state of $\hat{H}_0$? Is it a degenerate level?

**25.2.3**  What is the energy of the first excited state of $\hat{H}_0$? What is the degeneracy $d$ of this level? We shall denote by $\mathcal{E}^1$ the corresponding eigenspace of $\hat{H}_0$, of dimension $d$.

**25.2.4**  Show that $\hat{H}_1$ can be written as

$$\hat{H}_1 = \frac{A}{\hbar^2} \sum_{n=1}^{N} \left( \frac{1}{2} (\hat{J}_+^n \hat{J}_-^{n+1} + \hat{J}_-^n \hat{J}_+^{n+1}) + \hat{J}_z^n \hat{J}_z^{n+1} \right). \qquad (25.4)$$

## 25.3 Energy Levels of the Chain

We now work in the subspace $\mathcal{E}^1$. We introduce the following notation

$$|n, \pm\rangle = |\sigma_1 = 0, \sigma_2 = 0, \cdots, \sigma_n = \pm 1, \sigma_{n+1} = 0, \cdots, \sigma_N = 0\rangle. \qquad (25.5)$$

Owing to the periodicity of the chain, we define $|N + 1, \pm\rangle \equiv |1, \pm\rangle$.

**25.3.1** Show that

$$\hat{H}_1|n, \pm\rangle = A(|n - 1, \pm\rangle + |n + 1, \pm\rangle) + |\psi_n\rangle, \qquad (25.6)$$

where $|\psi_n\rangle$ is orthogonal to the subspace $\mathcal{E}^1$.

Without giving the complete form of $|\psi_n\rangle$, give an example of one of its components, and give the energy of the eigenspace of $\hat{H}_0$ to which $|\psi_n\rangle$ belongs.

**25.3.2** Consider the circular permutation operator $\hat{T}$, and its adjoint $\hat{T}^\dagger$, defined by

$$\hat{T}|\sigma_1, \sigma_2, \cdots, \sigma_N\rangle = |\sigma_N, \sigma_1, \cdots, \sigma_{N-1}\rangle,$$
$$\hat{T}^\dagger|\sigma_1, \sigma_2, \cdots, \sigma_N\rangle = |\sigma_2, \sigma_3, \cdots, \sigma_N, \sigma_1\rangle. \qquad (25.7)$$

Write the action of $\hat{T}$ and $\hat{T}^\dagger$ on the states $|n, \pm\rangle$.

**25.3.3** Check that, in the subspace $\mathcal{E}^1$, $\hat{H}_1$ and $A(\hat{T} + \hat{T}^\dagger)$ have the same matrix elements.

**25.3.4** Show that the eigenvalues $\lambda_k$ of $\hat{T}$ are the $N$-th roots of unity (we recall that $N$ is assumed to be even):

$$\lambda_k = e^{-iq_k} \qquad q_k = -\pi + \frac{2k\pi}{N} \qquad k = 0, \cdots, N - 1. \qquad (25.8)$$

**25.3.5** We seek, in $\mathcal{E}^1$, the $2N$ eigenvectors $|q_k, \pm\rangle$ of $\hat{T}$, each corresponding to an eigenvalue $\lambda_k$. Each $|q_k, \pm\rangle$ is written

$$|q_k, \pm\rangle = \sum_n c_n(k) |n, \pm\rangle. \qquad (25.9)$$

(a) Write a recursion relation between the coefficients $c_n$.
(b) Show that

$$c_n(k) = \frac{1}{\sqrt{N}} e^{iq_k n} \qquad (25.10)$$

is a solution of this recursion relation.

(c) Show that the states $|q_k, \pm\rangle$ defined using (25.9) and (25.10) are orthonormal.
(d) Show that the vectors $|q_k, \pm\rangle$ are also eigenvectors of $\hat{T}^\dagger$ and $\hat{T} + \hat{T}^\dagger$, and give the corresponding eigenvalues.
(e) Calculate the scalar product $\langle n, \epsilon | q_k, \epsilon' \rangle$ $(\epsilon, \epsilon' = \pm)$ and write the expansion of the states $|n, \pm\rangle$ in the basis $|q_k, \pm\rangle$.

**25.3.6** We treat the Hamiltonian $\hat{H}_1$ of Sect. 25.2 as a perturbation to $\hat{H}_0$. We limit ourselves to the first excited level of $\hat{H}_0$, and we want to calculate how the perturbation lifts the degeneracy of this level. We recall that, in the degenerate case, first order perturbation theory consists in diagonalizing the restriction of the perturbing Hamiltonian in the degenerate subspace of the dominant term $\hat{H}_0$.

(a) Explain why the results of question 25.3.5 above allow one to solve this problem.
(b) In first order perturbation theory, give the new energy levels which arise from the first excited state of $\hat{H}_0$, and the corresponding eigenstates.
(c) Draw qualitatively the energies $E(q_k)$ in terms of the variable $q_k$ which can be treated as a continuous variable, $q_k \in [-\pi, +\pi[$, if $N$ is very large. What is the degeneracy of each new energy level?

## 25.4  Vibrations of the Chain: Excitons

We now study the time evolution of the spin chain.

**25.4.1** Suppose that at time $t = 0$, the system is in the state

$$|\Psi(0)\rangle = \sum_{\epsilon=\pm} \sum_{k=0}^{N-1} \varphi_k^\epsilon |q_k, \epsilon\rangle \quad \text{with} \quad \sum_{\epsilon=\pm} \sum_{k=0}^{N-1} |\varphi_k^\epsilon|^2 = 1. \tag{25.11}$$

Setting $\omega = 2A/\hbar$, write the state $|\Psi(t)\rangle$ at a later time $t$.

**25.4.2** We assume that the initial state is $|\Psi(0)\rangle = |q_k, +\rangle$.

(a) Write the probability amplitude $\alpha_n(t)$ and the probability $P_n(t)$ of finding at time $t$ the $n$-th spin pointing upwards, i.e. $\sigma_n = +1$ and $\sigma_m = 0$ for $m \neq n$. Show that $P_n(t)$ is the same for all sites of the chain.
(b) The molecules of the chain are located at $x_n = na$, where $a$ is the lattice spacing. Show that the probability amplitude $\alpha_n(t)$ is equal to the value at $x = x_n$ of a monochromatic plane wave

$$\Psi_k(x, t) = Ce^{i(p(q)x - E(q)t)/\hbar}, \tag{25.12}$$

where $C$ is a constant, $q = q_k$, and $x$ is the abscissa along the chain. Express $p(q)$ in terms of $q$.

(c) Show that $\Psi_k(x, t)$ is an eigenstate of the momentum operator $\hat{p}_x = (\hbar/\mathrm{i})\partial/\partial x$ along the chain.

Show that the value of $p(q)$ ensures the periodicity of $\Psi_k(x, t)$, i.e. $\Psi_k(x + L, t) = \Psi_k(x, t)$, where $L = Na$ is the length of the chain.

(d) Show that, for $|q_k| \ll 1$, $\Psi_k(x, t)$ satisfies a Schrödinger equation for a particle of negative mass $m$, placed in a constant potential; give the value of $m$.

**25.4.3** In a more complete analysis, one can associate quasi-particles to the magnetic excitations of the chain. These quasi-particles, which we call "magnetic excitons", have an energy $E(q_k)$ and a momentum $p(q_k)$.

At very low temperatures, $T \approx 1.4$ K, the chain is in the ground state of $\hat{H}_0$. If low energy neutrons collide with it, they can create excitons whose energy and momentum can be determined by measuring the recoil of the neutrons. The experimental result for $E(q)$ as a function of $q \in [-\pi, 0]$ is given in Fig. 25.1.

(a) Deduce from that data approximate values for $D$ and $A$.

(b) What do you think of the approximation $D \gg A$ and of the comparison between theory and experiment? How could one improve the agreement between theory and experiment?

(c) Is it justified to assume that the chain is in its ground state when it is at thermal equilibrium at 1.4 K? We recall the Boltzmann factor: $N(E_2)/N(E_1) = \exp[-(E_2 - E_1)/kT]$, with $k = 8.6 \times 10^{-5}$ eV K$^{-1}$.

**25.4.4** Consider, at time $t = 0$, the state

$$|\Psi(0)\rangle = \sum_{k=0}^{N-1} \varphi_k |q_k, +\rangle \quad \text{with} \quad \sum_{k=0}^{N-1} |\varphi_k|^2 = 1. \tag{25.13}$$

We assume that $N \gg 1$, that the coefficients $\varphi_k$ have significant values only in a close vicinity of some value $k = k_0$, or, equivalently, $q \approx q_0$, and that, to a good

**Fig. 25.1** Experimental measurement of the excitation energy $E(q)$ as a function of $q$ between $-\pi$ and 0. The energy scale is in meV ($10^{-3}$ eV)

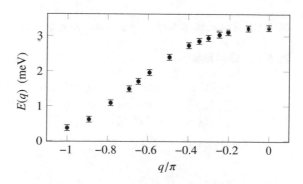

approximation, in this vicinity,

$$E(q) = E(q_0) + (q - q_0)u_0, \quad u_0 = \left.\frac{dE}{dq}\right|_{q=q_0}. \tag{25.14}$$

Show that the probability $P_n(t)$ of finding $\sigma_n = +1$ at time $t$ is the same as the probability $P_{n'}(t')$ of finding $\sigma_{n'} = +1$ at another time $t'$ whose value will be expressed in terms of $t$ and of the distance between the sites $n$ and $n'$.

Interpret the result as the propagation of a spin excitation wave along the chain. Calculate the propagation velocity of this wave and give its numerical value for $a = 0.7$ nm and $q_0 = -\pi/2$.

**25.4.5** We now assume that the initial state is $|\Psi(0)\rangle = |n = 1, +\rangle$.

(a) Write the probability $P_m(t)$ of finding $\sigma_m = +1$ at a later time $t$?
(b) Calculate the probabilities $P_1(t)$ and $P_2(t)$, in the case $N = 2$, and interpret the result.
(c) Calculate $P_1(t)$ in the case $N = 8$. Is the evolution of $P_1(t)$ periodic?
(d) For $N \gg 1$, one can convert the above sums into integrals. The probabilities are then $P_m(t) \approx |J_{m-1}(\omega t)|^2$ where the $J_n(x)$ are the Bessel functions. These functions satisfy $\sum |J_n(x)|^2 = 1$ and $J_n = (-)^n J_{-n}$.

For $x \gg 1$ we have $J_n(x) \approx \sqrt{\frac{2}{\pi x}} \cos(x - n\pi/2 - \pi/4)$ if $x > 2|n|/\pi$, and $J_n(x) \approx 0$ if $x < 2|n|/\pi$.

Which sites are appreciably reached by the probability wave at a time $t$ such that $\omega t \gg 1$?

(e) Interpret the result as the propagation along the chain of a probability amplitude (or wave). Calculate the propagation velocity and compare it with the result obtained in question 25.4.4).

## 25.5 Solutions

### 25.5.1 The Molecule CsFeBr$_3$

**25.5.1.1** The results are: $\hat{J}^2 : 2\hbar^2$, $\hat{J}_z : m\hbar; m = 1, 0, -1$.

**25.5.1.2** One has:

$$\begin{array}{ll}
J_+|1\rangle = 0 & J_-|1\rangle = \hbar\sqrt{2}|0\rangle \\
J_+|0\rangle = \hbar\sqrt{2}|1\rangle & J_-|0\rangle = \hbar\sqrt{2}|-1\rangle \\
J_+|-1\rangle = \hbar\sqrt{2}|0\rangle & J_-|-1\rangle = 0.
\end{array} \tag{25.15}$$

**25.5.1.3** The eigenstates are the states $|\sigma\rangle$. The state $|0\rangle$ corresponds to the eigenvalue $E = 0$, whereas $|+\rangle$ and $|-\rangle$, which are degenerate, correspond to $E = D$.

## 25.5.2 Spin–Spin Interactions in a Chain of Molecules

**25.5.2.1** It is straightforward to see that

$$\hat{H}_0|\sigma_1, \sigma_2 \cdots \sigma_N\rangle = D \sum_{n=1}^{N} (\sigma_n)^2 |\sigma_1 \cdots \sigma_N\rangle, \tag{25.16}$$

the corresponding eigenvalue being $E = D \sum \sigma_n^2$.

**25.5.2.2** The ground state of $\hat{H}_0$ corresponds to all the $\sigma_n$ equal to zero, so that $E = 0$. This ground state is non-degenerate.

**25.5.2.3** The first excited state corresponds to all the $\sigma$'s being zero except one: $\sigma_n$ equal to $\pm 1$. The energy is $D$, and the degeneracy $2N$, since there are $N$ possible choices of the non-vanishing $\sigma_n$, and two values $\pm 1$ of $\sigma_n$.

**25.5.2.4** $J\pm = J_x \pm i J_y$. A direct calculation leads to the result.

## 25.5.3 Energy Levels of the Chain

**25.5.3.1** The action of the perturbing Hamiltonian on the basis states is, setting $\epsilon = \pm$,

$$\hat{H}_1|n, \epsilon\rangle = A \left(|n-1, \epsilon\rangle + |n+1, \epsilon\rangle\right)$$
$$+ A \sum_{n' \neq n} (|0, \cdots 0, \sigma_n = \epsilon, 0 \cdots 0, \sigma_{n'} = -1, \sigma_{n'+1} = +1, 0 \cdots 0\rangle$$
$$+ |0 \cdots 0, \sigma_n = \epsilon, 0 \cdots 0, \sigma_{n'} = +1, \sigma_{n'+1} = -1, 0 \cdots 0\rangle).$$

The vector $|\psi\rangle = |\sigma_1 = 1, \sigma_2 = -1, 0 \cdots 0, \sigma_n = \epsilon, 0 \cdots 0\rangle$ belongs to this latter set; it is an eigenvector of $\hat{H}_0$ with energy $3D$.

**25.5.3.2** The definition of $\hat{T}$, $\hat{T}^\dagger$ and $|n, \pm\rangle$ implies:

$$\hat{T}|n, \pm\rangle = |n+1, \pm\rangle; \quad \hat{T}^\dagger|n, \pm\rangle = |n-1, \pm\rangle. \tag{25.17}$$

**25.5.3.3**  We therefore obtain

$$A(\hat{T} + \hat{T}^\dagger)|n, \pm\rangle = A(|n - 1, \pm\rangle + |n + 1, \pm\rangle). \tag{25.18}$$

Since

$$\hat{H}_1|n, \pm\rangle = A(|n-1, \pm\rangle + |n+1, \pm\rangle) + |\psi_n\rangle \quad \text{where } \langle n', \pm|\psi_n\rangle = 0, \tag{25.19}$$

$\hat{H}_1$ and $A(\hat{T} + \hat{T}^\dagger)$ obviously have the same matrix elements in the subspace $\mathcal{E}^1$.

**25.5.3.4**  Since $\hat{T}^N = \hat{I}$, an eigenvalue $\lambda_k$ satisfies $\lambda_k^N = 1$, which proves that each eigenvalue is an $N$-th root of unity. Conversely, we will see in the following that each $N$-th root of unity is an eigenvalue.

**25.5.3.5  (a)**  The corresponding eigenvectors satisfy

$$|q_k, \pm\rangle = \sum_n c_n|n, \pm\rangle \quad \hat{T}|q_k, \pm\rangle = \lambda_k|q_k, \pm\rangle \tag{25.20}$$

therefore one has

$$\sum_n c_n|n + 1, \pm\rangle = \lambda_k \sum_n c_n|n, \pm\rangle. \tag{25.21}$$

Hence the recursion relation and its solution are

$$\lambda_k c_n = c_{n-1} \quad c_n = \frac{1}{\lambda_k^{n-1}}c_1 = e^{iq_k(n-1)}c_1. \tag{25.22}$$

**(b)**  The normalization condition $\sum_n |c_n|^2 = 1$ gives $N|c_1|^2 = 1$. If we choose $c_1 = e^{iq_k}/\sqrt{N}$, the eigenvectors are of unit norm and we recover the solution given in the text of the problem.

**(c)**  The scalar product of $|q_k, \epsilon\rangle$ and $|q_{k'}, \epsilon'\rangle$ is easily calculated:

$$\langle q_{k'}, \epsilon'|q_k, \epsilon\rangle = \delta_{\epsilon,\epsilon'}\frac{1}{N}\sum_n e^{2i\pi n(k-k')/N} = \delta_{\epsilon,\epsilon'}\,\delta_{k,k'}. \tag{25.23}$$

**(d)**  The vectors $|q_k, \pm\rangle$ are eigenvectors of $\hat{T}^\dagger$ with the complex conjugate eigenvalues $\lambda_k^*$. Therefore they are also eigenvectors of $\hat{T} + \hat{T}^\dagger$ with the eigenvalue $\lambda_k + \lambda_k^* = 2\cos q_k = -2\cos(2k\pi/N)$.

**(e)** From the definition of the vectors, we have

$$\langle n, \epsilon | q_k, \epsilon' \rangle = \frac{1}{\sqrt{N}} e^{iq_k n} \delta_{\epsilon\epsilon'} \tag{25.24}$$

and (directly or by using the closure relation)

$$|n, \pm\rangle = \frac{1}{\sqrt{N}} \sum_{k=0}^{N-1} e^{-iq_k n} |q_k, \pm\rangle. \tag{25.25}$$

**25.5.3.6 (a)** The restriction of $\hat{H}_1$ to the subspace $\mathcal{E}^1$ is identical to $A(\hat{T} + \hat{T}^\dagger)$. In $\mathcal{E}^1$, the operator $A(\hat{T} + \hat{T}^\dagger)$ is diagonal in the basis $|q_k, \pm\rangle$. Therefore the restriction of $\hat{H}_1$ is also diagonal in that basis.

**(b)** The energy levels are

$$E(q_k) = D + 2A \cos(q_k), \tag{25.26}$$

corresponding to the states $|q_k, \pm\rangle$.

**(c)** As far as degeneracies are concerned, there is a twofold degeneracy for all levels (the spin value may be $+1$ or $-1$). In addition, for all levels except $q = -\pi$ and $q = 0$, there is a degeneracy $q_k \leftrightarrow -q_k$ (symmetry of the cosine). Therefore, in general, the degeneracy is 4.

## 25.5.4 Vibrations of the Chain: Excitons

**25.5.4.1** At time $t$ the state of the chain is (cf. (25.26)):

$$|\Psi(t)\rangle = e^{-iDt/\hbar} \sum_\epsilon \sum_k \varphi_k^\epsilon e^{-i\omega t \cos q_k} |q_k, \epsilon\rangle. \tag{25.27}$$

**25.5.4.2** We now consider an initial state $|q_k, \pm\rangle$, evolving as $e^{-iE(q)t/\hbar}|q_k, \pm\rangle$.

**(a)** We therefore obtain an amplitude

$$\alpha_n(t) = \frac{1}{\sqrt{N}} e^{i(q_k n - E(q_k)t/\hbar)} \tag{25.28}$$

and a probability $P_n(t) = |\alpha_n|^2 = \frac{1}{N}$, which is the same on each site.

**(b)** In the expression

$$\alpha_n(t) = \frac{1}{\sqrt{N}} e^{i(q_k x_n/a - E(q_k)t/\hbar)}, \tag{25.29}$$

we see that $\alpha_n(t)$ is the value at $x = x_n$ of the function $\Psi_k(x, t) = \frac{1}{\sqrt{N}} \exp[i(px - Et)/\hbar]$ with $E(q) = D + \hbar\omega \cos q$ and $p(q) = \hbar q/a$.

**(c)** The function $\Psi_k(x)$ is an eigenstate of $\hat{p}_x$ with the eigenvalue $\hbar q_k/a$. Since $N$ is even, we obtain:

$$e^{iq_k L/a} = e^{iNq_k} = e^{2\pi ik} = 1, \tag{25.30}$$

which proves the periodicity of $\Psi_k$.

**(d)** For $|q_k| \ll 1$, $\cos q_k = 1 - q_k^2/2$. Therefore $E = E_0 + p^2/2m$ with

$$E_0 = D + 2A \quad \text{and} \quad m = -\frac{\hbar^2}{2Aa^2} = -\frac{\hbar}{\omega a^2}. \tag{25.31}$$

$\Psi_k$ then satisfies the wave equation

$$i\hbar\frac{\partial\psi}{\partial t} = -\frac{\hbar^2}{2m}\frac{\partial^2\psi}{\partial x^2} + E_0\psi, \tag{25.32}$$

which is a Schrödinger equation for a particle of negative mass (in solid state physics, this corresponds to the propagation of holes and in field theory, to the propagation of anti-particles).

**25.5.4.3 (a)** With the data of the figure which resemble *grosso modo* the $E(q)$ drawn in Fig. 25.2, one finds $D+2A \sim 3.2 \times 10^{-3}$ eV, and $D-2A \sim 0.4 \times 10^{-3}$ eV.

**Fig. 25.2** Variation of the energy $E(q)$ as function of $q$

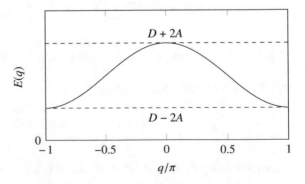

Therefore:

$$D \sim 1.8 \times 10^{-3} \, \text{eV} \quad A \sim 0.7 \times 10^{-3} \, \text{eV}. \tag{25.33}$$

**(b)** The approximation $D \gg A$ is poor. The theory is only meaningful to order $(A/D)^2 \sim 10\%$. Second order perturbation theory is certainly necessary to account quantitatively for the experimental curve which has a steeper shape than a sinusoid in the vicinity of $q = -\pi$.

**(c)** For $T = 1.4 \, \text{K}, kT \sim 1.2 \times 10^{-4} \, \text{eV}, \exp(-(D - 2A)/kT) \sim 0.04$. To a few %, the system is in its ground state.

**25.5.4.4** Approximating $E(q) = E(q_0) + (q - q_0)u_0$ in the vicinity of $q_0$, we obtain

$$\alpha_n(t) = \frac{1}{\sqrt{N}} e^{i(q_0 n - \omega_0 t)} \sum_k \varphi_k e^{i(q_k - q_0)(n - u_0 t/\hbar)}. \tag{25.34}$$

Since the global phase factor does not contribute to the probability, one has $P_n(t) = P_{n'}(t')$ with

$$t' = t + (n' - n) \frac{\hbar}{u_0}. \tag{25.35}$$

This corresponds to the propagation of a wave along the chain, with a group velocity

$$v_g = \frac{u_0 a}{\hbar} = \frac{a}{\hbar} \left. \frac{dE}{dq} \right|_{q=q_0} = -\frac{2aA}{\hbar} \sin q_0. \tag{25.36}$$

For $q_0 = -\pi/2$ and $a = 0.7 \, \text{nm}$, we find $v_g \sim 1500 \, \text{ms}^{-1}$. One can also evaluate $u_0 \sim 1.2 \, \text{meV}$ directly on the experimental curve, which leads to $v_g \sim 1300 \, \text{ms}^{-1}$.

**25.5.4.5** If $|\Psi(0)\rangle = |n = 1, +\rangle$, then $\varphi_k^+ = e^{-iq_k}/\sqrt{N}$ and $\varphi_k^- = 0$.

**(a)** The probability is $P_m(t) = |\langle m, +|\Psi(t)\rangle|^2$, where

$$\langle m, +|\Psi(t)\rangle = \frac{e^{-iDt/\hbar}}{N} \sum_k' e^{iq_k(m-1)} e^{-i\omega t \cos q_k}. \tag{25.37}$$

**(b) N = 2:**
There are two possible values for $q_k$ : $q_0 = -\pi$ and $q_1 = 0$. This leads to $P_1(t) = \cos^2 \omega t, P_2(t) = \sin^2 \omega t$. These are the usual oscillations of a two-state system, such as the inversion of the ammonia molecule.

**(c)  N = 8:**

| $q_k$ | $-\pi$ | $-\frac{3\pi}{4}$ | $-\frac{\pi}{2}$ | $-\frac{\pi}{4}$ | $0$ | $\frac{\pi}{4}$ | $\frac{\pi}{2}$ | $\frac{3\pi}{4}$ |
|---|---|---|---|---|---|---|---|---|
| $\cos(q_k)$ | $-1$ | $-\frac{1}{\sqrt{2}}$ | $0$ | $\frac{1}{\sqrt{2}}$ | $1$ | $\frac{1}{\sqrt{2}}$ | $0$ | $-\frac{1}{\sqrt{2}}$ |

The probability $P_1$ of finding the excitation on the initial site is

$$P_1(t) = \frac{1}{4}\left(\cos^2(\omega t/2) + \cos(\omega t/\sqrt{2})\right)^2. \tag{25.38}$$

The system is no longer periodic in time. There cannot exist $t \neq 0$ for which $P_1(t) = 1$, otherwise there would exist $n$ and $n'$ such that $\sqrt{2} = n'/n$.

**(d)**  Since $J_n(\omega t) \sim 0$ for $\omega t < 2|n|/\pi$, only sites for which $|m-1| < \pi\omega t/2$ are reached at time $t$. For large $\omega t$, the amplitude is the same for all sites of the same parity:

$$P_m(t) = \frac{2}{\pi\omega t}\cos^2\left(\omega t - (m-1)\frac{\pi}{2} - \frac{\pi}{4}\right). \tag{25.39}$$

We notice in particular that $P_m(t) + P_{m+1}(t) = 2/(\pi\omega t)$ is independent of $m$ and varies slowly with $t$.

**(e)**  The probability wave becomes delocalized very quickly on the chain ($\omega t >$ a few $\pi$). The edges of the region where the probability is non zero propagate in opposite directions with the velocity $v = \pi\omega a/2$. This is comparable with what we have found in 25.5.4.4 for a wave packet near $q = \pi/2$.

The experimental data displayed in this chapter were obtained by B. Dorner, D. Visser, U. Steigenberger, K. Kakurai and M. Steiner, *Magnetic excitations in the quasi one-dimensional antiferromagnetic singlet ground state system CsFeBr₃*, Z. Phys. **B 72**, 487 (1988).

# A Quantum Box

**26**

In recent years, it has become possible to devise quantum boxes (also called quantum dots) of nanometric dimensions, inside which the conduction electrons of a solid are confined at low temperatures. The ensuing possibility to control the energy levels of such devices leads to very interesting applications in micro-electronics and opto-electronics.

A quantum box is made of a material A on which another material B is deposited. A set of quantum boxes is shown on Fig. 26.1. The dots of In As (material B) are deposited on a substrate of Ga As (material A).

In this chapter, we are interested in the motion of an electron in a two-dimensional box. We note $-q$ the electric charge of the electron, and we neglect spin effects. We shall assume that in a solid, the dynamics of an electron is described by the usual Schrödinger equation where:

(i) the mass of the electron is replaced by an effective mass $\mu$,
(ii) the atoms of the materials A and B create an effective potential $V(x, y)$ which is slowly varying on the atomic scale.

## 26.1 Results on the One-Dimensional Harmonic Oscillator

Consider a particle of mass $\mu$ placed in the one-dimensional potential $V(x) = \mu\omega^2 x^2/2$. We recall the definition of the annihilation and creation operators $\hat{a}_x$ and $\hat{a}_x^\dagger$ of the oscillator in terms of the position and momentum operators $\hat{x}$ and $\hat{p}_x$

$$\hat{a}_x = \frac{1}{\sqrt{2}}\left(\hat{x}\sqrt{\frac{\mu\omega}{\hbar}} + i\frac{\hat{p}_x}{\sqrt{\hbar\mu\omega}}\right) \quad \hat{a}_x^\dagger = \frac{1}{\sqrt{2}}\left(\hat{x}\sqrt{\frac{\mu\omega}{\hbar}} - i\frac{\hat{p}_x}{\sqrt{\hbar\mu\omega}}\right). \tag{26.1}$$

© Springer Nature Switzerland AG 2019
J.-L. Basdevant, J. Dalibard, *The Quantum Mechanics Solver*,
https://doi.org/10.1007/978-3-030-13724-3_26

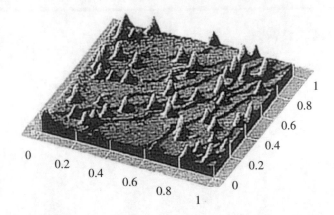

**Fig. 26.1** Picture of a set of quantum boxes obtained with a tunneling microscope. The side of the square is 1 μm long and the vertical scale will be studied below

The Hamiltonian of the system can be written as

$$\hat{H}_x = \frac{\hat{p}_x^2}{2\mu} + \frac{1}{2}\mu\omega^2\hat{x}^2 = \hbar\omega\left(\hat{n}_x + \frac{1}{2}\right) \quad \text{where} \quad \hat{n}_x = \hat{a}_x^\dagger\hat{a}_x. \tag{26.2}$$

We also recall that the eigenvalues of the number operator $\hat{n}_x$ are the non-negative integers. Noting $|n_x\rangle$ the eigenvector corresponding to the eigenvalue $n_x$, we have

$$\hat{a}_x^\dagger|n_x\rangle = \sqrt{n_x + 1}|n_x + 1\rangle \quad \hat{a}_x|n_x\rangle = \sqrt{n_x}|n_x - 1\rangle \tag{26.3}$$

**26.1.1** We recall that the ground state wave function is

$$\psi_0(x) = \left(\frac{\mu\omega}{\pi\hbar}\right)^{1/4}\exp\left(-\frac{\mu\omega x^2}{2h}\right). \tag{26.4}$$

What is the characteristic extension $\ell_0$ of the electron's position distribution in this state?

**26.1.2** The effective mass $\mu$ of the electron in the quantum box is $\mu = 0.07\,m_0$, where $m_0$ is the electron mass in vacuum. We assume that $\hbar\omega = 0.060$ eV, i.e. $\omega/(2\pi) = 1.45 \times 10^{13}$ Hz.

(a) Evaluate $\ell_0$ numerically.
(b) At a temperature of 10 Kelvin, how many levels of the oscillator are populated significantly?
(c) What is the absorption wavelength of radiation in a transition between two consecutive levels?

## 26.2   The Quantum Box

We assume that the effective two-dimensional potential seen by an electron in the quantum box is:

$$V(x, y) = \frac{1}{2}\mu\omega^2(x^2 + y^2).$$   (26.5)

We note $\hat{H}_0 = (\hat{p}_x^2 + \hat{p}_y^2)/2\mu + V(x, y)$ the Hamiltonian of the electron.

**26.2.1** We define the operators $\hat{a}_y, \hat{a}_y^\dagger$ and $\hat{n}_y$ in an analogous way to (26.1) and (26.2). Give a justification for the fact that the states $|n_x, n_y\rangle$, which are eigenstates of $\hat{n}_x$ and $\hat{n}_y$ with integer eigenvalues $n_x$ and $n_y$, form an eigenbasis of $\hat{H}_0$. Give the energy levels $E_N$ of $\hat{H}_0$ in terms of $n_x$ and $n_y$.

**26.2.2** What is the degeneracy $g_N$ of each level $E_N$ where $N = 0, 1, 2 \ldots$?

**26.2.3** Express the operator $\hat{L}_z = \hat{x}\hat{p}_y - \hat{y}\hat{p}_x$ as a function of the operators $\hat{a}_x, \hat{a}_x^\dagger, \hat{a}_y, \hat{a}_y^\dagger$.

**26.2.4** Write the action of $\hat{L}_z$ on the eigenstates $|n_x, n_y\rangle$ of $\hat{H}_0$. Do the states $|n_x, n_y\rangle$ have a well-defined angular momentum $L_z$?

**26.2.5** We are now interested in finding another eigenbasis of $\hat{H}_0$.

(a) Show that $\hat{H}_0$ and $\hat{L}_z$ commute. Interpret this result physically.
(b) We introduce the "left" and "right" annihilation operators as

$$\hat{a}_l = \frac{1}{\sqrt{2}}(\hat{a}_x + i\hat{a}_y) \quad \hat{a}_r = \frac{1}{\sqrt{2}}(\hat{a}_x - i\hat{a}_y)$$   (26.6)

and the corresponding creation operators $\hat{a}_l^\dagger, \hat{a}_r^\dagger$. Write the commutation relations of these four operators.
(c) Show that $\hat{n}_l = \hat{a}_l^\dagger\hat{a}_l$ and $\hat{n}_r = \hat{a}_r^\dagger\hat{a}_r$ commute. Using the values of the commutators $[\hat{a}_l, \hat{a}_l^\dagger]$ and $[\hat{a}_r, \hat{a}_r^\dagger]$, and following the same procedure as in the usual quantization of the harmonic oscillator, justify that the eigenvalues $n_l$ and $n_r$ of $\hat{n}_l$ and $\hat{n}_r$ are integers.

We assume that $\{\hat{n}_l, \hat{n}_r\}$ form a complete set of commuting observables in the problem under consideration, and we note $\{|n_l,\ n_r\rangle\}$ the corresponding eigenbasis.
(d) Write the expression of $\hat{n}_l$ and $\hat{n}_r$ in terms of $\hat{a}_x, \hat{a}_x^\dagger, \hat{a}_y$ and $\hat{a}_y^\dagger$. Deduce from that an expression of the operators $\hat{H}_0$ and $\hat{L}_z$ in terms of $\hat{n}_l$ and $\hat{n}_r$. Justify that the states $\{|n_l, n_r\rangle\}$ form an eigenbasis common to $\hat{H}_0$ and $\hat{L}_z$.

(e) We note $m\hbar$ and $E_N$ the eigenvalues of $\hat{L}_z$ and $\hat{H}_0$. What are the allowed values for the quantum number $m$ in a given energy level $E_N$?

(f) Represent the allowed couples of quantum numbers $(m, N)$ by points in the $(L_z, E_N)$ plane. Show that one recovers the degeneracy of the levels of $\hat{H}_0$.

**26.2.6** We consider the eigen-subspace of $\hat{H}_0$ generated by $|n_x = 1, n_y = 0\rangle$ and $|n_x = 0, n_y = 1\rangle$ as defined above. Write the eigenstates of $\hat{L}_z$ in this basis and give the corresponding eigenvalues.

---

## 26.3   Quantum Box in a Magnetic Field

One applies a uniform magnetic field $\boldsymbol{B}$, parallel to the $z$ axis, on the quantum box. This field derives from the vector potential $\boldsymbol{A}(\boldsymbol{r}) = -(yB/2)\boldsymbol{u}_x + (xB/2)\boldsymbol{u}_y$ where $\boldsymbol{u}_x$ and $\boldsymbol{u}_y$ are the unit vectors along the $x$ and $y$ directions. We assume that the Hamiltonian of the electron in the quantum box and in presence of the field is given by

$$\hat{H}_B = \frac{1}{2\mu}(\hat{p}_x + qA_x(\hat{\boldsymbol{r}}))^2 + \frac{1}{2\mu}(\hat{p}_y + qA_y(\hat{\boldsymbol{r}}))^2 + \frac{1}{2}\mu\omega^2(\hat{x}^2 + \hat{y}^2) \qquad (26.7)$$

with the usual canonical commutation relations $[\hat{x}, \hat{p}_x] = [\hat{y}, \hat{p}_y] = i\hbar$ and $[\hat{x}, \hat{p}_y] = [\hat{y}, \hat{p}_x] = 0$. We introduce the cyclotron angular frequency $\omega_c = qB/\mu$, $(\omega_c > 0)$.

**26.3.1** Expand $\hat{H}_B$ and show that one can always find a basis of the system for which both the *total energy* and the *angular momentum along z* are simultaneously well defined.

**26.3.2** We define $\Omega = \sqrt{\omega^2 + \omega_c^2/4}$. By redefining in a simple manner the operators of Sect. 26.2, give the energy levels $E_{n_l, n_r}$ of the system in terms of the two integers $n_l$ and $n_r$.

**26.3.3** For simplicity we subtract the zero point energy $\hbar\Omega$ from the energies $E_{n_l, n_r}$, and we set $\tilde{E}_{n_l, n_r} = E_{n_l, n_r} - \hbar\Omega$.

(a) Give the approximate expressions of the levels $\tilde{E}_{n_l, n_r}$ in the two limits of a weak and of a strong field, and give a definition of these regimes.

(b) Plot as a function of the magnetic field, the positions of the energy levels originating from the $N = 0, 1, 2$ levels in the absence of magnetic field.

(c) Show that two levels cross for a value of $B$ such that $\omega_c = \omega/\sqrt{2}$ and specify which states they correspond to.

**26.3.4** In the following we assume that $\omega_c < \omega/\sqrt{2}$. Using the values of $\omega$ and of the effective mass $\mu$ given in the first section of the problem, determine which values of the magnetic field correspond to this inequality.

**26.3.5** Show that the first three eigenstates of $\hat{H}_B$ have respective energies

$$E_0 = \hbar\Omega, \quad E_- = 2\hbar\Omega - \frac{\hbar\omega_c}{2}, \quad E_+ = 2\hbar\Omega + \frac{\hbar\omega_c}{2}. \tag{26.8}$$

We note $|u_0\rangle$, $|u_-\rangle$, $|u_+\rangle$ the three corresponding eigenstates. What is the value of the angular momentum $L_z$ in each of these three states?

## 26.4 Experimental Verification

One can study the energy levels of an electron in a quantum box by measuring the absorption spectrum of a light beam. Absorption peaks appear at the Bohr frequencies $(E_f - E_i)/h$ of the box, corresponding to excitations of the electron from an initial level $|u_i\rangle$ to a final one $|u_f\rangle$.

**26.4.1** At a temperature of 10 K, one observes that only the level $|u_i\rangle = |u_0\rangle$ contributes significantly to the absorption signal. Justify this fact.

**26.4.2** The experimental values of the frequencies of the first two absorption peaks of a quantum box are given in Fig. 26.2, according to the values of the applied magnetic field. Verify that the model developed above accounts for the slope of the curves for sufficiently large values of $B$, but that it fails to describe the behavior near $B = 0$.

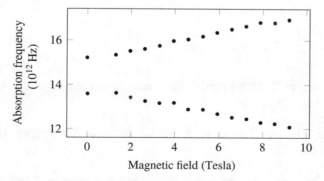

**Fig. 26.2** Frequencies $\nu = \omega/2\pi$ of the two first absorption peaks of a quantum box as a function of the magnetic field $B$

**26.4.3** Role of the $z$ Dimension.

The confinement along the $z$ direction can be simulated by an infinite one-dimensional square well potential of extension $D$ in this direction.

(a) Recall the energy levels of a particle of mass $\mu$ in a one-dimensional infinite well of width $D$. What is the energy difference between the two lowest lying states?
(b) Under what condition between $D$ and $\omega$ is it legitimate to consider that the motion along the $z$ axis is "frozen", and to restrict to the lowest lying levels of the harmonic motion in the $x$, $y$ plane?
(c) On the Fig. 26.1, the vertical ($z$) and horizontal ($xy$) scales are different. In order to neglect the motion along the $z$ direction, which of the two scales must be contracted? One can make use of the value of $l_0$ calculated before.

## 26.5   Anisotropy of a Quantum Box

In order to reproduce the positions of the absorption peaks for low values of the field (see Fig. 26.2), we assume that the confining potential in the quantum box is slightly anisotropic. We therefore replace the expression (26.5) of the confining potential by

$$V(x, y) = \frac{1}{2}\mu\omega^2(1 + \epsilon)x^2 + \frac{1}{2}\mu\omega^2(1 - \epsilon)y^2 \quad \text{with} \quad \epsilon \ll 1. \tag{26.9}$$

We treat the problem using perturbation theory. In what follows, we assume that the magnetic field is small ($\omega_c \ll \omega$). The Hamiltonian $\hat{H}_{B,\epsilon}$ can therefore be written as:

$$\hat{H}_{B,\epsilon} = \hat{H}_0 + \hat{W} \tag{26.10}$$

with

$$\hat{H}_0 = \frac{\hat{p}^2}{2\mu} + \frac{1}{2}\mu\omega^2(\hat{x}^2 + \hat{y}^2) \quad \text{and} \quad \hat{W} = \frac{\omega_c}{2}\hat{L}_z + \epsilon\frac{\mu\omega^2}{2}(\hat{x}^2 - \hat{y}^2). \tag{26.11}$$

In this expression, $\hat{W}$ is the perturbation, and we neglect terms of order $B^2$. We work in the basis $\{|n_x, n_y\rangle\}$.

**26.5.1** Using the potential given in (26.9), determine the energy levels of the electron for $B = 0$.

**26.5.2** Applying perturbation theory, evaluate to first order in $B$ and $\epsilon$ the energy shift of the ground state of the quantum box with respect to the value found in Sect. 26.2.

**26.5.3** We are now interested in the position of the two energy states originating from the first excited level of $\hat{H}_0$ found in Sect. 26.2.

(a) Write the restriction of the Hamiltonian $H_{B,\epsilon}$ in the basis $\{|n_x = 1, n_y = 0\rangle, |n_x = 0, n_y = 1\rangle\}$ of the corresponding subspace.

(b) Deduce the approximate values of the energies of interest $E_-(B, \epsilon)$ and $E_+(B, \epsilon)$ (with $E_- < E_+$).

(c) Give the expressions of the corresponding eigenstates $|u_-(B, \epsilon)\rangle$ and $|u_+(B, \epsilon)\rangle$. One can introduce the mixing angle $\alpha$ such that $\tan(2\alpha) = \omega_c/(\epsilon\omega)$.

**26.5.4** We now turn back to the experimental data of Fig. 26.2.

(a) Is the transition region between large values of $B$ and $B = 0$ correctly described by this model of an anisotropic quantum box?

(b) What value of the anisotropy $\epsilon$ can be deduced from the positions of the first two absorption peaks in the absence of a magnetic field?

---

## 26.6  Solutions

### 26.6.1  Results on the One-Dimensional Harmonic Oscillator

**26.6.1.1** The characteristic spatial extension of the position distribution is $\ell_0 = \sqrt{\hbar/(\mu\omega)}$. More precisely, $\ell_0/\sqrt{2}$ is the r.m.s. deviation of the probability law $|\psi_0(x)|^2$ for the position distribution.

**26.6.1.2 (a)** One finds $\ell_0 = 4.3$ nm.

**(b)** The ratio of populations in the first excited level $n = 1$ and the ground state $n = 0$ is given by the Boltzmann law

$$r = \exp(-\hbar\omega/(k_B T)) = \exp(-70) = 5 \times 10^{-31}, \tag{26.12}$$

which is negligible. The population of other excited levels is even smaller. Therefore, at $T = 10$ K, only the ground state is populated.

**(c)** One finds $\lambda = 2\pi c/\omega = 21\mu$m, which corresponds to an infrared radiation.

### 26.6.2  The Quantum Box

**26.6.2.1** The Hamiltonian is $\hat{H}_0 = \hbar\omega(\hat{n}_x + \hat{n}_y + 1)$, and the operators $\hat{n}_x$ and $\hat{n}_y$ commute. We can find an eigenbasis common to the two operators. If a function $\Psi(x, y)$ is an eigenfunction of $\hat{n}_x$, its $x$ dependence is completely determined (Hermite function of the variable $x\sqrt{\mu\omega/\hbar}$, corresponding to $|n_x\rangle$). Similarly, for the $y$ dependence. The set $\{\hat{n}_x, \hat{n}_y\}$ is therefore a CSCO, with eigenbasis $|n_x, n_y\rangle$. This basis is also an eigenbasis of $\hat{H}_0$, the eigenvalue corresponding to $|n_x, n_y\rangle$ is $E_N = \hbar\omega(N + 1)$ with $N = n_x + n_y$.

**26.6.2.2** The energy level $E_N = \hbar\omega(N + 1)$ corresponds to $N + 1$ possible couples for $(n_x, n_y)$ : $(N, 0), (N - 1, 1), \ldots, (0, N)$. The degeneracy is therefore $g_N = N + 1$.

**26.6.2.3** We replace $\hat{x}, \hat{p}_x, \hat{y}, \hat{p}_y$ by their values:

$$\hat{x} = \sqrt{\frac{\hbar}{2\mu\omega}}(\hat{a}_x^\dagger + \hat{a}_x) \quad \hat{p}_x = i\sqrt{\frac{\hbar\mu\omega}{2}}(\hat{a}_x^\dagger - \hat{a}_x) \quad \ldots \tag{26.13}$$

which leads to:

$$\hat{L}_z = \frac{i\hbar}{2}((\hat{a}_x^\dagger + \hat{a}_x)(\hat{a}_y^\dagger - \hat{a}_y) - (\hat{a}_y^\dagger + \hat{a}_y)(\hat{a}_x^\dagger - \hat{a}_x)) = i\hbar(\hat{a}_x \hat{a}_y^\dagger - \hat{a}_x^\dagger \hat{a}_y). \tag{26.14}$$

**26.6.2.4** Making use of the action of the creation and annihilation operators, one finds:

$$\hat{L}_z|n_x, n_y\rangle = i\hbar\left(\sqrt{n_x(n_y + 1)}|n_x - 1, n_y + 1\rangle - \sqrt{(n_x + 1)n_y}|n_x + 1, \ n_y - 1\rangle\right). \tag{26.15}$$

A state $|n_x, n_y\rangle$ is not, in general, an eigenstate of $\hat{L}_z$ and does not have a well defined angular momentum. The only exception is the state for which $n_x = n_y = 0$, and $\hat{L}_z|0, 0\rangle = 0$. It is a state with $L_z = 0$.

**26.6.2.5** We are now interested in finding another eigenbasis of $\hat{H}_0$.

**(a)** The commutator of $\hat{H}_0$ and $\hat{L}_z$ can be calculated using their expressions in terms of the creation and annihilation operators. We first calculate $[\hat{n}_x, \hat{a}_x^\dagger] = \hat{a}_x^\dagger$ and $[\hat{n}_x, \hat{a}_x] = -\hat{a}_x$, and the analogous relations for $y$. One obtains:

$$[\hat{H}_0, \hat{L}_z] = i\hbar^2\omega[\hat{n}_x + \hat{n}_y, \hat{a}_x \hat{a}_y^\dagger - \hat{a}_x^\dagger \hat{a}_y]$$

$$= i\hbar^2\omega([\hat{n}_x, \hat{a}_x]a_y^\dagger - [\hat{n}_x, \hat{a}_x^\dagger]\hat{a}_y + \hat{a}_x[\hat{n}_y, \hat{a}_y^\dagger] - \hat{a}_x^\dagger[\hat{n}_y, \hat{a}_y])$$

$$= i\hbar^2\omega\left(-\hat{a}_x a_y^\dagger - \hat{a}_x^\dagger \hat{a}_y + \hat{a}_x a_y^\dagger + \hat{a}_x^\dagger \hat{a}_y\right) = 0.$$

This result is the consequence of the rotation invariance around the $z$ axis of the potential $V(x, y)$. This can also be proven by using the polar coordinate expression: $\hat{L}_z = -i\hbar\frac{\partial}{\partial\phi}$, which commutes with the two contributions to $\hat{H}_0$ (kinetic and potential energy).

**(b)** There are four operators, and therefore six commutators to evaluate. We first remark that the two creation operators commute, and so do the two annihilation operators. Therefore:

$$[\hat{a}_l^\dagger, \hat{a}_r^\dagger] = 0, \quad [\hat{a}_l, \hat{a}_r] = 0. \tag{26.16}$$

A simple calculation leads to:

$$\left[\hat{a}_l, \hat{a}_l^\dagger\right] = \frac{1}{2}[\hat{a}_x + i\hat{a}_y, \hat{a}_x^\dagger - i\hat{a}_y^\dagger] = \frac{1}{2}[\hat{a}_x, \hat{a}_x^\dagger] + \frac{1}{2}[\hat{a}_y, \hat{a}_y^\dagger] = 1,$$

$$\left[\hat{a}_r, \hat{a}_r^\dagger\right] = \frac{1}{2}[\hat{a}_x - i\hat{a}_y, \hat{a}_x^\dagger + i\hat{a}_y^\dagger] = \frac{1}{2}[\hat{a}_x, \hat{a}_x^\dagger] + \frac{1}{2}[\hat{a}_y, \hat{a}_y^\dagger] = 1.$$

Finally, one finds:

$$\left[\hat{a}_l, \hat{a}_r^\dagger\right] = \frac{1}{2}[\hat{a}_x + i\hat{a}_y, \hat{a}_x^\dagger + i\hat{a}_y^\dagger] = \frac{1}{2}[\hat{a}_x, \hat{a}_x^\dagger] - \frac{1}{2}[\hat{a}_y, \hat{a}_y^\dagger] = 0,$$

$$\left[\hat{a}_r, \hat{a}_l^\dagger\right] = \frac{1}{2}[\hat{a}_x - i\hat{a}_y, \hat{a}_x^\dagger - i\hat{a}_y^\dagger] = \frac{1}{2}[\hat{a}_x, \hat{a}_x^\dagger] - \frac{1}{2}[\hat{a}_y, \hat{a}_y^\dagger] = 0.$$

Therefore, any "right" and "left" operators commute. The commutation relations between right (or left) creation and annihilation operators are the same as for usual one-dimensional creation and annihilation operators.

**(c)** The commutation between the hermitian operators $\hat{n}_l$ and $\hat{n}_r$ is obvious, since any left operator commutes with any right operator. The quantization of the one-dimensional harmonic oscillator is entirely based on the commutation relation $[\hat{a}, \hat{a}^\dagger] = 1$, which leads to the fact that the eigenvalues of $\hat{a}^\dagger\hat{a}$ are the non-negative integers. The same holds here for the operators $\hat{n}_l$ and $\hat{n}_r$, the corresponding eigenvalues are the couples of integers $(n_l, n_r)$. As it is suggested in the text, we assume for the moment that the common eigenbasis of $\hat{n}_l$ and $\hat{n}_r$ is unique, and we note $|n_l, n_r\rangle$ the eigenvector corresponding to the couple of eigenvalues $(n_l, n_r)$.

**(d)** One finds:

$$\hat{n}_l = \frac{1}{2}(\hat{a}_x^\dagger - i\hat{a}_y^\dagger)(\hat{a}_x + i\hat{a}_y) = \frac{1}{2}\left(\hat{n}_x + \hat{n}_y + i(\hat{a}_x^\dagger\hat{a}_y - \hat{a}_x\hat{a}_y^\dagger)\right)$$

$$= \frac{1}{2}\left(\frac{\hat{H}_0}{\hbar\omega} - 1 - \frac{\hat{L}_z}{\hbar}\right),$$

$$\hat{n}_r = \frac{1}{2}(\hat{a}_x^\dagger + i\hat{a}_y^\dagger)(\hat{a}_x - i\hat{a}_y) = \frac{1}{2}\left(\hat{n}_x + \hat{n}_y - i(\hat{a}_x^\dagger\hat{a}_y - \hat{a}_x\hat{a}_y^\dagger)\right)$$

$$= \frac{1}{2}\left(\frac{\hat{H}_0}{\hbar\omega} - 1 + \frac{\hat{L}_z}{\hbar}\right),$$

**Fig. 26.3** Allowed quantum numbers for the couple $L_z$, $E$

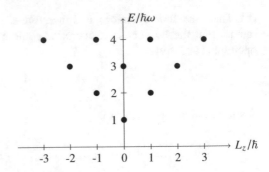

therefore

$$\hat{H}_0 = \hbar\omega(\hat{n}_l + \hat{n}_r + 1), \quad \hat{L}_z = \hbar(\hat{n}_r - \hat{n}_l). \tag{26.17}$$

The operators $\hat{H}_0$ and $\hat{L}_z$ can be expressed in terms of the sole operators $\hat{n}_l$ and $\hat{n}_r$. The eigenbasis common to $\hat{n}_l$ and $\hat{n}_r$ is therefore also a common eigenbasis of $\hat{H}_0$ and $\hat{L}_z$.

**(e)** The vector $|n_l, n_r\rangle$ is the eigenstate of $\hat{H}_0$ and $\hat{L}_z$, with eigenvalues $E = \hbar\omega(n_l + n_r + 1)$ and $L_z = \hbar(n_r - n_l)$. We therefore have $N = n_l + n_r$ and $m = n_r - n_l$. We recover the integer values for the orbital angular momentum, as expected. For a given $N$, the value of $m$ belongs to the set $\{-N, -N + 2, \ldots, N - 2, N\}$, therefore there are $N + 1$ possible values. We remark that in an energy level $E_N$, $m$ has the same parity as $N$. This comes from the parity invariance of the problem under consideration.

**(f)** The graphical representation is given on Fig. 26.3. On a given horizontal line corresponding to a given energy value, one finds $N + 1$ points, which correspond to the degeneracy of an energy level of $\hat{H}_0$ found previously. This justifies the fact that $\{\hat{n}_l, \hat{n}_r\}$ form a CSCO. If two different states would correspond to the same couple of eigenvalues $(n_l, n_r)$, the corresponding point of the diagram would be twofold degenerate, and the degeneracy of the energy level $E_N$ would be larger than $N + 1$.

**26.6.2.6** We must find in this subspace two eigenvectors of $\hat{L}_z$ corresponding to the two eigenvalues $\pm\hbar$. A first method for finding these eigenvectors consists in calculating the action of $\hat{L}_z$ on the vectors of the basis $\{|n_x, n_y\rangle\}$. Using the expression of $\hat{L}_z$ in terms of $\hat{a}_x, \hat{a}_y, \ldots$, one finds:

$$\hat{L}_z|n_x = 1, n_y = 0\rangle = i\hbar|n_x = 0, n_y = 1\rangle$$

$$\hat{L}_z|n_x = 0, n_y = 1\rangle = -i\hbar|n_x = 1, n_y = 0\rangle,$$

or the $2 \times 2$ matrix to diagonalize: $\begin{pmatrix} 0 & -i\hbar \\ i\hbar & 0 \end{pmatrix}$.

The eigenstates associated to the eigenvalues $\pm\hbar$ are therefore:

$$(|n_x = 1, n_y = 0\rangle \pm i|n_x = 0, n_y = 1\rangle)/\sqrt{2}. \tag{26.18}$$

Another method consists in starting from the ground state $|n_l = 0, n_r = 0\rangle$ and letting act on this state: (i) the operator $\hat{a}_r^\dagger$ in order to obtain the eigenvector of energy $2\hbar\omega$ and angular momentum $+\hbar$, (ii) the operator $\hat{a}_l^\dagger$ in order to obtain the eigenvector of energy $2\hbar\omega$ and angular momentum $-\hbar$. Of course, we recover the previous result.

### 26.6.3 Quantum Box in a Magnetic Field

**26.6.3.1** By expanding $\hat{H}_B$, one finds:

$$\hat{H}_B = \frac{\hat{p}_x^2 + \hat{p}_y^2}{2\mu} + \left(\frac{\mu\omega^2}{2} + \frac{q^2 B^2}{8}\right)(\hat{x}^2 + \hat{y}^2) + \frac{\omega_c \hat{L}_z}{2}. \tag{26.19}$$

**26.6.3.2** If we set $\Omega = \sqrt{\omega^2 + \omega_c^2/4}$, we can rewrite $\hat{H}_B = \hat{H}_0^{(\Omega)} + \omega_c \hat{L}_z/2$, where $\hat{H}_0^{(\Omega)}$ is the Hamiltonian of a two-dimensional oscillator of frequency $\Omega$:

$$\hat{H}_0^{(\Omega)} = \frac{\hat{p}_x^2 + \hat{p}_y^2}{2\mu} + \frac{\mu\Omega^2}{2}(\hat{x}^2 + \hat{y}^2). \tag{26.20}$$

One can then repeat the method of the previous section, by replacing $\omega$ by $\Omega$ in the definition of the operators $\hat{a}_x, \hat{a}_y, \ldots$ One constructs an eigenbasis common to $H_0^{(\Omega)}$ and $\hat{L}_z$, which we continue to note $\{|n_l, n_r\rangle\}$, the eigenvalues being $\hbar\Omega(n_l + n_r + 1)$ and $m\hbar$. Each vector $|n_l, n_r\rangle$ is also an eigenvector of $\hat{H}_B$, corresponding to the energy

$$\begin{aligned} E_{n_l, n_r} &= \hbar\Omega(n_l + n_r + 1) + \hbar\omega_c(n_r - n_l)/2 \\ &= \hbar\left(\Omega + \frac{\omega_c}{2}\right)n_r + \hbar\left(\Omega - \frac{\omega_c}{2}\right)n_l + \hbar\Omega. \end{aligned}$$

**26.6.3.3** (a) Two limiting regimes of the magnetic field can be considered, corresponding to the limits $\omega_c \ll \omega$ (very weak magnetic field) and $\omega_c \gg \omega$ (very strong magnetic field). In the first case, we have in first order in $B$:

$$\tilde{E}_{n_l, n_r} \simeq \hbar\omega(n_l + n_r) + \hbar\omega_c(n_r - n_l)/2, \tag{26.21}$$

which corresponds to a linear variation in $B$ of the $N + 1$ levels arising from the level $E_N$ in the absence of the field. The slope $(\hbar q B/(2\mu))(n_r - n_l)$ is different for each level, which means that there is no degeneracy if $B$ does not vanish.

**Fig. 26.4** Variation of the
energy levels $\tilde{E}_{n_l,n_r}$ arising
from $N = 0, 1, 2$ as a
function of the magnetic
field $B$

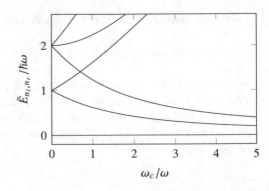

For strong fields, one finds

$$\Omega + \frac{\omega_c}{2} \simeq \omega_c, \quad \Omega - \frac{\omega_c}{2} \simeq \frac{\omega^2}{\omega_c} \ll \omega. \tag{26.22}$$

We therefore have

$$\tilde{E}_{n_l,n_r} \simeq \hbar\omega_c n_r \quad \text{if} \quad n_r \neq 0, \quad \text{and} \quad \tilde{E}_{n_l,0} \simeq \frac{\hbar\omega^2}{\omega_c} n_l. \tag{26.23}$$

For a non-vanishing $n_r$, the energy level increases linearly with $B$, the slope being proportional to $n_r$. For $n_r = 0$, the energy $\tilde{E}$ tends to zero as $1/B$.

**(b)** The energy levels $\tilde{E}_{n_l,n_r}$ corresponding to $N = 0, 1, 2$ are represented on Fig. 26.4.

**(c)** We notice on Fig. 26.4 that the levels $n_l = 2, n_r = 0$ and $n_l = 0, n_r = 1$ cross each other. The corresponding value of the field $B$ is given by the solution of the equation

$$\Omega + \frac{\omega_c}{2} = 2\left(\Omega - \frac{\omega_c}{2}\right), \tag{26.24}$$

or $3\omega_c = 2\Omega$, i.e. $\omega_c = \omega/\sqrt{2}$.

**26.6.3.4** The value of the field $B$ which corresponds to $\omega_c = \omega/\sqrt{2}$ is $\mu\omega/(q\sqrt{2}) \simeq 26$ T.

**26.6.3.5** If we assume that the field $B$ is smaller than 21 T, the three first energy levels of $\hat{H}_B$ correspond to $n_l = n_r = 0$ (ground state of energy $\hbar\Omega$), $n_l = 1, n_r = 0$ (energy $2\hbar\Omega - \hbar\omega_c/2$), and $n_l = 0, n_r = 1$ (energy $2\hbar\Omega + \hbar\omega_c/2$). These three states are eigenstates of $\hat{L}_z$ with the eigenvalues $0, -\hbar$ and $\hbar$ respectively.

## 26.6.4 Experimental Verification

**26.6.4.1** In the absence of a magnetic field, we saw above that only the level $n_x = n_y = 0$ is appreciably populated for $T = 10$ K. As the magnetic field increases, the energy splitting between the ground state and the first excited state diminishes but it stays much smaller than $k_B T$ if $\omega_c$ is less than $\omega/\sqrt{2}$. For $\omega_c = \omega/\sqrt{2}$, the splitting is $\hbar\omega/\sqrt{2}$. For that value, the ratio between the populations of the first excited state and the ground state is $r' = \exp(-49) = 3.7 \times 10^{-22}$.

Since only the ground state is populated, all the detectable absorption lines will occur from transition starting from this state.

**26.6.4.2** The first two absorption peaks correspond to the transitions $|u_0\rangle \leftrightarrow |u_-\rangle$ and $|u_0\rangle \leftrightarrow |u_+\rangle$. The corresponding frequencies $\nu\pm$ are such that

$$\nu_\pm = \frac{\Omega}{2\pi} \pm \frac{\omega_c}{4\pi}. \tag{26.25}$$

The domain in $B$ explored on the experimental figure of the text corresponds to values of $\omega_c$ which are small compared to $\omega$. We can therefore use the weak-field expansion obtained above in order to simplify this expression

$$\nu_\pm = \frac{\omega}{2\pi} \pm \frac{\omega_c}{4\pi}. \tag{26.26}$$

We therefore expect that the frequencies $\nu_\pm$ will vary linearly with $B$, the slopes being $\pm q/(4\pi\mu)$, and that the two straight lines will cross for a vanishing field at the frequency $\omega/(2\pi)$.

This linear variation of $\nu_\pm$ does appear on the figure for higher values of $B$ and the measured slope is close to the expected value $(2 \times 10^{11}$ Hz T$^{-1})$. However, the experimentally observed behavior for a very weak field does not correspond to our theoretical prediction. Instead of two lines of same frequency for $B = 0$, there is a finite difference $\nu_+ - \nu_-$.

**26.6.4.3** Role of the $z$ Dimension.

**(a)** The energy levels of an infinite square well of size $D$ are given by $E_n = \pi^2\hbar^2 n^2/(2\mu D^2)$, with $n$ positive integer, the corresponding eigenstates are the functions $\chi_n(z) \propto \sin(n\pi z/D)$. The splitting between the ground state and the first excited state is $\Delta E = 3\pi^2\hbar^2/(2\mu D^2)$.

**(b)** In order to consider that the motion along $z$ is "frozen", the energy splitting $\Delta E$ between the two first levels of the square well must be very large compared to $\hbar\omega$. If this condition is satisfied, the accessible states for the electron confined in the quantum box (in a reasonable domain of temperatures and exciting frequencies) will be simply combinations of the vectors $|n_x, n_y\rangle \otimes |\chi_0\rangle$. It is then legitimate to neglect the dynamics of the electron along $z$. If this condition is not satisfied, absorption lines can appear for frequencies near those presented on the experimental figure. They will correspond to the excitation of the motion along the $z$ axis. The condition that the $z$ motion be "frozen" is

$$\frac{3\pi^2\hbar^2}{2\mu D^2} \gg \hbar\omega, \quad \text{or equivalently} \quad D \ll \pi\ell_0. \tag{26.27}$$

(c)  In order for the harmonic approximation of the transverse motion to be valid, the transverse extension $\Delta L$ of the quantum box must be large compared to $\ell_0$. The condition obtained in the previous question $D \ll \pi\ell_0$, put together with $\ell_0 \ll \Delta L$, imposes that the box must have a very flat geometrical shape: the height $D$ along $z$ must be very small compared to its transverse extension in $xy$. We conclude that the vertical scale of the Fig. 26.1 is very dilated.

### 26.6.5  Anisotropy of a Quantum Box

**26.6.5.1**  For vanishing $B$, the Hamiltonian is $\hat{H}_x + \hat{H}_y$ with

$$\hat{H}_x = \frac{\hat{p}_x^2}{2\mu} + \frac{1}{2}\mu\omega^2(1+\epsilon)\hat{x}^2, \quad \hat{H}_y = \frac{\hat{p}_y^2}{2\mu} + \frac{1}{2}\mu\omega^2(1-\epsilon)\hat{x}^2. \tag{26.28}$$

One can find a common eigenbasis for $\hat{H}_x$ and $\hat{H}_y$, corresponding to products of Hermite functions in the variable $x\sqrt{\mu\omega(1+\epsilon)/\hbar}$ by Hermite functions in the variable $y\sqrt{\mu\omega(1-\epsilon)/\hbar}$. The corresponding eigenvalues are

$$\hbar\omega\sqrt{1+\epsilon}(n_x+1/2)+\hbar\omega\sqrt{1-\epsilon}(n_y+1/2) \simeq \hbar\omega(n_x+n_y+1)+\frac{\epsilon\hbar\omega}{2}(n_x-n_y) \tag{26.29}$$

where $n_x, n_y$ are non-negative integers.

**26.6.5.2**  To first order in $B$ and $\epsilon$, the shift of the ground state energy is given by the matrix element

$$\Delta E_{0,0} = \langle 0, 0|\hat{W}|0, 0\rangle = \frac{\omega_c}{2}\langle 0, 0|\hat{L}_z|0, 0\rangle + \frac{\epsilon\mu\omega^2}{2}\langle 0, 0|\hat{x}^2 - \hat{y}^2|0, 0\rangle. \tag{26.30}$$

The state $|0, 0\rangle$ is an eigenstate of $\hat{L}_z$ with eigenvalue 0. The first term in this sum therefore vanishes. By symmetry, we have $\langle 0, 0|\hat{x}^2|0, 0\rangle = \langle 0, 0|\hat{y}^2|0, 0\rangle$, which means that the second term also vanishes, to first order in $\epsilon$ and $B$.

**26.6.5.3  (a)**  We have already determined the matrix $\hat{L}_z$ in the basis under consideration in question 26.2.6. We must calculate the matrix elements of $\hat{x}^2$ and $\hat{y}^2$. In order to do that, the simplest is to use the expressions of $\hat{x}$ and $\hat{y}$ in terms of creation and annihilation operators. One has:

$$\hat{x}^2 = \frac{\hbar}{2\mu\omega}(\hat{a}_x + \hat{a}_x^\dagger)(\hat{a}_x + \hat{a}_x^\dagger), \tag{26.31}$$

which leads to

$$\langle 1, 0|\hat{x}^2|1, 0\rangle = \frac{\hbar}{2\mu\omega}\langle 1, 0|\hat{a}_x^\dagger\hat{a}_x + \hat{a}_x\hat{a}_x^\dagger|1, 0\rangle = \frac{\hbar}{2\mu\omega}(1 + 2) = \frac{3\hbar}{2\mu\omega},$$

$$\langle 1, 0|\hat{x}^2|0, 1\rangle = \langle 0, 1|\hat{x}^2|1, 0\rangle = 0,$$

$$\langle 0, 1|\hat{x}^2|0, 1\rangle = \frac{\hbar}{2\mu\omega}\langle 0, 1|\hat{a}_x\hat{a}_x^\dagger|0, 1\rangle = \frac{\hbar}{2\mu\omega},$$

where we have set $|0, 1\rangle \equiv |n_x = 0, n_y = 1\rangle$, etc. for simplicity. We obtain a similar result by exchanging the roles of $x$ and $y$. The restriction of $\hat{H}_{B,\epsilon}$ in the subspace of interest is therefore

$$[\hat{H}_{B,\epsilon}] = 2\hbar\omega + \frac{\hbar}{2}\begin{pmatrix} \epsilon\omega & -i\omega_c \\ i\omega_c & -\epsilon\omega \end{pmatrix}. \tag{26.32}$$

**(b)** The energy eigenvalues are obtained by diagonalizing this $2 \times 2$ matrix

$$E_\pm(B, \epsilon) = 2\hbar\omega \pm \frac{\hbar}{2}\sqrt{\epsilon^2\omega^2 + \omega_c^2}. \tag{26.33}$$

**(c)** Setting $\tan 2\alpha = \omega_c/(\epsilon\omega)$, the above matrix is written as:

$$[\hat{H}_{B,\epsilon}] = 2\hbar\omega + \frac{\hbar}{2}\sqrt{\epsilon^2\omega^2 + \omega_c^2}\begin{pmatrix} \cos 2\alpha & -i\sin 2\alpha \\ i\sin 2\alpha & -\cos 2\alpha \end{pmatrix}, \tag{26.34}$$

whose eigenvectors are

$$|u_-(B, \epsilon)\rangle = \begin{pmatrix} i\sin\alpha \\ \cos\alpha \end{pmatrix} \quad |u_+(B, \epsilon)\rangle = \begin{pmatrix} \cos\alpha \\ i\sin\alpha \end{pmatrix}. \tag{26.35}$$

**26.6.5.4 (a)** The variation of $E_\pm(B, \epsilon) - E_{0,0}$ with $B$ reproduces well the experimental observations. For large values of $B$ such that $\epsilon\omega \ll \omega_c$, we recover the linear variation with $B$ of the two transition frequencies. When $B$ tends to zero ($\omega_c \ll \epsilon\omega$), one finds two different Bohr frequencies corresponding respectively to the two transitions $n_x = n_y = 0 \to n_x = 0, n_y = 1$ and $n_x = n_y = 0 \to n_x = 1, n_y = 0$.

**(b)** When $B$ tends to zero, one finds experimentally that the limit of $(\nu_+ - \nu_-)/(\nu_+ + \nu_-)$ is of the order of 0.06. The theoretical prediction for this ratio is $\epsilon/2$. We therefore conclude that $\epsilon \simeq 0.12$.

## 26.6.6 Comments and References

Quantum boxes of semiconductors, a simple model of which has been examined here, have been the subject of many investigations over the last decades, both academic (Coulomb correlations) and applied (optronics). Here we have only considered electronic excitations, but collective modes in a lattice (phonons) also play an important role in the dynamics of quantum boxes and the two types of excitations can be strongly coupled. This is in contrast with the usual situation encountered in semiconductors, for which the coupling between the electrons and the phonons is weak.

The data presented here come from S. Hameau, Y. Guldner, O. Verzelen, R. Ferreira, G. Bastard, J. Zeman, A. Lemaître, and J. M. Gérard, *Strong Electron-Phonon Coupling Regime in Quantum Dots: Evidence for Everlasting Resonant Polarons*, Phys. Rev. Lett. **83**, 4152 (1999).

# Colored Molecular Ions

# 27

Some pigments are made of linear molecular ions, along which electrons move freely. We derive here the energy levels of such a system and we show how this explains the observed color of the pigments.

Consider molecular ions of the chemical formula $(C_n H_{n+2})^-$, which can be considered as deriving from polyethylene molecules, such as hexatriene

$$CH_2=CH\text{-}CH=CH\text{-}CH=CH_2,$$

with an *even* number of carbon atoms, by removing a $CH^+$ group. In an ion of this type, the bonds rearrange themselves and lead to a *linear* structure of the following type:

$$(CH_2 \cdots CH \cdots CH \cdots CH \cdots CH_2)^-, \tag{27.1}$$

with an odd number $n$ of equally spaced carbon atoms separated by $d = 1.4\,\text{Å}$. In this structure, one can consider that the $n + 1$ electrons of the double bonds of the original polyethylene molecule move independently of one another in a one-dimensional infinite potential well of length $L_n = nd$:

$$
\begin{aligned}
V(x) &= +\infty &&\text{for} && x < 0 \ \ \text{or} \ \ x > L_n \\
&= 0 &&\text{for} && 0 < x \le L_n
\end{aligned}
\tag{27.2}
$$

Actually, one should write $L_n = (n - 1)d + 2b$ where $b$ represents the edge effects. Experimentally, the choice $b = d/2$ appears to be appropriate.

## 27.1 Hydrocarbon Ions

**27.1.1** What are the energy levels $\varepsilon_k$ of an electron in this potential?

© Springer Nature Switzerland AG 2019
J.-L. Basdevant, J. Dalibard, *The Quantum Mechanics Solver*,
https://doi.org/10.1007/978-3-030-13724-3_27

**27.1.2** Owing to the Pauli principle, at most two electrons can occupy the same energy level. What are the energies of the ground state $E_0$ and of the first excited state $E_1$ of the set of $n+1$ electrons? We recall that $\sum_{k=1}^{n} k^2 = n(n+1)(2n+1)/6$.

**27.1.3** What is the wavelength $\lambda_n$ of the light absorbed in a transition between the ground state and the first excited state? One can introduce the Compton wavelength of the electron: $\lambda_C = h/(m_e c) = 2.426 \times 10^{-2}\,\text{Å}$.

**27.1.4** Experimentally, one observes that the ions $n = 9, n = 11$ and $n = 13$ absorb blue light ($\lambda_9 \sim 4700\,\text{Å}$), yellow light ($\lambda_{11} \sim 6000\,\text{Å}$) and red light ($\lambda_{13} \sim 7300\,\text{Å}$), respectively. Is the previous model in agreement with this observation? Are the ions $n \leq 7$ or $n \geq 15$ colored?

## 27.2   Nitrogenous Ions

One can replace the central CH group by a nitrogen atom, in order to form ions of the type:

$$(CH_2 \cdots CH \cdots N \cdots CH \cdots CH_2)^-. \qquad (27.3)$$

The presence of the nitrogen atom does not change the distances between atoms but it changes the above square well potential. The modification consists in adding a small perturbation $\delta V(x)$, attractive and localized around the nitrogen atom:

$$\delta V(x) = 0 \qquad \text{for} \quad |x - \frac{L_n}{2}| > \alpha/2$$

$$= -V_0 \quad \text{for} \quad |x - \frac{L_n}{2}| \leq \alpha/2,$$

where $\alpha/d \ll 1$ and $V_0 > 0$.

**27.2.1** Using first order perturbation theory, give the variations $\delta\varepsilon_k$ of the energy level $\varepsilon_k$ of an electron in the well. For convenience, give the result to leading order in $\alpha/d$.

**27.2.2** Experimentally, one observes that, for the same value of $n$, the spectrum of the nitrogenous ions (27.3) is similar to that of the ions (27.1) but that the wavelengths $\lambda_n^N$ are systematically shorter (blue-shifted) if $n = 4p + 1$, and systematically longer (red-shifted) if $n = 4p+3$, than those $\lambda_n^0$ of the corresponding hydrocarbons (27.1). Explain this phenomenon and show that $\lambda_n^N$ and $\lambda_n^0$ are related by:

$$\frac{\lambda_n^0}{\lambda_n^N} = 1 - (-1)^{\frac{n+1}{2}} \gamma \frac{n}{n+2}, \tag{27.4}$$

where $\gamma$ is a parameter to be determined.

**27.2.3** The nitrogenous ion $n = 11$ absorbs red light ($\lambda_{11}^N \sim 6700\,\text{Å}$). Check that the ion $n = 9$ absorbs violet light ($\lambda_9^N \sim 4300\,\text{Å}$). What is the color of the nitrogenous ion $n = 13$?

**27.2.4** For sufficiently large $n$, if the nitrogen atom is placed not in the central site but on either of the two sites adjacent to the center of the chain, one observes the reverse effect, as compared to question 27.2.2. There is a red shift for $n = 4p + 1$ and a blue shift for $n = 4p + 3$. Can you give a simple explanation for this effect?

## 27.3 Solutions

### 27.3.1 Hydrocarbon Ions

**27.3.1.1** The energy levels are

$$\varepsilon_k = \frac{\pi^2 \hbar^2 k^2}{2m L_n^2} \quad k = 1, 2, \dots . \tag{27.5}$$

**27.3.1.2** The ground state energy of the $n + 1$ electrons is

$$E_0 = \frac{\pi^2 \hbar^2}{m L_n^2} \sum_{k=1}^{(n+1)/2} k^2 = \frac{\pi^2 \hbar^2}{24 m L_n^2}(n + 1)(n + 2)(n + 3). \tag{27.6}$$

The energy of the first excited state is

$$E_1 = E_0 + \frac{\pi^2 \hbar^2}{8m L_n^2}[(n + 3)^2 - (n + 1)^2] = E_0 + \frac{\pi^2 \hbar^2}{2m L_n^2}(n + 2). \tag{27.7}$$

**27.3.1.3** One has $h\nu = E_1 - E_0 = \pi^2 \hbar^2 (n+2)/(2m L_n^2)$. Since $\lambda = c/\nu$, we obtain an absorption wavelength

$$\lambda_n = \frac{8\,d^2}{\lambda_C} \frac{n^2}{(n + 2)}. \tag{27.8}$$

**27.3.1.4** From the general form $\lambda_n = 646.33\, n^2/(n+2)$, we obtain $\lambda_9 = 4760\,\text{Å}$, $\lambda_{11} = 6020\,\text{Å}$, $\lambda_{13} = 7280\,\text{Å}$, in good agreement with experiment.

For smaller $n$, the wavelengths $\lambda_7 = 3520\,\text{Å}$ and $\lambda_5 = 2310\,\text{Å}$ are in the ultraviolet part of the spectrum. The ions $n \leq 7$ do not absorb visible light and are thus not colored.

For $n \geq 15$, the wavelengths $\lambda_{15} = 8550\,\text{Å}$ and $\lambda_{17} = 9830\,\text{Å}$ are in the infrared region. These ions do not absorb visible light in transitions from the ground state to the first excited state. They are nevertheless colored because of transitions to higher excited states.

### 27.3.2 Nitrogenous Ions

**27.3.2.1** The normalized wave functions are $\psi_k(x) = \sqrt{2/L_n}\,\sin(k\pi x/L_n)$. One has

$$\delta\varepsilon_k = \int \delta V(x)|\psi_k(x)|^2 \mathrm{d}x = -V_0 \int_{L_n - \alpha/2}^{L_n + \alpha/2} |\psi_k(x)|^2 \mathrm{d}x. \qquad (27.9)$$

Setting $y = x - L_n/2$, one obtains

$$\delta\varepsilon_k = -\frac{2V_0}{L_n} \int_{-\alpha/2}^{+\alpha/2} \sin^2\left(\frac{k\pi}{2} + \frac{k\pi y}{nd}\right) \mathrm{d}y. \qquad (27.10)$$

There are two cases:

- $k$ even:

$$\delta\varepsilon_k = -\frac{2V_0}{L_n} \int_{-\alpha/2}^{+\alpha/2} \sin^2\left(\frac{k\pi y}{nd}\right) \mathrm{d}y, \quad \text{i.e.} \quad \delta\varepsilon_k = O((\alpha/d)^3). \qquad (27.11)$$

The perturbation is negligible.
- $k$ odd:

$$\delta\varepsilon_k = -\frac{2V_0}{L_n} \int_{-\alpha/2}^{+\alpha/2} \cos^2\left(\frac{k\pi y}{nd}\right) \mathrm{d}y. \qquad (27.12)$$

To first order in $\alpha/d$, we have $\delta\varepsilon_k = -2V_0\alpha/nd < 0$.

The exact formulas are:

$$\delta\varepsilon_k = -\frac{V_0}{L_n}\left[\alpha - (-1)^k \frac{L_n}{k\pi} \sin(\frac{k\pi\alpha}{L_n})\right]. \qquad (27.13)$$

The (single particle) energy levels corresponding to even values of $k$ are practically unaffected by the perturbation; only those with $k$ odd are shifted. This is simple to understand. For $k$ even, the center of the chain is a node of the wave function, and the integral defining $\delta\varepsilon_k$ is negligible. For $k$ odd, on the contrary, the center is an antinode, we integrate over a maximum of the wave function, and the perturbation is maximum.

**27.3.2.2** The perturbation to the excitation energy $E_1 - E_0$ is

$$\delta E = \delta\epsilon_{(n+3)/2} - \delta\epsilon_{(n+1)/2}. \tag{27.14}$$

- $(n+1)/2$ even, i.e. $n = 4p + 3$, $\delta\varepsilon_{(n+1)/2} = 0$,

$$\delta E = \delta\varepsilon_{(n+3)/2} = -\frac{2V_0\alpha}{nd} < 0. \tag{27.15}$$

- $(n+1)/2$ odd, i.e. $n = 4p + 1$, $\delta\varepsilon_{(n+3)/2} = 0$,

$$\delta E = -\delta\varepsilon_{(n+1)/2} = \frac{2V_0\alpha}{nd} > 0. \tag{27.16}$$

We can summarize these results in the compact form

$$E_1 - E_0 + \delta E = \frac{\pi^2\hbar^2}{2md^2}\frac{n+2}{n^2}\left(1 - (-1)^{\frac{n+1}{2}}\gamma\frac{n}{n+2}\right), \tag{27.17}$$

with $\gamma = 4V_0\alpha md/(\pi h)^2$. We therefore obtain the desired relation

$$\frac{\lambda_n^0}{\lambda_n^N} = 1 - (-1)^{\frac{n+1}{2}}\gamma\frac{n}{n+2}. \tag{27.18}$$

For $n = 4p + 1$, the perturbation increases the excitation energy, and decreases $\lambda_n$. For $n = 4p + 3$, it decreases the excitation energy, and increases $\lambda_n$.

**27.3.2.3** For the ion $n = 11$ one obtains the relation $(1 - 11\gamma/13) = 6000/6700$, therefore $\gamma \sim 0.12$ and $\lambda_9^N = 4330$ Å, in good agreement with experiment. One also obtains $\lambda_{13}^N = 6600$ Å, which absorbs red light and gives a green color to the corresponding pigment. Note that the presence of the nitrogen atom yields $\lambda_{13}^N \leq \lambda_{11}^N$ whereas $\lambda_{13}^0 > \lambda_{11}^0$.

**27.3.2.4** The distance between a node and an antinode of $\psi_k(x)$ is $\delta x = nd/(2k)$.

For $k = (n+1)/2$ and $k = (n+3)/2$ which are the states of interest, we will have respectively $\delta x = nd/(n+1)$ and $\delta x = nd/(n+3)$, i.e. $\delta x \sim d$ if $n$ is large. Consequently, if a wave function has a node at the center, it has an antinode in the vicinity of the two adjacent sites, and vice versa. The argument is therefore similar to the answer to questions 27.2.1 and 27.2.2, with the reverse effect. The lines are red-shifted if $n = 4p + 1$ and they are blue-shifted if $n = 4p + 3$.

**Comment** Many further details can be found in the article by John R. Platt, *The Chemical Bound and the Distribution of Electrons in Molecules*, Handbuch Der Physik, Volume XXXVII/2, Molecules II, p. 173, Springer-Verlag (1961). This article provides an extensive description of the applications of Quantum Mechanics in Chemistry.

# Hyperfine Structure in Electron Spin Resonance

**28**

Many molecular species, such as free radicals, possess an unpaired electron. The magnetic spin resonance of this electron, called electron spin resonance (ESR) as opposed to nuclear magnetic resonance, provides useful information about the electronic structure of the molecule, as we shall see in this chapter. We assume here the following:

1. Spin variables and space variables are independent, both for electrons and for nuclei; we are only interested in the former.
2. The spatial ground state of the unpaired electron is non-degenerate, and one can neglect the effect of a magnetic field on its wave function.
3. We only take into account the following magnetic spin interactions: (a) the Zeeman interaction of spin magnetic moments with an external field $B$, and (b) the hyperfine interaction between the outer electron and the nuclei.
4. For a given nucleus in the molecule, the hyperfine interaction has the form $\hat{H}_{HF} = (A/\hbar^2)\hat{S}\cdot\hat{I} = (A/4)\hat{\sigma}_e\cdot\hat{\sigma}_n$ where $\hat{S} = \hbar\hat{\sigma}_e/2$ is the electron spin and $\hat{I} = \hbar\hat{\sigma}_n/2$ is the nuclear spin; $\hat{\sigma}_e$ and $\hat{\sigma}_n$ are the Pauli matrices which act respectively in the Hilbert spaces of the electron and of the nucleus. The constant $A$ is given by

$$A = -\frac{2}{3}\mu_0\gamma_e\gamma_n\hbar^2|\psi(\mathbf{r}_n)|^2, \tag{28.1}$$

where $\mu_0 = 1/\epsilon_0 c^2$ is the magnetic susceptibility of vacuum, $\gamma_e$ and $\gamma_n$ are the gyromagnetic factors of the electron and of the nucleus under consideration, and $\psi(\mathbf{r}_n)$ is the value of the electron wave function at the position $\mathbf{r}_n$ of this nucleus.
5. In all the problem, the system is considered to be in a constant uniform magnetic field $B$ directed along the $z$ axis. For simplicity, we set $A = \hbar a$, $\omega_e = -\gamma_e B$, $\omega_n = -\gamma_n B$ and $\eta = (\omega_e - \omega_n)/2$.

© Springer Nature Switzerland AG 2019
J.-L. Basdevant, J. Dalibard, *The Quantum Mechanics Solver*,
https://doi.org/10.1007/978-3-030-13724-3_28

The numerical values of gyromagnetic ratios are

$$
\begin{aligned}
\text{electron:} \quad & \gamma_e/(2\pi) = -28.024 \, \text{GHzT}^{-1}, \\
\text{proton:} \quad & \gamma_p/(2\pi) = +42.574 \, \text{MHzT}^{-1}.
\end{aligned}
\tag{28.2}
$$

## 28.1 Hyperfine Interaction with One Nucleus

**28.1.1** We first consider a species where the nuclei do not possess a magnetic moment, so that there is no hyperfine interaction.

Write the Zeeman interaction Hamiltonian of the electron with the magnetic field $B$.

What are the energy levels of the system?

What is the value of the frequency that can excite the system? Give its numerical value for a magnetic field of 1 Tesla.

**28.1.2** We now assume that the molecule has one spin-1/2 nucleus. We note the (factorized) eigenbasis common to $\hat{S}_z$ and $\hat{I}_z$ as $\{|\sigma_e; \sigma_n\rangle\}$ with $\sigma_e = \pm 1$ and $\sigma_n = \pm 1$.

(a) Write the complete spin Hamiltonian.
(b) Calculate the action of $\boldsymbol{\sigma}_e \cdot \boldsymbol{\sigma}_n$ on the vectors of the basis $\{|\sigma_e; \sigma_n\rangle\}$.
(c) Write the matrix form of the Hamiltonian in this basis, and calculate its eigenvalues.

**28.1.3** From now on, we assume that the magnetic field $B$ is strong, in the sense that $|\omega_e| \gg |a|$.

(a) Give the approximate form of the eigenvalues to first order in $a/\eta$.
(b) Recover these results by first diagonalizing the electron Zeeman Hamiltonian, and by treating the other terms, i.e. the nuclear Zeeman Hamiltonian and the hyperfine interaction, in first order perturbation theory. What are the corresponding eigenstates (to zeroth order in $a/\eta$)?
(c) One can show that the transitions that an electromagnetic field can induce occur only between states which differ by the value of a *single* spin (for instance, the transitions $|+; -\rangle \to |-; +\rangle$ are forbidden). Under these conditions, what are the observable transition frequencies, knowing that all transitions which are not forbidden actually occur? Classify these transitions in two sets corresponding respectively to *nuclear* and to *electronic* spin transitions.
(d) Calculate these frequencies numerically for the hydrogen atom in a field $B = 1$ T. We recall that, in this case, $A/(2\pi\hbar) \simeq 1.420 \, \text{GHz}$.

## 28.2    Hyperfine Structure with Several Nuclei

We now assume that the molecule has $N$ protons in hydrogen atoms located on sites $r_1, \ldots, r_N$, whose spins are denoted $\hat{I}_1, \ldots, \hat{I}_N$.

The Hilbert space of spin degrees of freedom is of dimension $2^{N+1}$. It is spanned by the set:

$$\{|\sigma_e; \sigma_1, \sigma_2, \ldots, \sigma_N\rangle\} \equiv \{|\sigma_e\rangle \otimes |\sigma_1\rangle \otimes |\sigma_2\rangle \otimes \cdots \otimes |\sigma_N\rangle\} \tag{28.3}$$

with $\sigma_e = \pm 1$ and $\sigma_k = \pm 1$, $k = 1, \ldots, N$. This set is an orthonormal eigenbasis common to the $z$ projection of the spin observables $\hat{S}_z$ and $\hat{I}_{kz}, k = 1, \ldots, N$, of the $N + 1$ particles.

**28.2.1** Let $A_k = \hbar a_k$ be the hyperfine constant of proton $k$. Write the expression for the spin Hamiltonian of the system (we recall that the magnetic nucleus-nucleus interaction is neglected).

**28.2.2** Show that the restriction of this Hamiltonian to each eigen-subspace of $\hat{S}_z$ is diagonal.

**28.2.3** Assuming that the field is strong, calculate the eigenvalues in first order perturbation theory, and the corresponding eigenstates.

**28.2.4** What are the observable *electron* spin transition frequencies? How many lines corresponding to these frequencies should the spectrum display in principle?

**28.2.5** What is the number of lines and the multiplicity of each of them (i.e. the number of transitions at the same frequency) if all the protons are equivalent, i.e. if all the $|\psi(r_k)|^2$, and therefore the coefficients $a_k$, are equal?

**28.2.6** What is the number of lines and their multiplicities, if there exist two sets of equivalent protons, one with $p$ protons corresponding to the constant $a_p$, the other with $q = N - p$ protons, corresponding to the constant $a_q$?

## 28.3    Experimental Results

Experimentally, one measures the positions and the intensities of the absorption lines in the microwave region. An absorption line appears as a peak in the absorbed intensity $\alpha(\nu)$ as a function of the frequency, whose qualitative shape is shown in Fig. 28.1.

It can be shown that the intensity of an absorption peak at a given frequency is proportional to the number of transitions (*multiplicity* of the line) which can occur at that frequency. For experimental convenience, one fixes the frequency of the

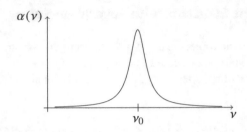

**Fig. 28.1** Typical shape of an ESR absorption curve as a function of the frequency

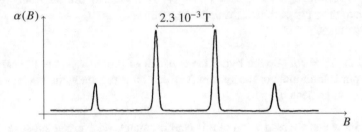

**Fig. 28.2** Microwave spectrum of the radical $^\bullet$CH$_3$

microwave at a given value, and one varies the magnetic field $B$. This results in an absorption curve $\alpha(B)$.

**28.3.1** Figure 28.2 shows a sketch of the spectrum of the free radical $^\bullet$CH$_3$ (methyl). The carbon nucleus does not possess any magnetic moment; only the protons of the hydrogen atoms give rise to hyperfine interactions.

(a) Interpret this spectrum qualitatively. Explain the number of lines and their relative intensities. How many different coefficients $a_k$ are there?
(b) Give the value of $a_k/(2\pi)$. Calculate the value of $|\psi(r_k)|^2$ for the unpaired electron in this molecule. It is convenient to express the result in terms of $|\psi(0)|^2_{\text{Hydrogen}} = 1/(\pi a_1^3)$ where $a_1$ is the Bohr radius of hydrogen.

**28.3.2** Answer the same questions for the spectrum of $CH_3 - ^\bullet COH - COO^-$ (the radical ion of lactic acid) shown in Fig. 28.3. Neither the oxygen nor the carbon nuclei carry magnetic moments. The only hyperfine interaction arises, again, from the protons of the hydrogen atoms.

**Fig. 28.3** Microwave spectrum of the radical $CH_3 - {}^\bullet COH - COO^-$

## 28.4 Solutions

### 28.4.1 Hyperfine Interaction with One Nucleus

**28.4.1.1** The magnetic Hamiltonian is $\hat{H} = -\hbar \gamma_e B \hat{\sigma}_{ez}/2$, hence the energy levels $E\pm = \mp \hbar \gamma_e B/2$ corresponding to the states $|\pm\rangle$. The transition frequency is given by $h\nu = E_+ - E_- = \hbar \omega_e$, $\nu = \omega_e/(2\pi)$. For $B = 1$ T, $\nu = 28.024$ GHz.

**28.4.1.2 (a)** The full Hamiltonian, including the hyperfine interaction, is

$$\hat{H} = -\gamma_e B \hat{S}_z - \gamma_n B \hat{I}_z + \frac{A}{\hbar^2} \hat{S} \cdot \hat{I} = \frac{\hbar \omega_e}{2} \hat{\sigma}_{ez} + \frac{\hbar \omega_n}{2} \hat{\sigma}_{nz} + \frac{\hbar a}{4} \hat{\sigma}_e \cdot \hat{\sigma}_n.$$

**(b)** The action of $\hat{\sigma}_e \cdot \hat{\sigma}_n$ on the basis states is:

$$\hat{\sigma}_e \cdot \hat{\sigma}_n |+; +\rangle = |+; +\rangle \qquad \hat{\sigma}_e \cdot \hat{\sigma}_n |+; -\rangle = 2|-; +\rangle - |+; -\rangle,$$

$$\hat{\sigma}_e \cdot \hat{\sigma}_n |-; -\rangle = |-; -\rangle \qquad \hat{\sigma}_e \cdot \hat{\sigma}_n |-; +\rangle = 2|+; -\rangle - |-; +\rangle.$$

**(c)** Hence the $4 \times 4$ matrix representation of the Hamiltonian

$$\hat{H} = \frac{\hbar}{4} \begin{pmatrix} a + 2(\omega_e + \omega_n) & 0 & 0 & 0 \\ 0 & 4\eta - a & 2a & 0 \\ 0 & 2a & -4\eta - a & 0 \\ 0 & 0 & 0 & a - 2(\omega_e + \omega_n) \end{pmatrix}, \qquad (28.4)$$

where the rows and columns are ordered as $|+; +\rangle, |+; -\rangle, |-; +\rangle, |-; -\rangle$. Hence the eigenstates and the corresponding eigenvalues:

$$|+; +\rangle \longrightarrow \frac{\hbar}{4}(a + 2(\omega_e + \omega_n)) \qquad |-; -\rangle \longrightarrow \frac{\hbar}{4}(a - 2(\omega_e + \omega_n)) \qquad (28.5)$$

and from the diagonalization of the $2 \times 2$ matrix between $|+; -\rangle$ and $|-; +\rangle$

$$\cos \phi \, |+; -\rangle + \sin \phi \, |-; +\rangle \longrightarrow \frac{\hbar}{4}(-a + 2\sqrt{4\eta^2 + a^2})$$

$$\sin \phi \, |+; -\rangle - \cos \phi \, |-; +\rangle \longrightarrow \frac{\hbar}{4}(-a - 2\sqrt{4\eta^2 + a^2})$$

with

$$\tan \phi = \frac{a}{2\eta + \sqrt{4\eta^2 + a^2}}. \tag{28.6}$$

**28.4.1.3 (a)** If $\eta \gg a$, the eigenvectors and eigenvalues are, to lowest order,

$$|+; +\rangle \longrightarrow (\hbar/4)(a + 2(\omega_e + \omega_n)) \qquad |+; -\rangle \longrightarrow \sim (\hbar/4)(4\eta - a)$$
$$|-; -\rangle \longrightarrow (\hbar/4)(a - 2(\omega_e + \omega_n)) \qquad |-; +\rangle \longrightarrow \sim (\hbar/4)(-4\eta - a).$$

**(b)** In each subspace corresponding respectively to $\sigma_e = 1$ and $\sigma_e = -1$, the perturbation is diagonal (the non-diagonal terms couple $\sigma_e = +1$ and $\sigma_e = -1$). The $2 \times 2$ matrices to be considered are indeed

$$\langle +, \sigma_n | \hat{H} | +, \sigma_n' \rangle \quad \text{and} \quad \langle -, \sigma_n | \hat{H} | -, \sigma_n' \rangle. \tag{28.7}$$

Consider for instance $\langle +, \sigma_n | \hat{H} | +, \sigma_n' \rangle$. Since

$$\langle +, \sigma_n | \hat{S}_x | +, \sigma_n' \rangle = \langle +, \sigma_n | \hat{S}_y | +, \sigma_n' \rangle = 0, \tag{28.8}$$

only $\langle +|\hat{S}_z|+\rangle \langle \sigma_n | \hat{I}_z | \sigma_n' \rangle$ has to be considered, and it is diagonal. The eigenstates at zeroth order are therefore $|\sigma_e; \sigma_n\rangle$ and we recover the above results.

**(c)** Transitions:

(i) *Nuclear transitions*: $|\sigma_e; +\rangle \leftrightarrow |\sigma_e; -\rangle$, i.e.

$$|+; +\rangle \leftrightarrow |+; -\rangle \quad \Delta E = \hbar(\omega_n + a/2), \quad \nu = |\omega_n + a/2|/(2\pi)$$
$$|-; +\rangle \leftrightarrow |-; -\rangle \quad \Delta E = \hbar(\omega_n - a/2), \quad \nu = |\omega_n - a/2|/(2\pi).$$

(ii) *Electronic transitions*: $|+; \sigma_n\rangle \leftrightarrow |-; \sigma_n\rangle$, i.e.

$$|+; +\rangle \leftrightarrow |-; +\rangle \quad \Delta E = \hbar(\omega_e + a/2), \quad \nu = |\omega_e + a/2|/(2\pi)$$
$$|+; -\rangle \leftrightarrow |-; -\rangle \quad \Delta E = \hbar(\omega_e - a/2), \quad \nu = |\omega_e - a/2|/(2\pi).$$

**(d)** For $B = 1$ T, $\nu_n = 42.6$ MHz; $a/(2\pi) = A/(2\pi\hbar) = 1420$ MHz; $\nu_e = 28.024$ GHz. The nuclear transitions occur at $\nu_1 = 753$ MHz and $\nu_2 = 667$ MHz, the electronic transitions occur at $\nu_1 = 28.734$ GHz and $\nu_2 = 27.314$ GHz.

### 28.4.2 Hyperfine Structure with Several Nuclei

**28.4.2.1** The total Hamiltonian is

$$\hat{H} = \frac{\hbar\omega_e}{2}\hat{\sigma}_{ez} + \sum_{k=1}^{N} \frac{\hbar\omega_n}{2}\hat{\sigma}_{kz} + \sum_{k=1}^{N} \frac{A_k}{4}\hat{\boldsymbol{\sigma}}_e \cdot \hat{\boldsymbol{\sigma}}_k. \tag{28.9}$$

**28.4.2.2** The restriction of $\hat{H}$ to a subspace corresponding to the eigenvalue $\hbar\sigma_e/2$ of $\hat{S}_{ez}(\sigma_e = \pm)$ can be written:

$$\hat{H}_{\sigma_e} = \frac{\hbar\omega_e}{2}\sigma_e + \sum_{k=1}^{N} \left(\frac{\hbar\omega_n}{2} + \frac{A_k\sigma_e}{4}\right)\hat{\sigma}_{kz}. \tag{28.10}$$

The operators $\hat{H}_+$ and $\hat{H}_-$ are diagonal in the basis $\{|\sigma_1, \sigma_2, \ldots, \sigma_N\rangle\}$.

**28.4.2.3** First order perturbation theory consists in diagonalizing the perturbing Hamiltonian $\sum_{k=1}^{N}(\hbar\omega_n/2)\hat{\sigma}_{kz} + \sum_{k=1}^{N}(A_k/4)\hat{\boldsymbol{\sigma}}_e \cdot \hat{\boldsymbol{\sigma}}_k$ in each eigen-subspace of the dominant term $\hbar\omega_e\hat{\sigma}_{ez}/2$. This is automatically satisfied. Therefore,

$$\sigma_e = +1 : E^+_{\sigma_1\ldots\sigma_N} = \frac{\hbar\omega_e}{2} + \sum_k \frac{\hbar(2\omega_n + a_k)}{4}\sigma_k, \text{ state } |+; \sigma_1, \ldots, \sigma_N\rangle,$$

$$\sigma_e = -1 : E^-_{\sigma_1\ldots\sigma_N} = -\frac{\hbar\omega_e}{2} + \sum_k \frac{\hbar(2\omega_n - a_k)}{4}\sigma_k, \text{ state } |-; \sigma_1, \ldots, \sigma_N\rangle.$$

**28.4.2.4** There are $2^N$ transitions $|+; \sigma_1, \ldots, \sigma_N\rangle \leftrightarrow |-; \sigma_1, \ldots, \sigma_N\rangle$ corresponding to the $2^N$ possible choices for the set $\{\sigma_k\}$. The corresponding frequencies are

$$\Delta\nu_{\sigma_1\ldots\sigma_N} = \frac{1}{2\pi}\left|\omega_e + \sum_k a_k\sigma_k/2\right|. \tag{28.11}$$

**28.4.2.5** If all $a_k$ are equal to $a$, we have

$$\Delta\nu = \frac{1}{2\pi}\left|\omega_e + a\sum_k \sigma_k/2\right| = \frac{1}{2\pi}|\omega_e + Ma/2|, \tag{28.12}$$

with $M = \sum \sigma_k = N, N - 2, \ldots, -N + 2, -N$, i.e. $N + 1$ absorption lines. There are $C_N^{(N-M)/2}$ transitions which have the same frequency and contribute to each line. The relative intensities of the lines will therefore be proportional to the binomial coefficients $C_N^{(N-M)/2}$. The splitting between two adjacent lines is $a$.

**28.4.2.6** If $p$ equivalent protons correspond to the coupling constant $A_p$, and $q = N - p$ correspond to $A_q$, then

$$\Delta\nu = \frac{1}{2\pi}\left|\omega_e + \frac{a_p}{2}\sum_{i=1}^{p}\sigma_i + \frac{a_q}{2}\sum_{j=1}^{q}\sigma_j\right| = \frac{1}{2\pi}\left|\omega_e + M_p\frac{a_p}{2} + M_q\frac{a_q}{2}\right|.$$

(28.13)

There are $p + 1$ values of $M_p : p, p - 2, \ldots, -p$, and $q + 1$ values of $M_q : M_q = q, q - 2, \ldots, -q$. The total number of lines is $(p + 1)(q + 1)$, and the multiplicity of a line corresponding to a given couple $(M_p, M_q)$ is $C_p^{(p-M_p)/2}C_q^{(q-M_q)/2}$.

### 28.4.3 Experimental Results

**28.4.3.1** The experimental results confirm the above analysis.

**(a)** For $^\bullet CH_3$ there are 4 equally spaced lines of relative intensities $1 : 3 : 3 : 1$. This is in perfect agreement with the fact that the three protons of $^\bullet CH_3$ are obviously equivalent. All the $A_k$ coefficients are equal.

**(b)** For a fixed $\omega$, one gets by considering two consecutive lines, for instance the center lines: $a/2 - \gamma_e B_1 = -a/2 - \gamma_e B_2$ so that $a = \gamma_e(B_1 - B_2)$. We deduce $\nu = |a|/2\pi = 65 \text{ MHz} = |A_k|/2\pi\hbar$, and

$$\pi a_1^3|\psi(r_k)|^2 = |\psi(r_k)|^2/|\psi(0)|^2_{\text{Hydrogen}} = 65/1420 \sim 0.045. \tag{28.14}$$

In the radical $^\bullet CH_3$, the probability that the outer electron is on top of a proton is smaller by a factor $3 \times 0.045 = 0.135$ than in the hydrogen atom.

**28.4.3.2** In the case of $CH_3 - {}^\bullet COH - COO^-$, there are four dominant lines, each of which is split into two. This agrees with the fact that, in the molecule $CH_3 - {}^\bullet COH - COO^-$, the 3 protons of the $CH_3$ group are equivalent and have the same hyperfine constant $a_1$ whereas the proton of the $^\bullet COH$ group has a different constant $a_2$ which is noticeably smaller than $a_1$.

A calculation similar to the previous one gives $|\psi(r_k)|^2/|\psi(0)|^2_{\text{Hydrogen}} \sim 0.034$ for the protons of the $CH_3$ group, and $|\psi(r_k)|^2/|\psi(0)|^2_{\text{Hydrogen}} \sim 0.004$ for the proton of $^\bullet COH$.

# Probing Matter with Positive Muons

**29**

A very efficient technique for probing the structure of crystals consists in forming, inside the material, pseudo hydrogenic atoms made of an electron and a positive muon, and called *muonium*. This chapter is devoted to the study of the dynamics of muonium, both in vacuum and in a silicon crystal.

The positive muon is a spin-1/2 particle which has the same charge as the proton. The muon mass is considerably larger than the electron mass: $m_\mu/m_e = 206.77$. The muon is unstable and decays with a lifetime $\tau = 2.2\,\mu$s. Its use in probing the structure of crystals is based on the rotation of its spin, once a muonium atom is formed.

- It is possible to form muonium atoms in a quantum state such that, at $t = 0$, the spin state of the $\mu^+$ is known.
- Using a technique of particle physics, one can measure its spin state at a later time $t$.
- The rotation of the muon spin can be related to the hyperfine structure of the 1s level of muonium.

Therefore, muonium is a local probe, sensitive to electric and magnetic fields in its vicinity. In this way, one can obtain information on the structure of the medium by methods analogous to magnetic resonance experiments.

In the first part of the chapter, we sketch the principle of the method by studying muonium in vacuum. When the method was first applied to a silicon crystal, in 1973, the results seemed anomalous. We shall see in the second section how these results were understood, in 1978, as being due to the anisotropy of crystalline media.

Throughout this chapter, the muon will be considered as stable. For simplicity, we set

$$\hat{\boldsymbol{\mu}}_{\mu^+} \equiv \hat{\boldsymbol{\mu}}_1 = \mu_1\hat{\boldsymbol{\sigma}}_1 \qquad \hat{\boldsymbol{\mu}}_e \equiv \hat{\boldsymbol{\mu}}_2 = \mu_2\hat{\boldsymbol{\sigma}}_2, \tag{29.1}$$

© Springer Nature Switzerland AG 2019
J.-L. Basdevant, J. Dalibard, *The Quantum Mechanics Solver*,
https://doi.org/10.1007/978-3-030-13724-3_29

where the $(x, y, z)$ components of $\hat{\sigma}_1$ and $\hat{\sigma}_2$ are the Pauli matrices.

Numerical values of interest are:

$$m_\mu c^2 = 105.66 \, \text{MeV} \qquad \mu_1/h = 67.5 \, \text{MHz T}^{-1}$$

$$m_e c^2 = 0.511 \, \text{MeV} \qquad \mu_2/h = -1.40 \times 10^4 \, \text{MHz T}^{-1}.$$

## 29.1  Muonium in Vacuum

Muonium is formed by slowing down a beam of $\mu^+$, prepared in a given spin state, in a thin metal foil. A sufficiently slow $\mu^+$ can capture an electron and form a hydrogen-like atom in an excited state. This atom falls into its ground state very quickly (in $\sim 10^{-9}$ s) and the muon's spin state *remains the same* during this process. Once it is formed, the muonium, which is electrically neutral, can diffuse outside the metal.

We assume that, at $t = 0$, the state of the muonium atom is the following:

- The muon spin is in the eigenstate $| + z \rangle \equiv |+\rangle$ of $\hat{\sigma}_{1z}$.
- The electron spin is in an arbitrary state $\alpha|+\rangle + \beta|-\rangle$, with $|\alpha|^2 + |\beta|^2 = 1$.
- The wave function $\Psi(r)$ of the system is the $1s$ wave function of the hydrogen-like system, $\psi_{100}(r)$.

Just as for the hyperfine structure of hydrogen, we work in the 4 dimensional Hilbert space corresponding to the spin variables of the electron and the muon. In this Hilbert space, the spin–spin interaction Hamiltonian is

$$\hat{H} = E_0 - \frac{2}{3} \frac{\mu_0}{4\pi} |\psi_{100}(0)|^2 \hat{\boldsymbol{\mu}}_1 \cdot \hat{\boldsymbol{\mu}}_2 = E_0 + \frac{A}{4} \hat{\sigma}_1 \cdot \hat{\sigma}_2, \tag{29.2}$$

where the indices 1 and 2 refer respectively to the muon and to the electron, and where $E_0 = -m_r c^2 \alpha^2/2$, with $m_r$ being the reduced mass of the $(e, \mu)$ system.

**29.1.1** Write the matrix representation of the Hamiltonian $\hat{H}$ in the basis $\{|\sigma_{1z}, \sigma_{2z}\rangle, \sigma_{iz} = \pm\}$.

**29.1.2** Knowing the value of $A$ in the hydrogen atom: $A/h = 1420 \, \text{MHz}$, calculate $A$ in muonium. We recall that $\mu_1 = q\hbar/(2m_\mu)$ for the muon, $\mu_2 = -q\hbar/(2m_e)$ for the electron, $\mu_p = 2.79 \, q\hbar/(2m_p)$ for the proton, where $q$ is the unit charge and $m_p = 1836.1 \, m_e$.

**29.1.3** Write the general form of an eigenstate of $\hat{\sigma}_{1z}$ with eigenvalue $+1$: (i) in the basis $\{|\sigma_{1z}, \sigma_{2z}\rangle\}$; (ii) in the eigenbasis of $\hat{H}$.

**29.1.4** We assume that, at $t = 0$, the system is in a state $|\psi(0)\rangle$ of the type defined above. Calculate $|\psi(t)\rangle$ at a later time.

**29.1.5** (a) Show that the operators $\hat{\pi}_\pm = (1 \pm \hat{\sigma}_{1z})/2$ are the projectors on the eigenstates of $\hat{\sigma}_{1z}$ corresponding to the eigenvalues $\pm 1$.
(b) Calculate for the state $|\psi(t)\rangle$ the probability $p(t)$ that the muon spin is in the state $|+\rangle$ at time $t$. Write the result in the form

$$p(t) = q \, p_+(t) + (1 - q) \, p_-(t) , \tag{29.3}$$

where $p_+$(or $p_-$) is the probability obtained if the electron is initially in the eigenstate of $\sigma_{2z}$ with eigenvalue $+1$(or $-1$), and where $q$ is a probability, as yet undefined.

**29.1.6** In practice, the electronic spins are unpolarized. A rigorous treatment of the problem then requires a statistical description in terms of a density operator. To account for this nonpolarization in a simpler way, we shall set heuristically that the *observed* probability $\overline{p}(t)$ corresponds to $q = 1/2$ in the above formula.
   Using this prescription, give the complete expression for $\overline{p}(t)$.

---

## 29.2   Muonium in Silicon

We now form muonium in a silicon crystal sufficiently thick that the muonium does not escape. The muonium stops in an interstitial position inside the crystal lattice, the nearest atoms forming a plane hexagonal mesh around it. The global effect of the interactions between the atoms of the crystal and the muonium atom is to break the spherical symmetry of the spin–spin interaction, but to preserve the rotational symmetry around the $z$ axis perpendicular to the plane of the mesh.
   We therefore consider the Hamiltonian

$$\hat{H} = E_0 + \frac{A'}{4}\hat{\sigma}_1 \cdot \hat{\sigma}_2 + D \, \hat{\sigma}_{1z} \, \hat{\sigma}_{2z} , \tag{29.4}$$

where the constant $A'$ may differ from $A$ since the presence of neighboring atoms modifies the Coulomb potential and, therefore, the wave function at the origin. The constants $A'$ and $D$ will be determined experimentally; their sign is known: $A' > 0$, $D < 0$.

**29.2.1** Calculate the spin energy levels and the corresponding eigenstates of the muonium trapped in the silicon crystal.

**29.2.2** We now reconsider the spin rotation experiment with the following modifications:

- Initially the $\mu_+$ spin is now in the eigenstate $|+x\rangle$ of $\sigma_x$.
- We want to know the probability of finding the $\mu_+$ spin in this same eigenstate $|+x\rangle$ at time $t$.

One can proceed as in question 29.1.5:

(a) Calculate in the $\{|\sigma_1, \ \sigma_2\rangle\}$ basis the states $|\psi_+(t)\rangle$ and $|\psi_-(t)\rangle$ which are initially eigenstates of $\hat{\sigma}_{2z}$ ($\hat{\sigma}_{2z}$ is the projection of the *electron* spin along the $z$ axis).

(b) Evaluate $\langle \psi_\epsilon(t)|\hat{\sigma}_{1x}|\psi_\epsilon(t)\rangle$, where $\epsilon = \pm$.

(c) Consider the projector $\hat{\pi}_x = (1+\hat{\sigma}_{1x})/2$, and deduce from the previous question the probabilities $p_\pm(t)$.

(d) Calculate the measured probability $\overline{p}(t) = (p_+(t) + p_-(t))/2$.

**29.2.3** Comparison with experiment. Present day technology in data processing allows one to determine not $p(t)$ itself, but a quantity which is easier to deal with, the characteristic function $g(\omega) = \mathrm{Re}(f(\omega))$ where

$$f(\omega) = \frac{1}{\tau} \int_0^\infty \overline{p}(t) e^{-t/\tau} e^{i\omega t} \, dt \qquad (29.5)$$

is the Fourier transform of $\overline{p}(t) e^{-t/\tau}/\tau$. In this expression, the factor $e^{-t/\tau}/\tau$ is due to the finite lifetime of the $\mu^+$ ($\tau \sim 2.2\,\mu$s). We recall that

$$\frac{1}{\tau} \int_0^\infty e^{-t/\tau} e^{i\omega t} \, dt = \frac{1}{1 - i\omega\tau}. \qquad (29.6)$$

(a) Figure 29.1a shows the distribution $g(\omega)$ as measured in the conditions of question 29.2.2. Check that this data is compatible with the results found in question 29.2.2, and deduce from the data the values of $A'/h$ and $D/h$ (we recall that $D < 0$).

(b) Figure 29.1b is obtained by a slight modification of the previous experiment. Can you tell what modification has been made? How can one evaluate the position of the third peak, in terms of the constants of the problem?

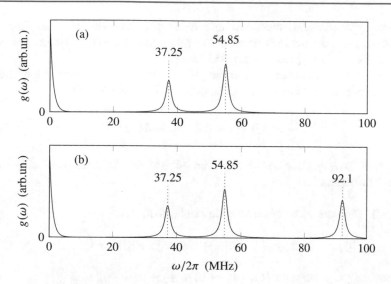

**Fig. 29.1** Sketch of the variations of the quantity $g(\omega)$, defined in the text, with the frequency $\nu = \omega/(2\pi)$. (a) In the conditions described in question 29.2.2, and (b) in another experimental configuration

## 29.3 Solutions

### 29.3.1 Muonium in Vacuum

**29.3.1.1** The Hamiltonian is

$$\hat{H} = E_0 + \frac{A}{4}(\hat{\sigma}_{1x}\hat{\sigma}_{2x} + \hat{\sigma}_{1y}\hat{\sigma}_{2y} + \hat{\sigma}_{1z}\hat{\sigma}_{2z}). \tag{29.7}$$

The matrix representation is therefore

$$\hat{H} = \begin{pmatrix} E_0 + A/4 & 0 & 0 & 0 \\ 0 & E_0 - A/4 & A/2 & 0 \\ 0 & A/2 & E_0 - A/4 & 0 \\ 0 & 0 & 0 & E_0 + A/4 \end{pmatrix}, \tag{29.8}$$

where the elements are ordered as: $|++\rangle$, $|+-\rangle$, $|-+\rangle$, $|--\rangle$.

**29.3.1.2** The constant $A$ is related to its value in the hydrogen atom by

$$\frac{A}{A_H} = \frac{|\psi(0)|^2}{|\psi(0)|_H^2} \frac{\mu_1}{\mu_p} = \frac{|\psi(0)|^2}{|\psi(0)|_H^2} \frac{m_p}{m_\mu} \frac{1}{2.79}. \tag{29.9}$$

In first approximation, muonium and hydrogen have similar sizes and wave functions, since the muon is much heavier than the electron. Therefore we obtain $A \simeq A_H(m_p/2.79\, m_\mu)$ and $A/h \simeq 4519$ MHz.

The reduced mass correction to the value of the wave function at the origin is straightforward to calculate. It is of the order of 1% and it leads to

$$\frac{A}{h} = 4519\,(1 - 0.0126) = 4462\,\text{MHz}. \qquad (29.10)$$

This value is very close to the observed 4463 MHz. the difference being due to relativistic effects.

**29.3.1.3** The state under consideration can be written as

$$|\psi\rangle = |+\rangle \otimes (\alpha|+\rangle + \beta|-\rangle) \text{ with } |\alpha|^2 + |\beta|^2 = 1. \qquad (29.11)$$

Equivalently, one can write it as $|\psi\rangle = \alpha|++\rangle + \beta|+-\rangle$.

The eigenbasis of $\hat{H}$ consists in the common eigenstates of the total spin operators $\hat{S}^2$ and $\hat{S}_z$:

$$\text{triplet states: } \begin{cases} |++\rangle \\ (|+-\rangle + |-+\rangle)/\sqrt{2} \\ |--\rangle \end{cases} \qquad (29.12)$$

and

$$\text{singlet state: } (|+-\rangle - |-+\rangle)/\sqrt{2}. \qquad (29.13)$$

Therefore, one also has the representation

$$|\psi\rangle = \alpha|1, 1\rangle + \frac{\beta}{\sqrt{2}}(|1, 0\rangle + |0, 0\rangle), \qquad (29.14)$$

where the only constraint on $\alpha$ and $\beta$ is $|\alpha|^2 + |\beta|^2 = 1$.

**29.3.1.4** We start from $|\psi(0)\rangle = |\psi\rangle$ as defined above. The energy levels and the corresponding eigenstates are known:

$$\text{triplet states } E_T = E_0 + A/4 \qquad \text{singlet state} \qquad E_S = E_0 - 3A/4. \qquad (29.15)$$

At time $t$ the state is:

$$|\psi(t)\rangle = e^{-iE_0t/\hbar}\left[ e^{-iAt/4\hbar}\left(\alpha|1, 1\rangle + \frac{\beta}{\sqrt{2}}|1, 0\rangle\right) + \frac{\beta}{\sqrt{2}}e^{i3At/4\hbar}|0, 0\rangle\right]. \qquad (29.16)$$

**29.3.1.5 (a)** It is straightforward to check that $\hat{\pi}_\pm$ are projectors:

$$\hat{\pi}_+|+\rangle = |+\rangle \qquad \hat{\pi}_+|-\rangle = 0 \qquad \hat{\pi}_-|-\rangle = |-\rangle \qquad \hat{\pi}_-|+\rangle = 0. \qquad (29.17)$$

**(b)** The probability to find the muon spin in the state $|+\rangle$ at time $t$ is by definition

$$p(t) = \|\hat{\pi}_+|\psi(t)\rangle\|^2 = \langle\psi(t)|\hat{\pi}_+|\psi(t)\rangle. \qquad (29.18)$$

Using

$$\hat{\pi}_+|1, 1\rangle = |1, 1\rangle$$

$$\hat{\pi}_+|1, 0\rangle = \hat{\pi}_+|0, 0\rangle = \frac{1}{\sqrt{2}}|+-\rangle$$

$$\hat{\pi}_+|1, -1\rangle = 0,$$

we obtain

$$\hat{\pi}_+|\psi(t)\rangle = e^{-i(E_0+A/4)t/\hbar}\left[\alpha|++\rangle + \frac{\beta}{2}\left(1 + e^{iAt/\hbar}\right)|+-\rangle\right]. \qquad (29.19)$$

Squaring the norm of $\hat{\pi}_+|\psi(t)\rangle$, we get

$$p(t) = |\alpha|^2 + |\beta|^2 \cos^2(At/(2\hbar)). \qquad (29.20)$$

There is a periodic modulation of the probability to observe the muon spin aligned with the positive $z$ axis, which can be interpreted as a rotation of the muon spin with frequency $\nu = A/h$.

- The probability $p_+(t)$ corresponds to the initial state $|\psi(0)\rangle = |++\rangle$. This is a stationary state so that $p(t) \equiv p_+(t) = 1$ in this case.
- The probability $p_-(t)$ corresponds to the initial state $|\psi(0)\rangle = |+-\rangle = (|1, 0\rangle + |0, 0\rangle)/\sqrt{2}$. There is in this case an oscillation with a 100% modulation between $|+-\rangle$ and $|-+\rangle$, so that $p(t) \equiv p-(t) = \cos^2(At/2(\hbar))$.

Therefore the result can be cast in the form suggested in the text:

$$p(t) = q\, p_+(t) + (1 - q)\, p_-(t), \qquad (29.21)$$

with $q = |\alpha|^2$.

**29.3.1.6** When the electronic spins are unpolarized, we obtain, following the assumption of the text:

$$\overline{p}(t) = \frac{3}{4} + \frac{1}{4}\cos(At/\hbar). \tag{29.22}$$

**Note** The rigorous way to treat partially polarized systems is based on the density operator formalism. In the present case the density operator for the unpolarized electron is:

$$\rho_2 = \frac{1}{2}(|+\rangle\langle+| + |-\rangle\langle-|), \tag{29.23}$$

so that the initial density operator for the muonium is:

$$\rho(0) = \frac{1}{2}|++\rangle\langle++| + \frac{1}{2}|+-\rangle\langle+-|$$

$$= \frac{1}{2}|1, 1\rangle\langle1, 1| + \frac{1}{4}(|1, 0\rangle\langle1, 0| + |1, 0\rangle\langle0, 0| + |0, 0\rangle\langle1, 0| + |0, 0\rangle\langle0, 0|).$$

The density operator at time $t$ is then given by:

$$\rho(t) = \frac{1}{2}|1, 1\rangle\langle1, 1|$$

$$+ \frac{1}{4}\left(|1, 0\rangle\langle1, 0| + e^{-iAt/\hbar}|1, 0\rangle\langle0, 0| + e^{iAt/\hbar}|0, 0\rangle\langle1, 0| + |0, 0\rangle\langle0, 0|\right)$$

hence the probability:

$$\overline{p}(t) = \langle+, +|\rho(t)|+, +\rangle + \langle+ - |\rho(t)| + -\rangle$$

$$= \frac{1}{2} + \frac{1}{4}\left(\frac{1}{2} + e^{-iAt/\hbar}\frac{1}{2} + e^{iAt/\hbar}\frac{1}{2} + \frac{1}{2}\right)$$

$$= \frac{3}{4} + \frac{1}{4}\cos(At/\hbar).$$

## 29.3.2 Muonium in Silicon

**29.3.2.1** In the factorized basis $\{|\sigma_1, \sigma_2\rangle\}$, the Hamiltonian is written as

$$\hat{H} = E_0 + \begin{pmatrix} A'/4+ D & 0 & 0 & 0 \\ 0 & -A'/4- D & A'/2 & 0 \\ 0 & A'/2 & -A'/4- D & 0 \\ 0 & 0 & 0 & A'/4+ D \end{pmatrix}. \tag{29.24}$$

This Hamiltonian is diagonal in the eigenbasis $\{|S, m\rangle\}$ of the total spin. A simple calculation shows that the eigenvalues and eigenvectors are

$$E_1 = E_4 = E_0 + A'/4 + D \ : \ |1, 1\rangle \text{ and } |1, -1\rangle \tag{29.25}$$

and

$$E_2 = E_0 + A'/4 - D \ : \ |1, 0\rangle, \qquad E_3 = E_0 - 3A'/4 - D \ : \ |0, 0\rangle. \tag{29.26}$$

**29.3.2.2 (a)** The initial states $|\psi_+(0)\rangle$ and $|\psi_-(0)\rangle$ are easily obtained in the factorized basis as

$$|\psi_+(0)\rangle = |+x\rangle \otimes |+\rangle = (|++\rangle + |-+\rangle)/\sqrt{2}$$
$$|\psi_-(0)\rangle = |+x\rangle \otimes |-\rangle = (|+-\rangle + |--\rangle)/\sqrt{2}.$$

They can be written in the total spin basis $\{|S, m\rangle\}$ as

$$|\psi_+(0)\rangle = \frac{1}{\sqrt{2}}|1, 1\rangle + \frac{1}{2}(|1, 0\rangle - |0, 0\rangle),$$

$$|\psi_-(0)\rangle = \frac{1}{\sqrt{2}}|1, -1\rangle + \frac{1}{2}(|1, 0\rangle + |0, 0\rangle).$$

Writing $\omega_i = -E_i/\hbar$, we find at time $t$:

$$|\psi_+(0)\rangle = \frac{e^{i\omega_1 t}}{\sqrt{2}}|1, 1\rangle + \frac{e^{i\omega_2 t}}{2}|1, 0\rangle - \frac{e^{i\omega_3 t}}{2}|0, 0\rangle$$

$$|\psi_-(0)\rangle = \frac{e^{i\omega_4 t}}{\sqrt{2}}|1, -1\rangle + \frac{e^{i\omega_2 t}}{2}|1, 0\rangle + \frac{e^{i\omega_3 t}}{2}|0, 0\rangle,$$

which can now be converted in the factorized basis:

$$|\psi_+(t)\rangle = \frac{e^{i\omega_1 t}}{\sqrt{2}}|++\rangle + \frac{e^{i\omega_2 t} - e^{i\omega_3 t}}{2\sqrt{2}}|+-\rangle + \frac{e^{i\omega_2 t} + e^{i\omega_3 t}}{2\sqrt{2}}|-+\rangle$$

$$|\psi_-(t)\rangle = \frac{e^{i\omega_4 t}}{\sqrt{2}}|--\rangle + \frac{e^{i\omega_2 t} + e^{i\omega_3 t}}{2\sqrt{2}}|+-\rangle + \frac{e^{i\omega_2 t} - e^{i\omega_3 t}}{2\sqrt{2}}|-+\rangle .$$

**(b)** Since $\hat{\sigma}_{1x}|\sigma_1, \sigma_2\rangle = |-\sigma_1, \sigma_2\rangle$, the matrix elements $\langle\psi_\pm(t)|\hat{\sigma}_{1x}|\psi_\pm(t)\rangle$ are equal to:

$$\langle\psi_+(t)|\hat{\sigma}_{1x}|\psi_+(t)\rangle = \frac{1}{2}\text{Re }(e^{-i\omega_1 t}(e^{i\omega_2 t} + e^{i\omega_3 t}))$$

$$= \frac{1}{2}\left(\cos\frac{2Dt}{h} + \cos\frac{(A'+2D)t}{h}\right)$$

$$\langle\psi_-(t)|\hat{\sigma}_{1x}|\psi_-(t)\rangle = \frac{1}{2}\text{Re }[e^{-i\omega_4 t}(e^{i\omega_2 t} + e^{i\omega_3 t})].$$

Since $\omega_1 = \omega_4$, the two quantities are equal.

**(c)** The desired probabilities are

$$p_\pm(t) = \|\hat{\pi}_{+x}|\psi_\pm(t)\rangle\|^2 = \langle\psi_\pm(t)|\hat{\pi}_{+x}|\psi_\pm(t)\rangle \tag{29.27}$$

or, equivalently,

$$p_\pm(t) = \langle\psi_\pm(t)|\frac{1}{2}(1 + \hat{\sigma}_{1x})|\psi_\pm(t)\rangle = \frac{1}{2} + \frac{1}{2}\langle\psi_\pm(t)|\hat{\sigma}_{1x}|\psi_\pm(t)\rangle. \tag{29.28}$$

Using the result obtained above, we get:

$$p_\pm(t) = \frac{1}{2} + \frac{1}{4}\left(\cos\frac{2Dt}{\hbar} + \cos\frac{(A'+2D)t}{\hbar}\right). \tag{29.29}$$

**(d)** Since $p_+(t) = p_-(t)$, the result for $\overline{p}(t)$ is simply:

$$\overline{p}(t) = \frac{1}{2} + \frac{1}{4}\left(\cos\frac{2Dt}{\hbar} + \cos\frac{(A'+2D)t}{\hbar}\right). \tag{29.30}$$

**29.3.2.3 Comparison with experiment.** In practice, the time $t$ corresponds to the decay of the $\mu^+$, with the emission of a positron $e^+$ and two neutrinos. The positron is sufficiently energetic and leaves the crystal. It is emitted preferentially in the muon spin direction. One therefore measures the direction where the positron is emitted as a function of time. For $N_0$ incoming muons, the number of positrons emitted in the $x$ direction is $dN(t) = N_0\overline{p}(t)e^{-t/\tau}dt/\tau$, where $\tau$ is the muon lifetime.

A convenient way to analyse the signal, and to extract the desired frequencies, consists in taking the Fourier transform of the above signal. Defining

$$f_0(\omega) = \frac{1}{\tau}\int_0^\infty e^{(i\omega - 1/\tau)t}dt = \frac{1}{1 - i\omega\tau}, \tag{29.31}$$

one obtains

$$f(\omega) = \frac{1}{2} f_0(\omega) + \frac{1}{8} \left[ f_0 \left( \omega - \frac{2D}{\hbar} \right) + f_0 \left( \omega + \frac{2D}{\hbar} \right) \right]$$
$$+ \frac{1}{8} \left[ f_0 \left( \omega - \frac{A' + 2D}{\hbar} \right) + f_0 \left( \omega + \frac{A' + 2D}{\hbar} \right) \right].$$

The function $\text{Re}(f_0(\omega))$ has a peak at $\omega = 0$ whose half width is $1/\tau$, which corresponds to $100\,\text{kHz}$.

(a) The curve of Fig. 29.1 is consistent with this observation. Besides the peak at $\omega = 0$, we find two peaks at $\omega_1 = -2D/\hbar$ and $\omega_2 = (A' + 2D)/\hbar$. Assuming that $D$ is negative, which can be confirmed by a more thorough analysis, one obtains

$$2D/h = -37.25\,\text{MHz} \qquad \text{and} \qquad A'/h = 92.1\,\text{MHz}. \tag{29.32}$$

(b) In general, one expects to see peaks at all frequencies $\omega_i - \omega_j$, and in particular at $\omega_2 - \omega_3 = -A'/\hbar$. In order to observe the corresponding peak, one must measure the $\mu^+$ spin projection along a direction which is not orthogonal to the $z$ axis. This leads to a term in $\cos(\omega_2 - \omega_3)t$ in $\overline{p}(t)$, which appears in Fig. 29.1.

### 29.3.3 References

More information on the interactions between muons and matter can be found in B. D. Patterson, A. Hintermann, W. Kündig, P. F. Meier, F. Waldner, H. Graf, E. Recknagel, A. Weidinger, and Th. Wichert, *Anomalous Muonium in Silicon* Phys. Rev. Lett. 40, 1347 (1978) and Bruce D. Patterson, *Muonium states in semiconductors*, Rev. Mod. Phys. 60, 69 (1988).

# Quantum Reflection of Atoms from a Surface

**30**

This chapter deals with the reflection of very slow hydrogen atoms from a surface of liquid helium. In particular, we estimate the sticking probability of the atoms onto the surface. This sticking proceeds via the excitation of a surface wave, called a ripplon. We show that this probability must vanish at low temperatures, and that, in this limit, the reflection of the atoms on the surface is specular.

In all the chapter, the position of a particle is defined by its coordinates $r = (x, y)$ in a horizontal plane, and its altitude $z$. The altitude $z = 0$ represents the position of the surface of the liquid He bath at rest. The wave functions $\psi(r, z)$ of the H atoms are normalized in a rectangular box of volume $L_x L_y L_z$. We write $m$ for the mass of a H atom ($m = 1.67 \ 10^{-27}$ kg).

## 30.1 The Hydrogen Atom–Liquid Helium Interaction

Consider a H atom above a liquid He bath at rest (cf. Fig. 30.1). We model the H-liquid He interaction as the sum of pairwise interactions between the H atom at point $(R, Z)(Z > 0)$, and the He atoms at $(r, z)$, with $z < 0$:

$$V_0(Z) = n \int d^2r \int_{-\infty}^{+\infty} dz \, U(\sqrt{(R - r)^2 + (Z - z)^2}) \, \Theta(-z) \tag{30.1}$$

where $n$ is the number of He atoms per unit volume, and $\Theta$ is the Heaviside function.

**30.1.1** We recall the form of the Van der Waals potential:

$$U(d) = -\frac{C_6}{d^6}, \tag{30.2}$$

which describes the long distance interaction between a H atom and a He atom separated by a distance $d$. Show that the long distance potential between the H atom

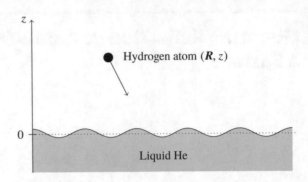

**Fig. 30.1** A hydrogen atom incident on a liquid helium bath. The oscillations of the surface of the liquid are studied in Sect. 30.2

and the liquid He bath is of the form:

$$V_0(Z) = -\frac{\alpha}{Z^3}. \tag{30.3}$$

Express $\alpha$ in terms of $C_6$ and $n$.

**30.1.2** Experimentally, one finds $\alpha = 1.9 \times 10^{-2}$eV Å$^3$. At what distance from the surface does gravity become larger than the Van der Waals force? In what follows, we shall neglect the gravitational force.

**30.1.3** Show that the eigenstates of the Hamiltonian which describes the motion of the H atom are of the form $|\mathbf{k}_\perp, \phi_\sigma\rangle$, where $\mathbf{k}_\perp$ represents a plane running wave propagating in the plane $Oxy$, i.e. parallel to the surface of the liquid He, and where $\phi_\sigma$ is an eigenstate of the Hamiltonian which describes the motion along the $z$ axis:

$$\langle \mathbf{R},\ Z|\mathbf{k}_\perp,\ \phi_\sigma\rangle = \frac{1}{\sqrt{L_x L_y}} e^{i(k_x X + k_y Y)} \phi_\sigma(Z). \tag{30.4}$$

**30.1.4** We want to evaluate the number of bound states of the motion along the $z$ axis in the potential:

$$V_0(Z) = -\frac{\alpha}{Z^3} \quad \text{if} \quad Z > z_{\min}$$

$$V_0(Z) = +\infty \quad \text{if} \quad Z \le z_{\min}.$$

We shall use the WKB approximation.

(a) Justify the shape of this potential.
(b) What is the continuity condition for the wave function at $Z = z_{\min}$?

(c) Show that the quantization condition for a motion with turning points $z_{min}$ and $b$ is

$$\int_{z_{min}}^{b} k(Z)\, dZ = \left(n + \frac{3}{4}\right) \pi \tag{30.5}$$

with $n$ integer $\geq 0$.

(d) Infer the order of magnitude of the number of bound states as a function of $z_{min}$ and $\alpha$. What is the domain of validity of the result?

(e) The parameter $z_{min}$ for the surface of liquid He is of the order of 2 Å. How many bound states does one expect for the motion along the $z$ axis?

(f) Experimentally, one finds that there is a single bound state H–liquid He, whose energy is $E_0 = -8.6 \times 10^{-5}$ eV. Compare this result with the WKB prediction. This unique bound state in the $z$-axis motion will be denoted $\phi_0$ in the rest of the chapter.

## 30.2 Excitations on the Surface of Liquid Helium

The general dispersion relation for waves propagating on the surface of a liquid is

$$\omega_q^2 = gq + \frac{A}{\rho_0} q^3 \quad \text{with } q = |\boldsymbol{q}|, \tag{30.6}$$

where $\omega_q$ and $\boldsymbol{q} = (q_x, q_y)$ are, respectively, the frequency and the wave vector of the surface wave, $g$ is the acceleration of gravity and $A$ and $\rho_0$ represent the surface tension and the mass density of the liquid.

**30.2.1** Discuss the nature of the surface waves (capillary waves or gravity waves) according to the value of the wavelength $\lambda = 2\pi/q$. Perform the numerical application in the case of liquid He: $\rho_0 = 145$ kg m$^{-3}$, $A = 3.5 \times 10^{-4}$ Jm$^{-2}$.

**30.2.2** Hereafter, we are only interested in waves for which $\hbar\omega_q \simeq |E_0|$. Show that these are always capillary waves and give their wavelengths. In what follows we shall use the simpler dispersion relation $\omega_q^2 = (A/\rho_0)q^3$.

**30.2.3** In order to quantize these surface waves, we introduce the bosonic operators $r_q$ and $r_q^\dagger$ corresponding to the annihilation and the creation of an excitation quantum. These elementary excitations are called *ripplons*. The Hamiltonian which describes these excitations is[1]:

$$H_S = \sum_q^{q_{max}} \hbar\omega_q \hat{r}_q^\dagger \, \hat{r}_q. \qquad (30.7)$$

The altitude $h(r)$ of the liquid surface at point $r = (x, y)$ becomes a two-dimensional scalar field operator:

$$\hat{h}(r) = \sum_q^{q_{max}} h_q(r_q^\dagger \, e^{-iq\cdot r} + r_q \, e^{iq\cdot r}) \quad \text{with} \quad h_q = \sqrt{\frac{\hbar q}{2\rho_0\omega_q L_x L_y}}. \qquad (30.8)$$

Evaluate, at zero temperature, the r.m.s. altitude $\Delta h$ of the position of the surface. We recall that, in two dimensions, the conversion of a discrete summation into an integral proceeds via:

$$\sum_q \rightarrow \frac{L_x L_y}{4\pi^2} \int d^2q. \qquad (30.9)$$

Numerical application: $q_{max} = 0.5 \, \text{Å}^{-1}$

## 30.3   Quantum Interaction Between H and Liquid He

We now investigate the modifications to the H–liquid He potential arising from the possible motion of the surface of the liquid helium bath. In order to do so, we replace the coupling considered above by

$$V(R, Z) = n \int d^2r \int_{-\infty}^{+\infty} dz \, U(\sqrt{(R - r)^2 + (Z - z)^2}) \, \Theta(\hat{h}(r) - z). \qquad (30.10)$$

**30.3.1** Expand $V(R, Z)$ to first order in $\hat{h}$ and interpret the result.

**30.3.2** Replacing $\hat{h}(r)$ by its expansion in terms of operators $\hat{r}_q, \hat{r}_q^\dagger$, cast $V(R, Z)$ in the form:

$$V(R, Z) = V_0(Z) + \sum_q (h_q e^{-iq\cdot R} \, V_q(Z) r_q^\dagger + h.c.) \qquad (30.11)$$

---

[1]The summation over $q$ is limited to $q < q_{max}$ where $q_{max}$ is of the order of a fraction of an inverse Ångstrom. For larger values of $q$, hence smaller wavelengths, the description of the vicinity of the surface in terms of a fluid does not hold any longer.

with

$$V_q(Z) = n \int d^2r \, e^{-iq \cdot r} \, U(\sqrt{r^2 + Z^2}) \,. \tag{30.12}$$

**30.3.3** Introducing the creation operators $\hat{a}_{k,\sigma}^\dagger$ and the annihilation operators $\hat{a}_{k,\sigma}$ of a hydrogen atom in an eigenstate of the motion in the potential $V_0(Z)$, write in second quantization the total hydrogen–ripplon Hamiltonian to first order in $\hat{h}$.

## 30.4 The Sticking Probability

We consider a H atom in an asymptotically free state in the $z$ direction (i.e. behaving as $e^{\pm ik_\sigma z}$ as $z \to +\infty$). This state denoted $|k_\perp, \, \phi_\sigma\rangle$ has an energy

$$E_i = \frac{\hbar^2}{2m}(k_\perp^2 + k_\sigma^2). \tag{30.13}$$

We now calculate the probability that this atom sticks on the surface, which is assumed here at zero temperature.

**30.4.1** How does the matrix element $\langle \phi_0 | V_q | \phi_\sigma \rangle$ vary with the size of the normalization box? We assume in the following that this matrix element is proportional to $k_\sigma$ if $k_\sigma$ is sufficiently small, and we introduce $M(q)$ such that

$$\langle \phi_0 | V_q | \phi_\sigma \rangle = \frac{\hbar k_\sigma}{\sqrt{2mL_z}} M(q). \tag{30.14}$$

All following results will be expressed in terms of $M(q)$ .

**30.4.2** Using Fermi's Golden Rule, define a probability per unit time for an atom to stick on the surface. In order to do so, one will define properly:

(a) the continuum of final states;
(b) the conditions imposed by energy conservation. For simplicity, we shall assume that the incident energy $E_i$ is negligible compared to the bound state energy $E_0$. Show that the emitted ripplon has a wave vector $q$ such that $|q| = q_0$ with:

$$\hbar \sqrt{\frac{A}{\rho_0} q_0^{3/2} + \frac{\hbar^2 q_0^2}{r2m}} = |E_0|; \tag{30.15}$$

(c) the density of final states.

**30.4.3** Express the flux of incident atoms in terms of $\hbar$, $k_\sigma$, $m$ and $L_z$.

**30.4.4** Write the expression for the probability that the hydrogen atom sticks on the surface of the liquid helium bath in terms of $\hbar$, $q_0$, $A$, $\rho_0$, $k_\sigma$ and $M(q)$. Check that this probability is independent of the normalization volume $L_x L_y L_z$.

**30.4.5** How does this probability vary with the energy of the incident hydrogen atoms?

**30.4.6** Describe qualitatively how one should modify the above treatment if the liquid helium bath is not at zero temperature.

## 30.5   Solutions

### 30.5.1  The Hydrogen Atom–Liquid Helium Interaction

**30.5.1.1** We use cylindrical coordinates, assuming that the H atom is at $R = 0$. The potential $V_0(Z)$ takes the form

$$
V_0(Z) = n \int d^2r \int_{-\infty}^{0} dz \, U(\sqrt{r^2 + (Z-z)^2})
$$

$$
= -nC_6 \int_{-\infty}^{0} dz \int_{0}^{\infty} dr \frac{2\pi r}{(r^2 + (Z-z)^2)^3}
$$

$$
= -\frac{\pi}{2} nC_6 \int_{-\infty}^{0} \frac{dz}{(Z-z)^4} = -\frac{\pi nC_6}{6Z^3}.
$$

Therefore

$$
V_0(Z) = -\frac{\alpha}{Z^3} \quad \text{with} \quad \alpha = \frac{\pi nC_6}{6}. \tag{30.16}
$$

**30.5.1.2** The force which derives from $V_0(Z)$ has modulus

$$
F(Z) = \frac{3\alpha}{Z^4}. \tag{30.17}
$$

We have $3\alpha/Z_g^4 = Mg$ for $Z_g = (3\alpha/(Mg))^{1/4}$. The numerical application yields $Z_g = 0.86\,\mu\text{m}$ which is very large on the atomic scale. For all the relevant H–liquid He distances, which are between 0.1 and 1 nm, gravity can be neglected.

**30.5.1.3** The Hamiltonian can be split as $\hat{H} = \hat{H}_\perp + \hat{H}_Z$, where

$$
\hat{H}_\perp = \frac{\hat{p}_x^2}{2m} + \frac{\hat{p}_y^2}{2m} \quad \text{and} \quad \hat{H}_Z = \frac{\hat{p}_z^2}{2m} + V_0(\hat{Z}). \tag{30.18}
$$

These two Hamiltonians commute and the eigenbasis of the total Hamiltonian $\hat{H}$ is factorized as a product $|k_\perp, \phi_\sigma\rangle$ of (i) the eigenstates of $\hat{H}_\perp$, where $k_\perp$ represents the wave vector of a plane running wave propagating in the $(x, y)$ plane, and (ii) the eigenstates $\phi_\sigma$ of $\hat{H}_Z$ which describes the motion along the $z$ axis.

**30.5.1.4 (a)** For $Z \leq z_{min}$, the overlap of the electron wave functions of the H and He atoms causes a repulsion between these atoms, which is modeled here by a hard core potential. For $Z \gg z_{min}$, the Van der Waals forces are dominant.

**(b)** For $Z \leq z_{min}$, the wave function $\phi(Z)$ is such that $\phi(Z) = 0$. Since $\phi(Z)$ is continuous, we have $\phi(z_{min}) = 0$.

**(c)** For a turning point $b$, the WKB eigenfunction of energy $E$ has the following form in the allowed region $(E > V_0(Z))$ :

$$\phi(Z) = \frac{C}{\sqrt{k(Z)}} \cos\left(\int_Z^b k(Z')\,dZ' - \frac{\pi}{4}\right), \tag{30.19}$$

where $C$ is a normalization constant and where

$$\hbar k(Z) = \sqrt{2m(E - V_0(Z))}. \tag{30.20}$$

Imposing the condition $\phi(z_{min}) = 0$ yields

$$\int_{z_{min}}^b k(Z')\,dZ' - \frac{\pi}{4} = \left(n + \frac{1}{2}\right)\pi, \text{ i.e. } \int_{z_{min}}^b k(Z')\,dZ' = \left(n + \frac{3}{4}\right)\pi \tag{30.21}$$

with $n$ a positive integer.

**(d)** If the WKB method were exact, the number of bound states would be

$$n = 1 + \text{Int}\left(\int_{z_{min}}^\infty \frac{k(Z')}{\pi}\,dZ' - \frac{3}{4}\right), \tag{30.22}$$

where Int denotes the integer part and where $k(Z)$ is calculated for a zero energy $E$. As usual for the WKB method, the accuracy of this expression is good if the number of bound states is large. We can take in this case: $n \simeq \pi^{-1} \int_{z_{min}}^\infty k(Z')\,dZ'$ with $\hbar k(Z) = \sqrt{2m\alpha/Z^3}$, which yields

$$n \simeq \frac{2}{\pi\hbar}\sqrt{\frac{2m\alpha}{z_{min}}}. \tag{30.23}$$

(e) The above formula yields $n \simeq 1.36$. We therefore expect a number of bound states close to 1, say between 0 and 2.

(f) The experimental result compares favorably with the WKB. prediction, but it is beyond the validity of the WKB approximation to give a correct expression for $\phi_0(Z)$ .

## 30.5.2 Excitations on the Surface of Liquid Helium

**30.5.2.1** The two terms of the dispersion relation are equal if $q = \sqrt{g\rho_0/A}$ or, equivalently, for a wavelength

$$\lambda = 2\pi \sqrt{\frac{A}{g\rho_0}}. \tag{30.24}$$

Numerically, one obtains $\lambda = 3$ mm. Therefore, we observe capillary waves ($\omega_q^2 \simeq Aq^3/\rho_0$) for $\lambda \ll 3$ mm, and gravity waves ($\omega_q^2 \simeq gq$) for $\lambda \gg 3$ mm. For $\lambda = 3$ mm the corresponding energy is $\hbar_{\omega_q}=1.3 \times 10^{-13}$ eV.

**30.5.2.2** For an energy such that $|E_0| \gg 10^{-13}$ eV, we are therefore in the regime of capillary waves, with the wavelength:

$$\lambda = \frac{2\pi}{q} = 2\pi \left( \frac{Ah^2}{\rho_0 E_0^2} \right)^{1/3}. \tag{30.25}$$

The numerical value is $\lambda = 33$ Å.

**30.5.2.3** We have

$$\Delta h^2 = \langle \hat{h}^2 \rangle - \langle \hat{h} \rangle^2 = \langle \hat{h}^2 \rangle = \sum_q h_q^2 \langle r_q r_q^\dagger \rangle = \sum_q h_q^2. \tag{30.26}$$

Converting this into an integral, we obtain

$$\Delta h^2 = \frac{L_x L_y}{4\pi^2} \int \frac{\hbar q}{2\rho_0 \omega_q L_x L_y} \, \mathrm{d}^2 q = \frac{\hbar}{4\pi \sqrt{A\rho_0}} \int_0^{q_{max}} \sqrt{q} \, \mathrm{d}q$$

$$= \frac{\hbar}{6\pi} \sqrt{\frac{q_{max}^3}{A\rho_0}} = \frac{\hbar \omega_{max}}{6\pi A},$$

which yields $\Delta h = 0.94$ Å.

### 30.5.3 Quantum Interaction Between H and Liquid He

**30.5.3.1** Using the fact that $\Theta'(z) = \delta(z)$, we can write $\Theta(-z + \hat{h}(r)) \simeq \Theta(-z) + \hat{h}(r)\delta(z)$ since the $\delta$ function is even. Therefore, we obtain

$$V(\boldsymbol{R}, \ Z) \simeq V_0(Z) + n \int \mathrm{d}^2 r \, U(\sqrt{(\boldsymbol{R} - \boldsymbol{r})^2 + Z^2}) \, \hat{h}(r). \tag{30.27}$$

In this expression, the second term describes the interaction with the "additional" or "missing" atoms on the surface as compared to the equilibrium position $z = 0$.

**30.5.3.2** Replacing $\hat{h}(r)$ by its expansion we obtain

$$V(\boldsymbol{R}, \ Z) \simeq V_0(Z) + n \int \mathrm{d}^2 r \, U(\sqrt{(\boldsymbol{R} - \boldsymbol{r})^2 + Z^2}) \sum_q h_q(\hat{r}_q^\dagger e^{-i q \cdot r} + \hat{r}_q e^{i q \cdot r}). \tag{30.28}$$

Considering the term $r_q^\dagger$ and setting $\boldsymbol{r}' = \boldsymbol{r} - \boldsymbol{R}$, we obtain in a straightforward manner

$$V(\boldsymbol{R}, \ Z) = V_0(Z) + \sum_q (h_q e^{-i q \cdot \boldsymbol{R}} V_q(Z)\hat{r}_q^\dagger + \text{h.c.}), \tag{30.29}$$

with

$$V_q(Z) = n \int \mathrm{d}^2 r' \, e^{-i q \cdot r'} \, U(\sqrt{r'^2 + Z^2}). \tag{30.30}$$

**30.5.3.3** The Hamiltonian is the sum of the "free" Hamiltonians $\hat{H}_{\text{at}} = \dfrac{P^2}{2M} + V_0(Z)$ and $\hat{H}_{\text{S}}$, and the coupling term found above. One has

$$\hat{H}_{\text{at}} = \sum_{k,\sigma} E_{k,\sigma} \hat{a}_{k,\sigma}^\dagger \hat{a}_{k,\sigma} \qquad \hat{H}_{\text{S}} = \sum_q \hbar \omega_q \hat{r}_q^\dagger \hat{r}_q. \tag{30.31}$$

The coupling term becomes

$$\sum_{k,\sigma} \sum_{k',\sigma'} \sum_q h_q \hat{a}_{k,\sigma}^\dagger \hat{a}_{k',\sigma'} \hat{r}_q^\dagger \langle k, \ \phi_\sigma | e^{-i q \cdot \boldsymbol{R}} V_q(Z) | k', \ \phi_{\sigma'} \rangle + \text{h.c.} \tag{30.32}$$

The matrix element is

$$\begin{aligned}
\langle k, \ \phi_\sigma | e^{-i q \cdot \boldsymbol{R}} V_q(Z) | k', \ \phi_{\sigma'} \rangle &= \langle k | e^{-i q \cdot \boldsymbol{R}} | k' \rangle \, \langle \phi_\sigma | V_q(Z) | \phi_{\sigma'} \rangle \\
&= \delta_{k', k+q} \, \langle \phi_\sigma | V_q(Z) | \phi_{\sigma'} \rangle.
\end{aligned}$$

We end up with the total hydrogen–ripplon Hamiltonian to first order in $\hat{h}$:

$$\hat{H} = \sum_{k,\sigma} E_{k,\sigma} \hat{a}^\dagger_{k,\sigma} \hat{a}_{k,\sigma} + \sum_q \hbar\omega_q \hat{r}^\dagger_q r_q$$

$$+ \sum_{q,k,\sigma,\sigma'} h_q \hat{a}^\dagger_{k,\sigma} \hat{a}_{k+q,\sigma'} \hat{r}^\dagger_q \langle \phi_\sigma | V_q(Z) | \phi_{\sigma'} \rangle + \text{h.c.} .$$

In the $(x, y)$ plane, the momentum is conserved owing to the translation invariance of the problem. This can be seen directly on the form of the coupling

$$\hat{a}^\dagger_{k,\sigma} \hat{a}_{k+q,\sigma'} \hat{r}^\dagger_q, \tag{30.33}$$

which annihilates a H atom with momentum $\hbar(k + q)$ term, and creates a H atom with momentum $\hbar k$ and a ripplon with momentum $\hbar q$.

### 30.5.4  The Sticking Probability

**30.5.4.1**  We have by definition

$$\langle \phi_0 | V_q | \phi_\sigma \rangle = \int \phi_0^*(Z) V_q(Z) \phi_\sigma(Z) \mathrm{d}Z. \tag{30.34}$$

Since $|\phi_\sigma\rangle$ is an asymptotically free state, it is normalized in a segment of length $L_z$. Therefore its amplitude varies as $L_z^{-1/2}$. Since $|\phi_0\rangle$ is a localized state which does not depend on $L_z$, we find

$$\langle \phi_0 | V_q | \phi_\sigma \rangle \propto \frac{1}{\sqrt{Lz}}. \tag{30.35}$$

The fact that this matrix element is proportional to $k_\sigma$ in the limit of small incident momenta is more subtle. The positions $Z$ contributing to the matrix element are close to zero, since the bound state $\phi_0(Z)$ is localized in the vicinity of the He surface. Therefore only the values of $\phi_\sigma(Z)$ around $Z = 0$ are relevant for the calculation of the integral. For the $Z^{-3}$ potential between the H atom and the He surface, one finds that the amplitude of $\phi_\sigma$ in this region is proportional to $k_\sigma$, hence the result. Such a linear dependance can be recovered analytically by replacing the $Z^{-3}$ potential by a square well, but is out of reach of the WKB approximation, which would predict a dependance in $\sqrt{k_\sigma}$ for the amplitude around $Z = 0$ of $\phi_\sigma$. The reason for this discrepancy is that the potential in $-\alpha Z^{-3}$ is too stiff for the WKB to be valid for the calculation of $\phi_\sigma$ at distances larger than $m\alpha/\hbar^2$.

**30.5.4.2** We start with the initial state $k_\perp$, $\phi_\sigma$. If the atom sticks to the surface, the final state along the $z$ axis is $|\phi_0\rangle$. The sticking proceeds via the emission of a ripplon of momentum $\hbar q$ and a change of the transverse momentum $\hbar k_\perp \to \hbar k_\perp - \hbar q$.

**(a)** The continuum of final states is characterized by the vector $q$:

$$|k_\perp, \phi_\sigma\rangle \to |k_\perp - q, \phi_0\rangle \otimes |q\rangle. \tag{30.36}$$

**(b)** Energy conservation gives $E_i = E_f$ with:

$$E_i = \frac{\hbar^2(k_\sigma^2 + k_\perp^2)}{2m} \qquad E_f = E_0 + \frac{\hbar^2(k_\perp - q)^2}{2m} + \hbar\omega_q. \tag{30.37}$$

We suppose that $E_i$ is negligible compared to the bound state energy $E_0$. Therefore $\hbar^2(k_\perp \cdot q)/m \sim \sqrt{|E_0|\hbar^2 k_\perp^2/(2m)}$ is also very small compared to $|E_0|$, and we obtain:

$$\frac{\hbar^2 q^2}{2m} + \hbar\omega_q \simeq |E_0|. \tag{30.38}$$

This equation, in addition to the dispersion relation for ripplons, determines the modulus $q_0$ of $q$:

$$\hbar\sqrt{\frac{\Lambda}{\rho_0}}q_0^{3/2} + \frac{\hbar^2 q_0^2}{2m} = |E_0|. \tag{30.39}$$

**(c)** A variation $\delta E$ of the final state energy corresponds to a variation $\delta q$ such that

$$\left(\frac{\hbar^2 q_0}{m} + \frac{3\hbar}{2}\sqrt{\frac{\Lambda q_0}{\rho_0}}\right)\delta q = \delta E. \tag{30.40}$$

The number of states $\delta^2 n$ in a domain $\delta^2 q$ is:

$$\delta^2 n = \frac{L_x L_y}{4\pi^2}\delta^2 q = \frac{L_x L_y}{4\pi^2}q_0\delta q\delta\theta. \tag{30.41}$$

After integrating over $\delta\theta$, we obtain:

$$\rho(E_f) = \frac{L_x L_y}{\pi}\frac{m q_0}{2\hbar^2 q_0 + 3m\hbar\sqrt{\Lambda q_0/\rho_0}}. \tag{30.42}$$

**30.5.4.3** The number of atoms which cross a plane of altitude $Z$ in the direction $Z < 0$ during a time interval $dt$ is $v_z dt/(2L_z) = \hbar k_\sigma dt/(2mL_z)$ . The flux is therefore:

$$\Phi_\sigma = \frac{\hbar k_\sigma}{2mL_z}. \tag{30.43}$$

**30.5.4.4** The sticking probability is the ratio of the probability per unit time, given by Fermi's Golden Rule, and the incident flux:

$$P = \frac{2\pi}{\hbar} |\langle k_\perp, \ \phi_\sigma | V | k_\perp - q, \ \phi_0, \ q \rangle|^2 \ \rho(E_f) \frac{2mL_z}{\hbar k_\sigma}. \tag{30.44}$$

This reduces to

$$P = \frac{m k_\sigma |M(q)|^2}{3Am + 2\hbar\sqrt{A\rho_0 q_0}}. \tag{30.45}$$

**30.5.4.5** $P$ varies as $k_\sigma \propto \sqrt{E}$. At very small energies, the sticking probability goes to zero and the H atoms bounce elastically on the liquid He surface.

**30.5.4.6** If the liquid helium bath is not at zero temperature, other processes can occur, in particular a sticking process accompanied by the stimulated emission of a ripplon. One must therefore take into account the number $n_{q_0}$ of thermal ripplons.

## 30.5.5 Comments and References

The model studied in this chapter is inspired from the work of D.S. Zimmerman and A.J. Berlinsky, *The sticking probability for hydrogen atoms on the surface of liquid* $^4He$, Can. J. Phys. **61**, 508 (1983). Thorough experimental studies of this process are presented by J.J. Berkhout, O.J. Luiten, J.D. Setija, T.W. Hijmans, T. Mizusaki, and J.T.M. Walraven, *Quantum reflection: Focusing of hydrogen atoms with a concave mirror*, Phys. Rev. Lett. **63**, 1689 (1989) and J.M. Doyle, J.C. Sandberg, I.A. Yu, C.L. Cesar, D. Kleppner, and T.J. Greytak, *Hydrogen in the submillikelvin regime: Sticking probability on superfluid* $^4He$, Phys. Rev. Lett. **67**, 603 (1991).

Quantum reflection has also been observed for atoms originating from a Bose–Einstein condensate by T. A. Pasquini, M. Saba, G.-B. Jo, Y. Shin, W. Ketterle, D. E. Pritchard, T. A. Savas, and N. Mulders, *Low Velocity Quantum Reflection of Bose-Einstein Condensates*, Phys. Rev. Lett. **97**, 093201 (2006). It is currently considered as a promising tool to confine antimatter particles in boxes made out of normal matter, without having these antiparticles annihilating on the walls of the box, see e.g. A. Yu. Voronin, P. Froelich, and V. V. Nesvizhevsky, *Gravitational quantum states of antihydrogen*, Phys. Rev. A 83, 032903 (2011), P. Pérez et al., *The GBAR antimatter gravity experiment*, Hyperfine Interactions **233**, 21 (2015).

# Part IV

# Appendix

# Appendix: Memento of Quantum Mechanics 31

In the following pages we recall the basic definitions, notations and results of quantum mechanics.

## 31.1 Principles

### 31.1.1 Hilbert Space

The first step in treating a quantum physical problem consists in identifying the appropriate Hilbert space to describe the system. A Hilbert space is a complex vector space, with a Hermitian scalar product. The vectors of the space are called kets and are noted $|\psi\rangle$. The scalar product of the ket $|\psi_1\rangle$ and the ket $|\psi_2\rangle$ is noted $\langle\psi_2|\psi_1\rangle$. It is linear in $|\psi_1\rangle$ and antilinear in $|\psi_2\rangle$ and one has: $\langle\psi_1|\psi_2\rangle = (\langle\psi_2|\psi_1\rangle)^*$.

### 31.1.2 Definition of the State of a System; Pure Case

The state of a physical system is completely defined at any time $t$ by a vector of the Hilbert space, normalized to 1, noted $|\psi(t)\rangle$. Owing to the superposition principle, if $|\psi_1\rangle$ and $|\psi_2\rangle$ are two possible states of a given physical system, any linear combination $|\psi\rangle \propto c_1|\psi_1\rangle + c_2|\psi_2\rangle$, where $c_1$ and $c_2$ are complex numbers, is a possible state of the system. These coefficients must be chosen such that $\langle\psi|\psi\rangle = 1$.

### 31.1.3 Measurement

To a given physical quantity $A$ one associates a self-adjoint (or Hermitian) operator $\hat{A}$ acting in the Hilbert space. In a measurement of the quantity $A$, the only possible results are the eigenvalues $a_\alpha$ of $\hat{A}$.

© Springer Nature Switzerland AG 2019
J.-L. Basdevant, J. Dalibard, *The Quantum Mechanics Solver*,
https://doi.org/10.1007/978-3-030-13724-3_31

Consider a system in a state $|\psi\rangle$. The probability $\mathcal{P}(a_\alpha)$ to find the result $a_\alpha$ in a measurement of $A$ is

$$\mathcal{P}(a_\alpha) = \left\| \hat{P}_\alpha |\psi\rangle \right\|^2, \tag{31.1}$$

where $\hat{P}_\alpha$ is the projector on the eigensubspace $\mathcal{E}_\alpha$ associated to the eigenvalue $a_\alpha$.

After a measurement of $\hat{A}$ which has given the result $a_\alpha$, the state of the system is proportional to $\hat{P}_\alpha |\psi\rangle$ (wave packet projection or reduction).

A single measurement gives information on the state of the system after the measurement has been performed. The information acquired on the state before the measurement is very "poor", i.e. if the measurement gave the result $a_\alpha$, one can only infer that the state $|\psi\rangle$ was not in the subspace orthogonal to $\mathcal{E}_\alpha$.

In order to acquire accurate information on the state before measurement, one must use $N$ independent systems, all of which are prepared in the same state $|\psi\rangle$ (with $N \gg 1$) If we perform $N_1$ measurements of $\hat{A}_1$ (eigenvalues $\{a_{1,\alpha}\}$), $N_2$ measurements of $\hat{A}_2$ (eigenvalues $\{a_{2,\alpha}\}$), and so on (with $\sum_{i=1}^p N_i = N$), we can determine the probability distribution of the $a_{i,\alpha}$, and therefore the $\| \hat{P}_{i,\alpha} |\psi\rangle \|^2$. If the $p$ operators $\hat{A}_i$ are well chosen, this determines unambiguously the initial state $|\psi\rangle$.

### 31.1.4 Evolution

When the system is not being measured, the evolution of its state vector is given by the Schrödinger equation

$$i\hbar \frac{d}{dt} |\psi\rangle = \hat{H}(t) |\psi(t)\rangle, \tag{31.2}$$

where the hermitian operator $\hat{H}(t)$ is the Hamiltonian, or energy observable, of the system at time $t$.

If we consider an isolated system, whose Hamiltonian is time-independent, the energy eigenstates of the Hamiltonian $|\phi_n\rangle$ are the solution of the time independent Schrödinger equation:

$$\hat{H}|\phi_n\rangle = E_n|\phi_n\rangle. \tag{31.3}$$

They form an orthogonal basis of the Hilbert space. This basis is particularly useful. If we decompose the initial state $|\psi(0)\rangle$ on this basis, we can immediately write its expression at any time as:

$$|\psi(0)\rangle = \sum_n \alpha_n |\phi_n\rangle \quad \rightarrow \quad |\psi(t)\rangle = \sum_n \alpha_n \, e^{-iE_n t/\hbar}|\phi_n\rangle. \tag{31.4}$$

The coefficients are $\alpha_n = \langle \phi_n | \psi(0) \rangle$, i.e.

$$|\psi(t)\rangle = \sum_n e^{-iE_n t/\hbar} |\phi_n\rangle\langle\phi_n|\psi(0)\rangle. \tag{31.5}$$

### 31.1.5 Complete Set of Commuting Observables (CSCO)

A set of operators $\{\hat{A}, \hat{B}, \ldots, \hat{X}\}$ is a CSCO if all of these operators commute and if their common eigenbasis $\{|\alpha, \beta, \ldots, \xi\rangle\}$ is unique (up to a phase factor).

In that case, after the measurement of the physical quantities $\{A, B, \ldots, X\}$, the state of the system is known unambiguously. If the measurements have given the values $\alpha$ for $A$, $\beta$ for $B$, $\ldots$, $\xi$ for $\hat{X}$, the state of the system is $|\alpha, \beta, \ldots, \xi\rangle$.

### 31.1.6 Entangled States

Consider a quantum system $S$ formed by two subsystems $S_1$ and $S_2$. The Hilbert space in which we describe $S$ is the tensor product of the Hilbert spaces $\mathcal{E}_1$ and $\mathcal{E}_2$ respectively associated with $S_1$ and $S_2$. If we note $\{|\alpha_m\rangle\}$ a basis of $S_1$ and $\{|\beta_n\rangle\}$ a basis of $S_2$, a possible basis of the global system is $\{|\alpha_m\rangle \otimes |\beta_n\rangle\}$.

Any state vector of the global system can be written as:

$$|\Psi\rangle = \sum_{m,n} C_{m,n} |\alpha_m\rangle \otimes |\beta_n\rangle. \tag{31.6}$$

If this vector can be written as $|\Psi\rangle = |\alpha\rangle \otimes |\beta\rangle$, where $|\alpha\rangle$ and $|\beta\rangle$ are vectors of $\mathcal{E}_1$ and $\mathcal{E}_2$ respectively, one calls it a factorized state.

In general an arbitrary state $|\Psi\rangle$ is not factorized: there are quantum correlations between the two subsystems, and $|\Psi\rangle$ is called an *Entangled state*.

### 31.1.7 Statistical Mixture and the Density Operator

If we have an incomplete information on the state of the system, for instance because the measurements are incomplete, one does not know exactly its state vector. The state can be described by a density operator $\hat{\rho}$ whose properties are the following:

- The density operator is hermitian and its trace is equal to 1.
- All the eigenvalues $\Pi_n$ of the density operator are non-negative. The density operator can therefore be written as

$$\hat{\rho} = \sum_n \Pi_n |\phi_n\rangle\langle\phi_n|, \tag{31.7}$$

where the $|\phi_n\rangle$ are the eigenstates of $\hat{\rho}$ and the $\Pi_n$ can be interpreted as a probability distribution. In the case of a pure state, all eigenvalues vanish except one which is equal to 1.

- The probability to find the result $a_\alpha$ in a measurement of the physical quantity $A$ is given by

$$\mathcal{P}(a_\alpha) = \mathrm{Tr}\left(\hat{P}_\alpha \hat{\rho}\right) = \sum_n \Pi_n \langle \phi_n | \hat{A} | \phi_n \rangle. \tag{31.8}$$

The state of the system after the measurement is $\hat{\rho}' \propto \hat{P}_\alpha \hat{\rho} \hat{P}_\alpha$.

- As long as the system is not measured, the evolution of the density operator is given by

$$i\hbar \frac{d}{dt}\hat{\rho}(t) = [\hat{H}(t), \hat{\rho}(t)]. \tag{31.9}$$

## 31.2   General Results

### 31.2.1 Uncertainty Relations

Consider $2N$ physical systems which are identical and independent, and are all prepared in the same state $|\psi\rangle$ (we assume $N \gg 1$). For $N$ of them, we measure a physical quantity $A$, and for the $N$ others, we measure a physical quantity $B$. The rms deviations $\Delta a$ and $\Delta b$ of the two series of measurements satisfy the inequality

$$\Delta a \, \Delta b \geq \frac{1}{2} \left| \langle \psi | [\hat{A}, \hat{B}] | \psi \rangle \right|. \tag{31.10}$$

### 31.2.2 Ehrenfest Theorem

Consider a system which evolves under the action of a Hamiltonian $\hat{H}(t)$, and an observable $\hat{A}(t)$. The expectation value of this observable evolves according to the equation:

$$\frac{d}{dt}\langle a \rangle = \frac{1}{i\hbar}\langle \psi | [\hat{A}, \hat{H}] | \psi \rangle + \langle \psi | \frac{\partial \hat{A}}{\partial t} | \psi \rangle. \tag{31.11}$$

In particular, if $\hat{A}$ is time-independent and if it commutes with $\hat{H}$, the expectation value $\langle a \rangle$ is a constant of the motion.

## 31.3   The Particular Case of a Point-Like Particle; Wave Mechanics

### 31.3.1 The Wave Function

For a point-like particle for which we can neglect possible internal degrees of freedom, the Hilbert space is the space of square integrable functions (written in mathematics as $L^2(R^3)$).

The state vector $|\psi\rangle$ is represented by a wave function $\psi(r)$. The quantity $|\psi(r)|^2$ is the probability density to find the particle at point $r$ in dimensional space. Its Fourier transform $\varphi(p)$:

$$\varphi(p) = \frac{1}{(2\pi\hbar)^{3/2}} \int e^{-i p \cdot r/\hbar} \psi(r) \, d^3 r \tag{31.12}$$

is the probability amplitude to find that the particle has a momentum $p$.

### 31.3.2 Operators

Among the operators associated to usual physical quantities, one finds:

- The position operator $\hat{r} \equiv (\hat{x}, \hat{y}, \hat{z})$, which consists in multiplying the wave function $\psi(r)$ by $r$.
- The momentum operator $\hat{p}$ whose action on the wave function $\psi(r)$ is the operation $-i\hbar\nabla$.
- The Hamiltonian, or energy operator, for a particle placed in a potential $V(r)$:

$$\hat{H} = \frac{\hat{p}^2}{2M} + V(\hat{r}) \quad \rightarrow \quad \hat{H}\psi(r) = -\frac{h^2}{2M}\nabla^2\psi(r) + V(r)\psi(r), \tag{31.13}$$

where $M$ is the mass of the particle.

### 31.3.3 Continuity of the Wave Function

If the potential $V$ is continuous, the eigenfunctions of the Hamiltonian $\psi_\alpha(r)$ are continuous and so are their derivatives. This remains true if $V(r)$ is a step function: $\psi$ and $\psi'$ are continuous where $V(r)$ has discontinuities.

In the case of infinitely high potential steps, (for instance $V(x) = +\infty$ for $x < 0$ and $V(x) = 0$ for $x \geq 0$), $\psi(x)$ is continuous and vanishes at the discontinuity of $V(\psi(0) = 0)$, while its first derivative $\psi'(x)$ is discontinuous.

In one dimension, it is interesting to consider potentials which are Dirac distributions, $V(x) = g\delta(x)$. The wave function is continuous and the discontinuity of its derivative is obtained by integrating the Schrödinger equation around the center of the delta function [$\psi'(0_+) - \psi'(0_-) = (2Mg/\hbar^2)\psi(0)$ in our example].

### 31.3.4 Position-Momentum Uncertainty Relations

Using the above general result, one finds:

$$[\hat{x}, \hat{p}_x] = i\hbar \quad \rightarrow \quad \Delta x \, \Delta p_x \geq \hbar/2, \tag{31.14}$$

and similar relations for the $y$ and $z$ components.

## 31.4    Angular Momentum and Spin

### 31.4.1 Angular Momentum Observable

An angular momentum observable $\hat{\boldsymbol{J}}$ is a set of three operators $\{\hat{J}_x, \hat{J}_y, \hat{J}_z\}$ which satisfy the commutation relations

$$[\hat{J}_x, \hat{J}_y] = i\hbar \, \hat{J}_z, \quad [\hat{J}_y, \hat{J}_z] = i\hbar \, \hat{J}_x, \quad [\hat{J}_z, \hat{J}_x] = i\hbar \, \hat{J}_y. \tag{31.15}$$

The orbital angular momentum with respect to the origin $\hat{\boldsymbol{L}} = \hat{\boldsymbol{r}} \times \hat{\boldsymbol{p}}$ is an angular momentum observable.

The observable $\hat{J}^2 = \hat{J}_x^2 + \hat{J}_y^2 + \hat{J}_z^2$ commutes with all the components $\hat{J}_i$. One can therefore find a common eigenbasis of $\hat{J}^2$ and one of the three components $\hat{J}_i$. Traditionally, one chooses $i = z$.

### 31.4.2 Eigenvalues of the Angular Momentum

The eigenvalues of $\hat{J}^2$ are of the form $\hbar^2 j(j+1)$ with $j$ integer or half integer. In an eigensubspace of $\hat{J}^2$ corresponding to a given value of $j$, the eigenvalues of $\hat{J}_z$ are of the form

$$\hbar m, \quad \text{with } m \in \{-j, -j+1, \ldots, j-1, j\} \quad (2j+1 \text{ values}). \tag{31.16}$$

The corresponding eigenstates are noted $|\alpha, j, m\rangle$, where $\alpha$ represents the other quantum numbers which are necessary in order to define the states completely. The states $|\alpha, j, m\rangle$ are related to $|\alpha, j, m \pm 1\rangle$ by the operators $\hat{J}_\pm = \hat{J}_x \pm i\hat{J}_y$:

$$\hat{J}_\pm|\alpha, j, m\rangle = \sqrt{j(j+1) - m(m \pm 1)} \, |\alpha, j, m \pm 1\rangle. \tag{31.17}$$

### 31.4.3 Orbital Angular Momentum of a Particle

In the case of an orbital angular momentum, only integer values of $j$ and $m$ are allowed. Traditionally, one notes $j = \ell$ in this case. The common eigenstates $\psi(\mathbf{r})$ of $\hat{L}^2$ and $\hat{L}_z$ can be written in spherical coordinates as $R(r)\, Y_{\ell,m}(\theta, \varphi)$, where the radial wave function $R(r)$ is arbitrary and where the functions $Y_{\ell,m}$ are the spherical harmonics, i.e. the harmonic functions on the sphere of radius one. The first are:

$$Y_{0,0}(\theta, \varphi) = \frac{1}{\sqrt{4\pi}}, \quad Y_{1,0}(\theta, \varphi) = \sqrt{\frac{3}{4\pi}} \cos\theta,$$

$$Y_{1,1}(\theta, \varphi) = -\sqrt{\frac{3}{8\pi}} \sin\theta\, e^{i\varphi}, \quad Y_{1,-1}(\theta, \varphi) = \sqrt{\frac{3}{8\pi}} \sin\theta\, e^{-i\varphi}.$$

### 31.4.4 Spin

In addition to its angular momentum, a particle can have an intrinsic angular momentum called its Spin. The spin, which is noted traditionally $j = s$, can take half-integer as well as integer values.

The electron, the proton, the neutron are spin $s = 1/2$ particles, for which the projection of the intrinsic angular momentum can take either of the two values $m\hbar$ : $m = \pm 1/2$. In the basis $|s = 1/2, m = \pm 1/2\rangle$, the operators $\hat{S}_x, \hat{S}_y, \hat{S}_z$ have the matrix representations:

$$\hat{S}_x = \frac{\hbar}{2} \begin{pmatrix} 0 & 1 \\ 1 & 0 \end{pmatrix}, \quad \hat{S}_y = \frac{\hbar}{2} \begin{pmatrix} 0 & -i \\ i & 0 \end{pmatrix}, \quad \hat{S}_z = \frac{\hbar}{2} \begin{pmatrix} 1 & 0 \\ 0 & -1 \end{pmatrix}. \tag{31.18}$$

### 31.4.5 Addition of Angular Momenta

Consider a system $S$ made of two subsystems $S_1$ and $S_2$, of angular momenta $\hat{\boldsymbol{J}}_1$ and $\hat{\boldsymbol{J}}_2$. The observable $\hat{\boldsymbol{J}} = \hat{\boldsymbol{J}}_1 + \hat{\boldsymbol{J}}_2$ is an angular momentum observable. In the subspace corresponding to given values $j_1$ and $j_2$ (of dimension $(2j_1 + 1) \times (2j_2 + 1)$), the possible values for the quantum number $j$ corresponding to the total angular momentum of the system $\hat{\boldsymbol{J}}$ are:

$$j = |j_1 - j_2|, \; |j_1 - j_2| + 1, \; \cdots, \; j_1 + j_2, \tag{31.19}$$

with, for each value of $j$, the $2j + 1$ values of $m$: $m = -j, -j + 1, \cdots, j$. For instance, adding two spins $1/2$, one can obtain an angular momentum 0 (singlet state $j = m = 0$) and three states of angular momentum 1 (triplet states $j = 1, m = 0, \pm 1$).

The relation between the factorized basis $|j_1, m_1\rangle \otimes |j_2, m_2\rangle$ and the total angular momentum basis $|j_1, j_2 ; j, m\rangle$ is given by the Clebsch-Gordan coefficients:

$$|j_1, j_2; j, m\rangle = \sum_{m_1 m_2} C^{j,m}_{j_1,m_1; j_2,m_2} |j_1, m_1\rangle \otimes |j_2, m_2\rangle. \tag{31.20}$$

## 31.5   Exactly Soluble Problems

### 31.5.1  The Harmonic Oscillator

For simplicity, we consider the one-dimensional problem. The harmonic potential is written $V(x) = m\omega^2 x^2/2$. The natural length and momentum scales are

$$x_0 = \sqrt{\frac{\hbar}{m\omega}}, \quad p_0 = \sqrt{\hbar m\omega}. \tag{31.21}$$

By introducing the reduced operators $\hat{X} = \hat{x}/x_0$ and $\hat{P} = \hat{p}/p_0$, the Hamiltonian is:

$$\hat{H} = \frac{\hbar\omega}{2} \left( \hat{P}^2 + \hat{X}^2 \right), \quad \text{with } [\hat{X}, \hat{P}] = i. \tag{31.22}$$

We define the creation and annihilation operators $\hat{a}^\dagger$ and $\hat{a}$ by:

$$\hat{a} = \frac{1}{\sqrt{2}} \left( \hat{X} + i\hat{P} \right), \quad \hat{a}^\dagger = \frac{1}{\sqrt{2}} \left( \hat{X} - i\hat{P} \right), \quad [\hat{a}, \hat{a}^\dagger] = 1. \tag{31.23}$$

One has

$$\hat{H} = \hbar\omega \left( \hat{a}^\dagger \hat{a} + 1/2 \right). \tag{31.24}$$

The eigenvalues of $\hat{H}$ are $(n + 1/2)\hbar\omega$, with $n$ non-negative integer. These eigenvalues are non-degenerate. The corresponding eigenvectors are noted $|n\rangle$. We have:

$$\hat{a}^\dagger |n\rangle = \sqrt{n+1}|n+1\rangle \tag{31.25}$$

and

$$\hat{a}|n\rangle = \sqrt{n}|n-1\rangle \quad \text{if } n > 0, \tag{31.26}$$

$$= 0 \quad \text{if } n = 0. \tag{31.27}$$

The corresponding wave functions are the Hermite functions. The ground state $|n = 0\rangle$ is given by:

$$\psi_0(x) = \frac{1}{\pi^{1/4}\sqrt{x_0}} \exp(-x^2/2x_0^2). \tag{31.28}$$

Higher dimension harmonic oscillator problems are deduced directly from these results.

### 31.5.2 The Coulomb Potential (Bound States)

We consider the motion of an electron in the electrostatic field of the proton. We note $\mu$ the reduced mass ($\mu = m_e m_p/(m_e + m_p) \simeq m_e$) and we set $e^2 = q^2/(4\pi\epsilon_0)$. Since the Coulomb potential is rotation invariant, we can find a basis of states common to the Hamiltonian $\hat{H}$, to $\hat{L}^2$ and to $\hat{L}_z$. The bound states are characterized by the 3 quantum numbers $n, \ell, m$ with:

$$\psi_{n,\ell,m}(\boldsymbol{r}) = R_{n,\ell}(r) \, Y_{\ell,m}(\theta, \varphi), \tag{31.29}$$

where the $Y_{\ell,m}$ are the spherical harmonics. The energy levels are of the form

$$E_n = -\frac{E_I}{n^2} \quad \text{with} \quad E_I = \frac{\mu e^4}{2\hbar^2} \simeq 13.6 \, \text{eV}. \tag{31.30}$$

The principal quantum number $n$ is a positive integer and $\ell$ can take all integer values from 0 to $n - 1$. The total degeneracy (in $m$ and $\ell$) of a given energy level is $n^2$ (we do not take spin into account). The radial wave functions $R_{n,\ell}$ are of the form:

$$R_{n,\ell}(r) = r^\ell P_{n,\ell}(r) \exp(-r/(na_1)), \quad \text{with} \quad a_1 = \frac{\hbar^2}{\mu e^2} \simeq 0.53 \, \text{Å}. \tag{31.31}$$

$P_{n,\ell}(r)$ is a polynomial of degree $n - \ell - 1$ called a Laguerre polynomial. The length $a_1$ is the Bohr radius. The ground state wave function is $\psi_{1,0,0}(\boldsymbol{r}) = e^{-r/a_1}/\sqrt{\pi a_1^3}$.

## 31.6   Approximation Methods

### 31.6.1 Time-Independent Perturbations

We consider a time-independent Hamiltonian $\hat{H}$ which can be written as $\hat{H} = \hat{H}_0 + \lambda \hat{H}_1$. We suppose that the eigenstates of $\hat{H}_0$ are known:

$$\hat{H}_0 |n, r\rangle = E_n |n, r\rangle, \quad r = 1, 2, \ldots, p_n \tag{31.32}$$

where $p_n$ is the degeneracy of $E_n$. We also suppose that the term $\lambda \hat{H}_1$ is sufficiently small so that it only results in small perturbations of the spectrum of $\hat{H}_0$.

**Non-degenerate Case** In this case, $p_n = 1$ and the eigenvalue of $\hat{H}$ which coincides with $E_n$ as $\lambda \to 0$ is given by:

$$\tilde{E}_n = E_n + \lambda \langle n|\hat{H}_1|n\rangle + \lambda^2 \sum_{k \neq n} \frac{|\langle k|\hat{H}_1|n\rangle|^2}{E_n - E_k} + O(\lambda^3). \tag{31.33}$$

The corresponding eigenstate is:

$$|\psi_n\rangle = |n\rangle + \lambda \sum_{k \neq n} \frac{\langle k|\hat{H}_1|n\rangle}{E_n - E_k} |k\rangle + O(\lambda^2) \tag{31.34}$$

**Degenerate Case** In order to obtain the eigenvalues of $\hat{H}$ at first order in $\lambda$, and the corresponding eigenstates, one must diagonalize the restriction of $\lambda \hat{H}_1$ to the subspace of $\hat{H}_0$ associated with the eigenvalue $E_n$, i.e. find the $p_n$ solutions of the "secular" equation:

$$\begin{vmatrix} \langle n, 1|\lambda\hat{H}_1|n, 1\rangle - \Delta E & \cdots & \langle n, 1|\lambda\hat{H}_1|n, p_n\rangle \\ \vdots & \langle n, r|\lambda\hat{H}_1|n, r\rangle - \Delta E & \vdots \\ \langle n, p_n|\lambda\hat{H}_1|n, 1\rangle & \cdots & \langle n, p_n|\lambda\hat{H}_1|n, p_n\rangle - \Delta E \end{vmatrix} = 0. \tag{31.35}$$

The energies to first order in $\lambda$ are $\tilde{E}_{n,r} = E_n + \Delta E_r, r = 1, \ldots, p_n$. In general, the perturbation is lifted (at least partially) by the perturbation.

### 31.6.2 Variational Method for the Ground State

Consider an arbitrary state $|\psi\rangle$ normalized to 1. The expectation value of the energy in this state is greater than or equal to the ground state energy $E_0$: $\langle\psi|\hat{H}|\psi\rangle \geq E_0$. In order to find an upper bound to $E_0$, one uses a set of trial wave functions which depend on a set of parameters, and one looks for the minimum of $\langle E\rangle$ for these functions. This minimum always lies above $E_0$.

---

### 31.7   Identical Particles

All particles in nature belong to one of the following classes:

- Bosons, which have integer spin. The state vector of $N$ identical bosons is totally symmetric with respect to the exchange of any two of these particles.

- Fermions, which have half-integer spin. The state vector of $N$ identical fermions is totally antisymmetric with respect to the exchange of any two of these particles.

Consider a basis $\{|n_i\rangle,\ i = 1, 2, \ldots\}$ of the one particle Hilbert space. Consider a system of $N$ identical particles, which we number arbitrarily from 1 to $N$.

(a) If the particles are bosons, the state vector of the system with $N_1$ particles in the state $|n_1\rangle$, $N_2$ particles in the state $|n_2\rangle$, etc., is:

$$|\Psi\rangle = \frac{1}{\sqrt{N!}} \frac{1}{\sqrt{N_1! N_2! \cdots}} \sum_P |1 : n_{P(1)}; 2 : n_{P(2)}; \ldots; N : n_{P(N)}\rangle,$$

(31.36)

where the summation is made on the $N!$ permutations of a set of $N$ elements.

(b) If the particles are fermions, the state corresponding to one particle in the state $|n_1\rangle$, another in the state $|n_2\rangle$, etc., is given by the Slater determinant:

$$|\Psi\rangle = \frac{1}{\sqrt{N!}} \begin{vmatrix} |1 : n_1\rangle & |1 : n_2\rangle & \cdots & |1 : n_N\rangle \\ |2 : n_1\rangle & |2 : n_2\rangle & \cdots & |2 : n_N\rangle \\ \vdots & \vdots & & \vdots \\ |N : n_1\rangle & |N : n_2\rangle & \cdots & |N : n_N\rangle \end{vmatrix}.$$

(31.37)

Since the state vector is antisymmetric, two fermions cannot be in the same quantum state (Pauli's exclusion principle). The above states form a basis of the $N$–fermion Hilbert space.

## 31.8 Time-Evolution of Systems

### 31.8.1 Rabi Oscillation

Consider a two-level system $|\pm\rangle$, of Hamiltonian $\hat{H}_0 = \hbar\omega_0|+\rangle\langle+|$. We couple these two states with a Hamiltonian $\hat{H}_1$:

$$\hat{H}_1 = \frac{\hbar\omega_1}{2} \left( e^{-i\omega t}|+\rangle\langle-| + e^{i\omega t}|-\rangle\langle+| \right).$$

(31.38)

We assume that the state of the system is $|-\rangle$ at time $t = 0$. The probability to find the system in the state $|+\rangle$ at time $t$ is:

$$P(t) = \frac{\omega_1^2}{\Omega^2} \sin^2(\Omega T/2) \quad \text{with} \quad \Omega^2 = (\omega - \omega_0)^2 + \omega_1^2.$$

(31.39)

## 31.8.2 Time-Dependent Perturbation Theory

We consider a system whose Hamiltonian is $\hat{H}(t) = \hat{H}_0 + \hat{H}_1(t)$. We assume the eigenstates $|n\rangle$ of $\hat{H}_0$ and the corresponding energies $E_n$ are known. At time $t = 0$, we assume that the system is in the eigenstate $|i\rangle$ of $\hat{H}_0$. To first order in $\hat{H}_1$, the probability amplitude to find the system in another eigenstate $|f\rangle$ at time $t$ is:

$$a(t) = \frac{1}{i\hbar} \int_0^t e^{i(E_f - E_i)t/\hbar} \langle f|\hat{H}_1(t')|i\rangle \, dt'. \qquad (31.40)$$

In the case of a time-independent perturbation $H_1$, the probability is:

$$P(t) = |a(t)|^2 = \frac{1}{\hbar^2} \left| \langle f|\hat{H}_1|i\rangle \right|^2 \frac{\sin^2(\omega t/2)}{(\omega/2)^2}, \qquad (31.41)$$

where we have set $\hbar\omega = E_f - E_i$.

## 31.8.3 Fermi's Golden Rule and Exponential Decay

Consider a system with an unperturbed Hamiltonian $\hat{H}_0$. Initially, the system is in an eigenstate $|i\rangle$ of energy $E_i$. We assume that this system is coupled to a continuum $\{|f\rangle\}$ of eigenstates of $\hat{H}_0$ by the time-independent perturbation $\hat{V}$. For simplicity, we assume that the matrix elements $\langle f|\hat{V}|i\rangle$ only depend on the energies $E_f$ of the states $|f\rangle$.

To lowest order in $\hat{V}$, this coupling results in a finite lifetime $\tau$ of the state $|i\rangle$: the probability to find the system in the state $|i\rangle$ at time $t > 0$ is $e^{-t/\tau}$ with:

$$\frac{1}{\tau} = \frac{2\pi}{\hbar} |\langle f|\hat{V}|i\rangle|^2 \, \rho(E_i). \qquad (31.42)$$

The matrix element $\langle f|\hat{V}|i\rangle$ is evaluated for a state $|f\rangle$ of energy $E_f = E_i$. The function $\rho(E)$ is the density of final states. For non relativistic particles ($E = p^2/2m$) or ultra-relativistic particles ($E = cp$, for instance photons), its values are respectively:

$$\rho_{\text{non rel.}}(E) = \frac{mL^3\sqrt{2mE}}{2\pi^2\hbar^3} \qquad \rho_{\text{ultra rel.}}(E) = \frac{L^3 E^2}{2\pi^2\hbar^3 c^3}. \qquad (31.43)$$

When the spin degree of freedom of the particle comes into play, this density of state must be multiplied by the number of possible spin states $2s + 1$, where $s$ is the spin of the particle. The quantity $L^3$ represents the normalization volume (and cancels identically with the normalization factors of the states $|i\rangle$ and $|f\rangle$). Consider an atomic transition treated as a two-level system, an excited state $|e\rangle$ and a ground state $|g\rangle$, separated by an energy $\hbar\omega$ and coupled via an electric dipole interaction. The lifetime $\tau$ of the excited state due to this spontaneous emission is given by:

$$\frac{1}{\tau} = \frac{\omega^3}{3\pi\epsilon_0\hbar c^3}\left|\langle e|\hat{\boldsymbol{D}}|g\rangle\right|^2, \tag{31.44}$$

where $\hat{\boldsymbol{D}}$ is the electric dipole operator.

## 31.9 Collision Processes

### 31.9.1 Born Approximation

We consider an elastic collision process of a non-relativistic particle of mass $m$ with a fixed potential $V(r)$. To second order in $V$, the elastic scattering cross-section for an incident particle in the initial momentum state $\boldsymbol{p}$ and the final momentum state $\boldsymbol{p}'$ is given by:

$$\frac{d\sigma}{d\Omega} = \left(\frac{m}{2\pi\hbar^2}\right)^2 |\tilde{V}(\boldsymbol{p} - \boldsymbol{p}')|^2, \quad \text{with} \quad \tilde{V}(\boldsymbol{q}) = \int e^{i\boldsymbol{q}\cdot\boldsymbol{r}/\hbar} V(r)\,d^3r. \tag{31.45}$$

**Example: the Yukawa potential** We consider

$$V(r) = g\,\frac{\hbar c}{r}\,e^{-r/a}, \tag{31.46}$$

which gives, writing $p = \hbar k$:

$$\frac{d\sigma}{d\Omega} = \left(\frac{2mgca^2}{\hbar}\right)^2 \frac{1}{\left(1 + 4a^2k^2\sin^2(\theta/2)\right)^2} \quad \text{(Born)}, \tag{31.47}$$

where $\theta$ is the scattering angle between $\boldsymbol{p}$ and $\boldsymbol{p}'$. The total cross-section is then:

$$\sigma(k) = \left(\frac{2mgca}{\hbar}\right)^2 \frac{4\pi a^2}{1 + 4k^2a^2} \quad \text{(Born)}. \tag{31.48}$$

In the case where the range $a$ of the potential tends to infinity, we recover the Coulomb cross section:

$$\frac{d\sigma}{d\Omega} = \left(\frac{g\hbar c}{4E}\right)^2 \frac{1}{\sin^4(\theta/2)} \quad \text{(exact)}, \tag{31.49}$$

where $E = p^2/(2m)$.

### 31.9.2 Scattering by a Bound State

We consider a particle $a$ of mass $m$ undergoing an elastic scattering on a system composed of $n$ particles $b_1, \ldots, b_n$. These $n$ particles form a bound state whose wave function is $\psi_0(r_1, \ldots, r_n)$. In Born approximation, the cross section is

$$\frac{d\sigma}{d\Omega} = \left(\frac{m}{2\pi\hbar^2}\right)^2 |\mathcal{V}(p - p')|^2 \quad \text{with} \quad \mathcal{V}(q) = \sum_j \tilde{V}_j(q) \, F_j(q). \tag{31.50}$$

The potential $V_j$ represents the interaction between particles $a$ and $b_j$. The form factor $F_j$ is defined by:

$$F_j(q) = \int e^{iq \cdot r_j/\hbar} |\psi_0(r_1, \ldots, r_j, \ldots, r_n)|^2 \, d^3r_1 \ldots d^3r_j \ldots d^3r_n. \tag{31.51}$$

In general, interference effects can be observed between the various $q$ contributing to the sum which defines $\mathcal{V}(q)$. In the case of a charge distribution, $\tilde{V}$ is the Rutherford amplitude, and the form factor $F$ is the Fourier transform of the charge density.

### 31.9.3 General Scattering Theory

In order to study the general problem of the scattering of a particle of mass $m$ by a potential $V(r)$, it is useful to determine the positive energy $E = \hbar^2 k^2/(2m)$ eigenstates of $\hat{H} = \hat{p}^2/(2m) + V(r)$ whose asymptotic form is

$$\psi_k(r) \underset{|r| \to \infty}{\sim} e^{ik \cdot r} + f(k, u, u') \frac{e^{ikr}}{r}. \tag{31.52}$$

This corresponds to the superposition of an incident plane wave $e^{ik \cdot r}$ and a scattered wave. Such a state is called a *stationary scattering state*. The scattering amplitude $f$ depends on the energy, on the incident direction $u = k/k$, and on the final direction $u' = r/r$. The differential cross section is given by:

$$\frac{d\sigma}{d\Omega} = |f(k, u, u')|^2. \tag{31.53}$$

The scattering amplitude is given by the implicit equation

$$f(k, u, u') = -\frac{m}{2\pi\hbar^2} \int e^{-ik' \cdot r'} V(r') \, \psi_k(r') d^3r' \quad \text{with} \quad k' = ku'. \tag{31.54}$$

We recover Born's approximation by choosing $\psi_k(r') \simeq e^{ik \cdot r}$.

### 31.9.4 Low Energy Scattering

When the wavelength of the incident particle $\lambda \sim k^{-1}$ is large compared to the range of the potential, the amplitude $f$ does not depend on $\boldsymbol{u}$ and $\boldsymbol{u}'$ (at least if the potential decreases faster than $r^{-3}$ at infinity). The scattering is isotropic. The limit $a_s = -\lim_{k \to 0} f(k)$ is called the scattering length.

# Author Index

# Subject Index

Absorption, 204, 297
Adiabatic approximation, 64
Angular momentum, 275
Annihilation, 53
Annihilation operator, 121, 138
Anomaly (electron), 37
Antineutrino, 61
Antiquark, 215
Approximation
   adiabatic, 64
   WKB, 316
Atom
   antihydrogen, 326
   cesium, 41, 253
   hydrogen, 51, 71, 315
   lithium, 224
   nitrogen, 79
   positronium, 51
   rubidium, 41, 232, 236
   silver, 23, 79
   sodium, 232
Atomic clocks, 41
Atomic fountain, 42

Band conduction, 259
Baryon, 215
Beam splitter, 109
Bell inequality, 92
Bethe–Bloch formula, 84
Bloch oscillations, 249
Bohr radius, 72, 298
Bose–Einstein condensate, 223, 235, 326
Bragg angle, 9
Bragg diffraction, 3
Bragg reflection, 259
Brillouin zone, 254

Cat (Schrödinger), 137
CERN, 61
Cesium atom, 41, 253
Chain (spins), 261
Condensate (Bose–Einstein), 223, 235, 326
Conductor, 259
Constant (fundamental), 45
Cooling (laser), 199
Correlated pair, 152
Correlation, 90, 101, 152
Creation operator, 121, 138
Cryptography (quantum), 151
Cyclotron, 40
Cyclotron motion, 183

Damping, 123, 134
Decay rate, 53
Decoherence, 168
Density operator, 199, 310
Detector
   non-destructive, 109
   Von Neumann, 164
Diffraction, 3, 19
Dimer, 3
Dipole force, 202
Dirac equation, 37
Dispersion relation, 317
Doppler cooling, 201
Double slit, 3
Drift (fundamental constants), 45

Ehrenfest theorem, 23, 39, 128
Einstein–Podolsky–Rosen (EPR) problem, 89
Electric dipole, 182, 199
Electromagnetic cavity, 119

© Springer Nature Switzerland AG 2019
J.-L. Basdevant, J. Dalibard, *The Quantum Mechanics Solver*,
https://doi.org/10.1007/978-3-030-13724-3

Printed in the United States
By Bookmasters